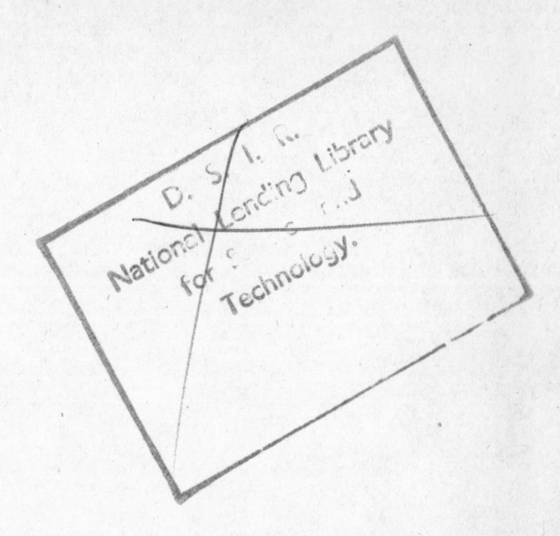
D. S. I. R.
National Lending Library
for Science and
Technology.

D1422410

A History of Electrical Engineering

TECHNOLOGY TODAY AND TOMORROW

Edited by P. F. R. Venables, Ph.D., B.Sc., F.R.I.C.
Principal of the College of Advanced Technology, Birmingham

*

A SOCIAL HISTORY OF ENGINEERING

W. H. G. Armytage, M.A., *Professor of Education in the University of Sheffield*

A HISTORY OF ELECTRICAL ENGINEERING

Percy Dunsheath, C.B.E., D.Sc., M.A.

A HISTORY OF MECHANICAL ENGINEERING

A. F. Burstall, D.Sc., Ph.D., *Professor of Mechanical Engineering and Director of the Stephenson Engineering Laboratories at King's College, University of Durham*

THE PROFESSIONAL TECHNOLOGIST

P. F. R. Venables, Ph.D., B.Sc., F.R.I.C.

A HISTORY OF ELECTRICAL ENGINEERING

PERCY DUNSHEATH, C.B.E.
M.A. (Cantab), D.Sc. (Eng.) London,
Hon.D.Eng. (Sheffield), Hon. LL.D. (London)
Past President, Institution of Electrical Engineers,
Past President, International Electrotechnical Commission

FABER AND FABER
24 Russell Square
London
1962.

First published in mcmlxii
by Faber and Faber Limited
24 *Russell Square, London, W.C.*1
Printed in Great Britain by
Western Printing Services Limited, Bristol

Acknowledgments

The author wishes to acknowledge help received from many quarters in the preparation of the book. Illustrations have been generously provided by industrial concerns as the individual acknowledgments will indicate, and special thanks are due to the assistance given by Dr. Follett, the Director of the Science Museum, and his staff, one of whom, Miss Weston, has made many valuable suggestions. The collection of historical material and confirmation of dates have been greatly facilitated by the willing help given by the Librarian and Staff of the Institution of Electrical Engineers. In particular Mr. H. Lansley has been tireless in his responses to my many calls on him. Friends in the Supply Industry have also been most helpful in producing information on early power stations, much of the information being channelled through Mr. P. A. Lingard of the London Electricity Board.

Dr. P. F. R. Venables, the General Editor of the Series, has taken considerable trouble in reading and advising about the book.

The many cases of help received from the light current, heavy current, and instrument sides of the electrical industry and from the electrical press are too numerous to permit a mention by name, but to all I express my most grateful thanks.

PERCY DUNSHEATH

Preface

In the preparation of a history of any branch of engineering an author is faced with a major problem in the relations between the science of the subject and the practical applications. Shall he confine himself strictly to engineering on the assumption that the underlying scientific principles constitute a separate subject, or shall he accept the two as being inseparable? In civil and mechanical engineering practical invention has usually preceded any theoretical analysis of the principles of operation; good examples are hydraulic dams and steam engines. In electrical engineering, on the other hand, as indeed to a lesser extent also in chemical engineering, evolution has followed an entirely different course. In these fields, apart from accidental observation as in Perkins' discovery of the mauve dye and Galvani's twitching frog's legs, engineering progress has been closely knit with observed phenomena resting on, and frequently suggested by, a deliberately assembled corpus of theoretical knowledge. Examples are not difficult to find—Fleming and Florey's production of penicillin and the intellectual structure built on the experimental results of Arago, Oersted and others by Faraday and Ampère.

In the introduction to his *History of the Institution of Electrical Engineers*, Mr. Rollo Appleyard, writing of the evolution of the Institution from the former Society of Telegraph Engineers and Electricians, says, 'It has stood at the confluence of the streams of academic and practical knowledge where for fifty years it has directed and safeguarded electrical progress. To describe its activities in their various aspects it is necessary to take note of the ever accelerating advance of electrical science and practice and of the men of mark by whose exertions the work has flourished and acquired form and vitality.'

So in a history of electrical engineering the underlying science cannot be divorced from its practical application and the following pages therefore tell the story of a remarkable composite growth over a century and a half in which the two have always been, and still are, inseparably associated.

Another practical problem has been one of chronology. The history of such a subject as electrical engineering does not follow a simple straightforward list of dates. The contributions of one man may continue over many decades during which period others pick up the same threads to weave completely different patterns. Many tributaries may flow simultaneously but reach the main stream at different times. This difficulty has been overcome by adopting a reasonable amount of 'leapfrogging' and, time and time again throughout the pages, it will be found that a particular interest or development is followed through beyond the date at which another, to be discussed later, makes its debut. To continue the river metaphor, the whole watershed has been kept under review at the same time as the course of the main stream has been traced.

Bearing in mind the broad cultural intention of this series of books, and as any history is essentially an account of the work of men, opportunity has been taken throughout of recording interesting biographical details of some of the many picturesque characters who have crossed the stage in this fascinating drama of electrical engineering. They include many of outstanding personality, courage, intellectual attainment and original thought. Their enthusiasm and application have contributed in no mean way to the progress of our civilization, and they well deserve individual recognition for their contributions to this remarkable story.

In the treatment of such a comprehensive subject, it is difficult to be certain of giving adequate recognition to the contributions made by men of all nationalities. In these pages, notwithstanding the space he has devoted to major contributions made by Britain, the author has aimed at giving full credit to the achievements of other countries. Where however he has failed in this respect, he must plead for tolerance through exigencies of time and space.

Finally a word on the title of the book. The range of modern electrical engineering is so vast that no history confined to a single volume could claim to deal adequately with the developments during recent decades. The early history of electrical engineering is largely an account of individual discoveries and inventions with which names and dates can be associated. As the field broadened the contributions became more and more the work of groups rather than of individuals. Moreover the complexity of electrical engineering increased and adequate treatment required a much longer story which made a complete account in one volume quite impossible. The nearer we approach the present day, the greater the problem, which can thus only be solved by the procedure adopted for covering recent decades by restricting the statement to a limited number of examples representing selected branches of the subject.

Two major examples will illustrate this need for curtailing the treatment. In radio engineering, for instance, the spate of invention which started just before the First World War resulted in a vast number of designs of both components and circuits. In his Presidential Address to the Institution of Electrical Engineers in 1946, the author gave a broad review of this explosive development, but since then hundreds of detailed papers have appeared on this one branch of electrical engineering.

Another example is the generation of electricity by hydro-electric stations. In this field, it is on the civil and mechanical engineering side that the most revolutionary development has taken place—the appearance of vast man-made lakes through the construction of huge concrete dams, and the design and manufacture of turbines of ever-increasing size and efficiency. The electrical side has involved the design and manufacture of large low-speed vertical-axis generators, but the rest of the equipment—switchgear, transformers and transmission lines—has all followed closely on normal practice. The scale of hydro-electric development has, however, already ensured it a major place in electrical history and, with the increasing scarcity of coal, it will in the future no doubt share the honours with atomic power in ensuring an adequate supply of electricity—and possibly, chapters for future histories of electrical engineering.

PERCY DUNSHEATH

Contents

Illustrations

PLATES

LINE DRAWINGS

CHAPTER I

Origins

The whole of Electrical Engineering is based on magnetic and electrical phenomena and no history of the subject can ignore the origins of these two groups, remote and sometimes uncertain as these origins may be. For many centuries man has observed magnetic effects in natural minerals found in the ground and electrical effects in lightning, the aurora borealis, St. Elmo's fire, the electric eel and the attraction of light objects by natural resins when rubbed.

Some of these observations have been put to practical use from the very earliest recorded times—the lodestone for navigation, the electric eel for medicinal purposes—so that, if electrical engineering is the practical application of electrical and magnetic science, there is a sense in which it has not only its roots in the remote past but actually existed as a human activity even in those far-off days. The two sides, magnetism and electricity, however, remained quite apart until the beginning of the nineteenth century when the discovery of the close relationship between them brought the two streams of thought together and opened the way to the establishment of their interrelation. The great surge forward on the foundation of electromagnetism made modern electrical engineering.

The records of magnetic effects date back to remotest antiquity. Mottelay opens his comprehensive and fascinating history of electricity and magnetism[1] with a statement that in the year 2637 B.C. the Chinese Emperor Hoang-ti constructed a chariot carrying a prominent female figure which always pointed to the south no matter in whatever direction the chariot was moving.

Sixteen centuries later we hear again of these 'south-seeking carts', Tcheou-Koung, a Chinese Minister of State, is said to have taught the use of the magnetic needle compass to ambassadors sent from Cochin China and to have given them an instrument called *tchi-nan*, meaning 'Chariot of the South'. On one side it turned towards the north and on the other side to the south, the better to direct them on their homeward voyage.

Several writers have supported the view that, in the ninth century B.C., navigation on land and sea was carried out with the aid of a floating needle, one of the most authentic accounts being written in the second century B.C. by Szu-ma-thsian, a great Chinese historian. King Solomon, son of David, is said to have employed the compass and indeed to have invented it, while certain verses in Homer's *Odyssey* are interpreted as evidence that the properties of the lodestone were understood and applied in his time. On the authority of Socrates we understand that Euripides referred to the natural magnetic ore as Magnesian stone or the Herculean, while many have considered the word magnet to have derived from its origin in Magnesia, a part of Asia Minor.

Attempts to account for the working of the elementary phenomena of magnetism, again, go back a long way. Lucretius (55 B.C.) for example, considered that the lodestone had hooks on its surface which engaged with rings on the surface of the attracted iron. This same poet, in his *De Rerum Natura*, vividly describes magnetic induction in iron by the lodestone in the following lines:

When without aid of hinges, links or springs,
A pendant chain we hold of steely rings,
Dropt from the stone; the stone the binding source,
Ring cleaves to ring, and owns magnetic force;
Those held superior those below maintain
Circle neath circle downward draws in vain.

Attraction of iron at a distance was also well known as is evident from the further lines:

The steel will move to seek the Stone's embrace
Or up or down or t'any other place.

Three centuries later the Chinese writer Koupho, referring to the attraction of iron by the lodestone, speaks of the 'breath of wind that promptly and mysteriously penetrates both bodies, uniting them imperceptibly with the rapidity of an arrow. It is incomprehensible.'

In A.D. 428 Saint Augustine, the early Christian writer, describes the attraction of a piece of iron lying on a silver dish by the lodestone underneath. In his *De Civitate Dei* he speaks of being thunderstruck by magnetic experiments which he witnessed. Speaking of his brother in the episcopate, Severus, Bishop of Milevis, he says (Dod's translation):

'He told me that Bathanarius, once Count of Africa, when the Bishop was dining with him, produced a magnet and held it under a silver plate on which he placed a bit of iron; then as he moved his hand with the magnet underneath the plate, the iron upon the plate moved about accordingly. The intervening silver was not affected at all, but precisely as the magnet was moved backward and forward below it, no matter how quickly, so was the iron attracted above. I have related what I have myself witnessed; I have related what I was told by one whom I trust as I trust my own eyes.'

In a Chinese dictionary completed in A.D. 121 there appears to be the first reference to the communication of magnetic polarity to an iron needle by rubbing with the lodestone and by striking it in a methodical manner.

During the Sung dynasty in China about A.D. 1000, the magnetization of iron by rubbing was certainly known for, we are told, fortune-tellers rubbed the needle with the lodestone to make it indicate the south. During the following two centuries, French sailors were rubbing needles upon the ugly brown stone called *mariniere* to produce the element for navigational compasses.

From the earliest times the outstanding practical application of magnetic effects has been of course in the field of navigation. Allowing for some uncertainty in the first references, it is possible that about 1000 B.C. the Chinese were finding their way across the boundless plains of Tartary with the aid of the compass.

There is nothing really authentic however to indicate sea navigation by the magnetic compass until the third century A.D. In a Chinese work Mung-khi-py-than soothsayers are recorded as using the needle floating on water and pointing to the south. They also suspended the needle on a thread in a place free from draughts and, although they were unaware of the fact that one end of the needle was attracted to the south, and one to the north—they thought that the difference was between needles—they did discover magnetic deviation.

During the next few hundred years this deviation was more and more observed and by the twelfth century, the knowledge of the compass and its application had spread, due to various travellers, to many countries. The famous letter of Peter Peregrinus, written in 1269, provides a remarkable account of the knowledge of the subject up to that time and gives a detailed specification for the construction of a mariner's compass enclosed in a case and complete with a 360-degree scale marked North, South, East and West.

Christopher Columbus naturally employed the magnetic compass on his famous voyage and was well aware of the deviation of the magnetic from the geographic pole. He was surprised to find, however, that the deviation changed as he sailed west. Washington Irving has left a striking account of the episode. He says:

'On the 13 of September (1492) in the evening being about two hundred leagues from the Island of Ferro (the smallest of the Canaries) Columbus, for the first time, noticed the variation of the needle.—He perceived about nightfall, that the needle, instead of pointing to the North Star, varied about half a point, or between five and six degrees to the north west, and still more on the following morning. Struck with this circumstance he observed it attentively for three days and found that the variation increased as he advanced. He at first made no mention of the phenomenon, knowing how ready his people were to take alarm, but it soon attracted the attention of the pilots, and filled them with consternation. It seemed as if the laws of nature were changing as they advanced and that they were entering into another world subject to unknown influences. They apprehended that the compass was about to lose its mysterious virtues; and without that guide what was to become of them in a vast and trackless ocean.'

To allay their terrors Columbus told them that the direction of the needle was really towards some remote point beyond the Pole Star. The deviation was not due to any failure of the compass but to the movement of the Pole Star! With this explanation and his great reputation as an astronomer their alarm subsided.

The practical and rapidly extending use of the mariner's compass by many navigators, including Vasco da Gama and Sebastian Cabot, led to an intensive study of terrestrial magnetism and in 1544, a Nuremburg clergyman named Hartmann discovered the phenomenon of 'dip' or inclination. By 1576 Robert Norman, 'a good seaman and ingenious artificer', had established a factory at Wapping for the manufacture of compass needles and, ignorant of Hartmann's priority, announced the discovery of 'dip' in his instrument. It was a serious problem for him because having constructed his instruments before magnetizing the needle he found it necessary to add a small weight to bring back the needle into a horizontal position. Norman provided an early example of the electrical engineer adding to fundamental knowledge. As the result of his practical manufac-

PLATE I

(*a*) Alessandro Volta (1745–1827)
(*Photo: Institution of Electrical Engineers*)

(*b*) Hans Christian Oersted (1771–1851)
(*Photo: Institution of Electrical Engineers*)

(*c*) André Marie Ampère (1775–1836)
(*Photo: Institution of Electrical Engineers*)

(*d*) Humphry Davy (1778–1829)
(*Photo: Royal Institution*)

Plate II

(*a*) Francis Ronalds (1788–1873)
(*Photo: H.M. Postmaster-General*)

(*b*) Michael Faraday (1791–1867)
(*Photo: Royal Institution*)

(*c*) Charles Wheatstone (1802–1875)
(*Photo: Institution of Electrical Engineers*)

(*d*) Karl Wilhelm Siemens (1823–1883)
(*Photo: Institution of Electrical Engineers*)

PLATE III

(*a*) William Thomson, Lord Kelvin (1824–1907) (*Photo: Annan*)

(*b*) Joseph Wilson Swan (1828–1914) (*Photo: A.E.I. Ltd.*)

(*c*) David Edward Hughes (1831–1900)

(*Photo: Institution of Electrical Engineers*)

(*d*) Rookes Evelyn Bell Crompton (1845–1940)
(*Photo: Institution of Electrical Engineers*)

Plate IV

(*a*) Thomas Alva Edison (1847–1931)
(*Photo: A.E.I. Ltd.*)

(*b*) Alexander Graham Bell (1847–1922)
(*Photo: Institution of Electrical Engineers*)

(*c*) John Hopkinson (1849–1898)
(*Photo: Institution of Electrical Engineers*)

(*d*) Oliver Heaviside (1850–1925)
(*Photo: Institution of Electrical Engineers*)

PLATE V

(*a*) Silvanus P. Thompson (1851–1916)
(*Photo: Institution of Electrical Engineers*)

(*b*) Elihu Thomson (1853–1937)
(*Photo: Institution of Electrical Engineers*)

(*c*) Charles Algernon Parsons (1854–1931)
(*Photo: C. A. Parsons & Co. Ltd.*)

(*d*) Heinrich Hertz (1857–1894)
(*Photo: Institution of Electrical Engineers*)

Plate VI

(*a*) Nikola Tesla (1857–1943)
(*Photo: Royal Institution*)

(*b*) Sebastian Ziani de Ferranti (1864–1930)
(*Photo: Institution of Electrical Engineers*)

(*c*) William Du Bois Duddell (1872–1917)
(*Photo: Institution of Electrical Engineers*)

(*d*) Guglielmo Marconi (1874–1937)
(*Photo: Institution of Electrical Engineers*)

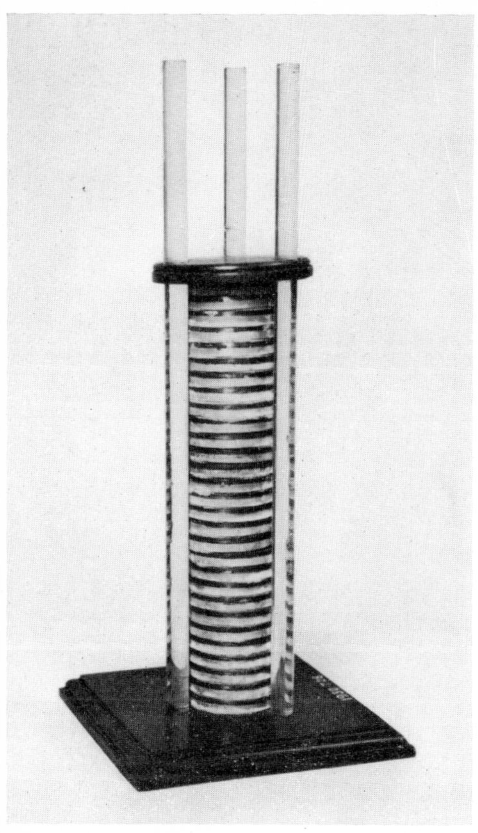

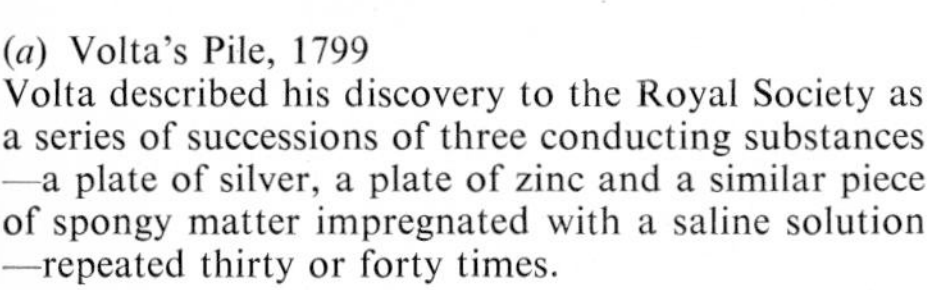

(*a*) Volta's Pile, 1799
Volta described his discovery to the Royal Society as a series of successions of three conducting substances —a plate of silver, a plate of zinc and a similar piece of spongy matter impregnated with a saline solution —repeated thirty or forty times.

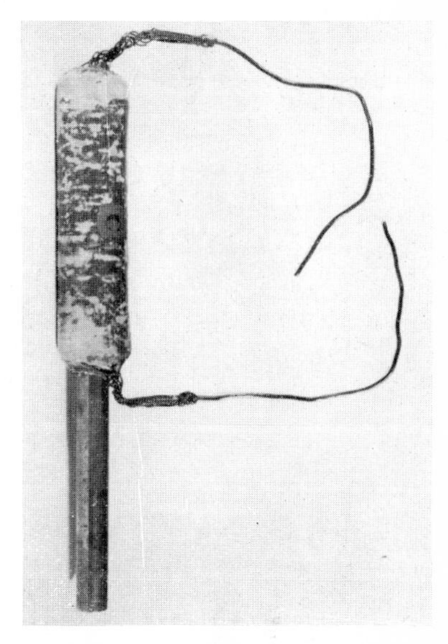

(*b*) Faraday's Original Coil and Magnet, 17th October, 1831
Faraday discovered that when he thrust a bar magnet into a helix of copper wire connected to a galvanometer he obtained a deflection of the needle.

(*Photo: Royal Institution*)

(*c*) Faraday's Original Ring, 29th August, 1831
Faraday produced the first transformer by winding on an iron ring six inches in diameter a number of coils of insulated wire. By connecting a battery to the coil on one side he discovered that "making" the current induced a momentary current in the other side which was connected to a galvanometer and "breaking" the current induced a momentary current in the reverse direction.

(*Photo: Royal Institution*)

Plate VIII

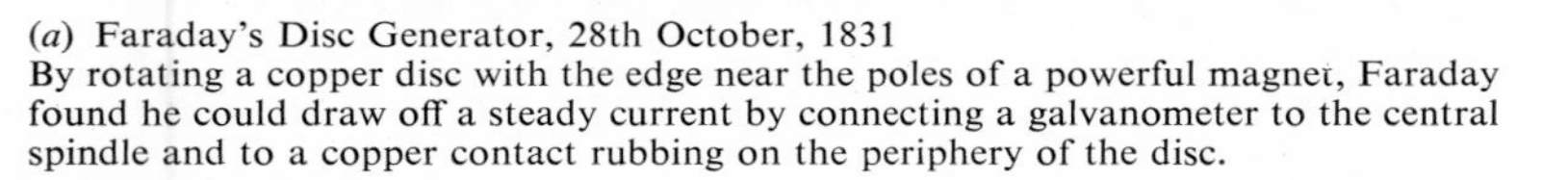

(*a*) Faraday's Disc Generator, 28th October, 1831
By rotating a copper disc with the edge near the poles of a powerful magnet, Faraday found he could draw off a steady current by connecting a galvanometer to the central spindle and to a copper contact rubbing on the periphery of the disc.

(*Photo: Science Museum*)

(*b*) Woolrich's Electromagnetic Generator, 1844
One of the first industrial generators used in the electro-plating works of Messrs Prime, at Birmingham. The heavy oak frame-work carries four horse-shoe magnets and the armature consists of eight bobbins rotated between the poles.

turing difficulty he made a 'dip circle' and determined the value of the magnetic inclination in London to be 71 degrees 50 minutes.

In the year A.D. 1600 William Gilbert of Colchester, who was Physician to Queen Elizabeth I, published a book *De Magnete*, the appearance of which was an epoch-making event in electrical progress. As its title suggests, the work dealt largely with the phenomena of magnetism but it also covered the complete range of electrical effects as known at that time, as well as adding much new knowledge through Gilbert's own experiments. To appreciate the significance of *De Magnete* as a factor in the foundations of electrical engineering, therefore, it is necessary to glance back over the years and see to what extent electrical phenomena had been observed alongside the magnetic development already described.

It is safe to assume that the first electrical effects to be noticed by man were the lightning flash and the aurora borealis. Their existence called for no deliberate act on his part and they occurred long before he had produced electrical charges himself, either fortuitously or by design. At first they and their effects were completely out of man's control, though several thousand years ago damage to buildings by lightning strokes was prevented by the nature of their construction. Several famous historic buildings, including the Temple of Juno and Solomon's Temple, had their roofs covered with metallic points—sword blades or sharp ornamental objects—with resulting immunity to damage by lightning, and there is no record that Solomon's Temple was ever struck by lightning during a period of a thousand years. The pipes which are known to have carried the roof water into caverns under the hill no doubt contributed to this result. There seems to be no evidence that protection of buildings was understood or deliberately adopted, although Pliny in his famous *Naturalis Historia* written during the first half of the first century A.D. asserts that the Etruscans had a secret method of drawing lightning from the clouds five centuries earlier and turning it aside in any desired direction.

It was not until the great Benjamin Franklin, American writer, philosopher and statesman, became interested in electrical phenomena that the idea of deliberate protection of buildings emerged. After his famous and extremely dangerous experiments of collecting electric charges by sending kites up into thunderclouds, Franklin, in 1750, conceived the idea of a lightning conductor and in his *Poor Richards Almanac* for 1753 he puts forward the proposal for the protection of buildings 'from mischief by thunder and lightning'.

St. Paul's Cathedral, which had been partially destroyed by lightning in 1561,[2] was first provided with lightning conductors in 1769. A few years later a futile dispute arose over the relative merits of pointed and blunt conductors. The majority of the members of the Royal Society accepted Franklin's view that points were the more effective, but owing to his participation in the American Revolution he was regarded as an enemy of England and his scientific views were the subject of disparagement. King George III, the patron of the Society, was persuaded to have the pointed conductors on Buckingham Palace replaced by ball-ended ones and suggested to Sir John Pringle, the President, that he should use his influence in favour of this preference. The result was the resignation of the President, an episode recalling more than one attempt, even in recent times, to put science in strings for political purposes.

Closely allied with lightning, two other atmospheric electrical effects have been observed for some centuries past, the polar lights—those in the north named aurora borealis by Pierre Gassendi in 1621 and those in the south named aurora australis by Ulloa, a Spanish mathematician in 1752—and St. Elmo's Fire observed by Italian sailors in the Mediterranean as early as the third century A.D. They noticed that light was emitted at night from the mastheads and rigging of their ships during dry stormy weather. In his record of his second voyage in 1493 Columbus wrote 'St. Elmo appeared on the top gallant masts with seven lighted tapers.'

Through these early years the electrical effects displayed by certain fishes were frequently described. Greek philosophers, including Aristotle and Plutarch, two thousand years ago, knew that the electric torpedo was capable of stunning its prey by an electric charge and over the next few centuries there were many references to the phenomenon. Pliny reported that a man could receive a shock by touching such a fish with a spear and the electrical properties of the *torpedo* and *gymnotus* were frequently proposed for the cure of human ailments, including gout and rheumatism.

The development of frictional or static electricity grew up over the centuries alongside that of magnetism and originated in the observation of the tiny crackling sparks produced by combing the hair in dry weather. For many centuries the fossil resin, amber, has been known to acquire the property of attracting light objects when heated and rubbed, but it was only in the year 1600 that any systematic study was made of such controllable electrical effects. In the second book of Gilbert's *De Magnete* he recorded a careful study of amber and distinguished between 'electrics' and 'non-electrics'. He spoke of electric force for the first time, electric attraction and the absence of poles in an electric such as were recognized in a magnet.

Other philosophers became interested and in an English translation of a work by the Belgian scientist Van Helmont, published in 1650, the following reference to static electricity appeared:
'the phansy of amber delights to allect strawes, chaffe, and other festucous bodies; by an attraction, we confesse, observe obscure and weake enough, yet sufficiently manifest and strong to attest an *Electricity* or attractive signature.'

Ten years later, that is in 1660, following a number of sporadic experiments, the first frictional electrical machine was constructed by Otto von Guericke of Magdeburg. He made a sulphur globe mounted on an axis which, when rotated against a cloth rubber pressed to its surface, emitted crackling sparks and evinced the well-known phenomenon of attracting light pieces of straw.

The first one to observe the electric light *in vacuo* appears to have been Jean Picard, French astronomer, who noticed a light inside the tube of a mercury barometer which he was carrying. Sir Isaac Newton in 1675 communicated to the Royal Society an observation that rubbed glass would also attract light bodies and moreover, that the glass showed a second kind of electricity on the side opposite to that on which it had been rubbed. He also noted the similarity between the electric spark and the flash of lightning.

The first half of the eighteenth century saw many discoveries and applications of electricity. Outstanding among these was the principle of conduction and insulation enunciated by Stephen Grey in 1720. By suspending a hempen line on silken threads he transmitted electric charges hundreds of feet. When metallic wire was substituted for the hempen cord, circuits up to several miles were made to carry the charge.

The French scientist Dufay seems to have been the first to have established the idea that electricity appeared in two distinct forms, *vitreous* and *resinous*, the former produced on glass, and certain other materials, and the latter on amber, silk, paper, etc. He also observed that each repels its own kind and attracts the other.

Frictional machines were developed to produce powerful charges and in 1745 the great step forward of storing the charge in a Leyden jar was taken. Several different experimenters are claimed as the discoverers of the Leyden jar and it does appear that the abilty to store a charge in a bottle—or jar-shaped container—was noticed at the end of 1745 and the beginning of 1746 by more than one independent observer. Popular opinion gives the credit definitely to Cunaeus working with Musschenbroek at Leyden. Musschenbroek, the more readily to prevent the escape of electricity from a conductor, employed a glass jar containing water with the conductor inserted through the neck. Cunaeus, attempting to remove the conductor which had been charged by a powerful friction machine, unexpectedly received through his body the full discharge from the inner conductor to the outside of the jar which he was holding in his hand.

The English scientist, Dr. Watson, made many experiments with the Leyden jar, established the idea of two coatings separated by the dielectric and spoke of 'plus' and 'minus' electricity. With others he also made up long circuits up to several miles and discharged the Leyden jar through them.

The tempo of investigation into static electrical phenomena increased rapidly towards the end of the eighteenth century when it dramatically resulted in the discovery of the steady electric current. Over a period of twenty-five years two names stand out prominently, Galvani and Volta, the former in connection with his observations on the contraction of the muscles in legs of dead frogs and the latter in the developing of the Voltaic Pile, the first device to produce a steady controllable current of electricity.

As early as 1678 Swammerdam, a celebrated Dutch scientist, carried out the first known experiment in the influence of electricity on animal nerve and muscle. Drawing a dissected muscle with a protruding nerve from a glass tube by means of a silver wire attached to the nerve and bringing the nerve into contact with a copper ring, the muscle was seen to contract. The report of the experiment appears in the account of his experiment published at Leipzig in 1752.

In 1762 another flood of light was thrown on the phenomenon by Sulzer, a Swiss philosopher. Using the words of Sabini from his *Nouvelle Theorie des Plaisirs* published in 1767: 'On taking two pieces of different metals—silver and zinc—and placing one of them above and the other underneath his tongue, he found that, so long as the metals did not make contact with each other, he felt nothing; but when the edges were brought together over the tip of his tongue, the moment contact took place, and as long as it lasted, he experienced an itching sensation and a taste resembling that of sulphate of iron.'

The significance of Sulzer's discovery was not appreciated by him and remained unrecognized until twenty-four years later when, in 1786, Galvani, a young Italian scientist, made his now famous discovery. Galvani, studying the effects of lightning discharges and discharges from electrical machines, found that they produced similar convulsions in the limbs of dead frogs. He also found that the effect could be produced without any other external agency than a pair of dissimilar metals. First of all a frog's leg hung on a copper hook fastened to an iron railing and blown into contact with the railing produced the twitching phenomenon. The result was also obtained from the two extremities of a bimetallic strip. He found copper and zinc very effective.

In the light of our present knowledge it seems strange that Galvani should have formulated and persisted in the idea that the source of the electricity lay in the nerves and muscles of the frog but his views were not shared by all his contemporaries. A great rivalry sprang up between him and the other Italian scientist Volta. Born in Como in 1745 and later in life a professor in the University of Pavia, Volta made many valuable contributions to the understanding and development of electrical phenomena. In 1775 he had invented the electrophorus by which a small charge of static electricity can be multiplied many times by the manipulation of mechanical apparatus. A dish of solid resin being electrified by rubbing or striking with warm flannel or a silk handkerchief has a metal disc provided with an insulating handle placed over it. The surface of the disc is discharged by touch while resting on the resin and it is then removed, bringing with it a charge which is held captive up to that point.

Between the rival Universities of Bologna and Pavia the ideas of Volta and Galvani were made the basis of serious factions and the dispute spread to other European countries where it raged for several years, as absurd a display of unscientific intransigence as that of the lightning conductors a little earlier.

This story of Galvani and the physiological effects of electrical discharge takes us into a curious chapter of history of electrical engineering in which the bodies of executed murderers were made to perform gruesome contortions by the application of electricity, but rather away from the origins. In this direction and more appropriate to the present introduction, it was Volta's reasoning arising from his observations of Galvani's experiments that led to the epoch-making discovery of the steady electric current.

There had already been some anticipation of the electric current and conduction in certain of the experiments made with static electricity and in observations on the effects of lightning. Lightning rods had been found to be heated by the discharge. In 1761 Ebenezer

Kinnersley, for instance, a friend of Benjamin Franklin and an English master in the College of Philadelphia, came very near to the conception of an electric current when he wrote: 'No heat is produced by passing shocks through a large wire but a small wire is heated red hot, expanded and melted.' Again in 1796, John Cuthbertson, an English instrument maker, showed that the discharge from a battery of fifteen Leyden jars containing 17 square feet of coated glass fused a 6-foot length of iron wire 1/150 of an inch in diameter.

Henry Cavendish, the famous English scientist who later gave his name to the well-known Cavendish Laboratory at Cambridge, found in 1772 that 'a saturated solution of sea-salt conducts seven hundred and twenty times better than fresh water, also that electricity experiences as much resistance in passing through a column of water one inch long as it does in passing through an iron wire of the same diameter four hundred million inches long'.

By the year 1780 Volta was deeply engaged in a wide range of electrical experiments and was particularly interested in Galvani's use of the lively muscular contraction of the frog's leg as a sensitive electroscope. Galvani had found that it was sufficient to touch the lumbar nerve and the muscles of the thigh with the ends of a wire to produce the contraction, a phenomenon which he attributed to 'animal electricity'.

Volta, after experimenting for many years, succumbed to the same explanation and in a published account of his work stated 'the evidence of many experiments well combined and accurately described (shows) that there exists a true and real animal electricity, that is to say, electricity excited by the living organs themselves.' He soon concluded, however, that the action of a circuit of two dissimilar metals was greatly superior to that of a homogeneous circuit in its effect on the muscle and in the course of these experiments he rediscovered the phenomenon observed by Sulzer twenty-five years before, whereby the contact of two dissimilar metals applied to the tongue produced a curious effect of taste. While making mental reservations on the existence of the effects produced he was soon convinced that they were not due to animal electricity but were produced in some way by the contact of dissimilar metals.

About this time Volta observed that this new electricity differed from the flow from a Leyden jar, it was a continuous phenomenon and he began to use the term 'current of electricity'. The world-wide controversy which was waged with ardour—though with courtesy—

between the views of Galvani and those of Volta began to wane when once it had been established that an electromotive force exists between two wet conductors in contact.

Volta found that adding a second pair of metals in series, separated by a moistened fibrous diaphragm, increased the effect and then, adding more and more, was led to the 'Voltaic Pile', a discovery which astonished the world. In a centenary lecture delivered in Como on 18 September 1899, Professor Auguste Righi said 'Alessandro Volta's noble intellect shone nowhere so brightly as in his researches on contact electricity. A lively imagination, controlled at all times by the coolest judgment, a serene spirit in the face of difficulties, which compelled him to modify his ideas; a penetrating sagacity in contriving experiments, and unequalled skill in their execution, sane reasoning power in the interpretation and collation of facts, and in drawing conclusions from them—these were the salient characteristics of his lofty mind.' In the same address Professor Righi referred to Volta's Pile as producing with extremely simple means many of the effects of the discharge of electrical bodies and obtaining entirely new and unexpected results. Anticipating a famous dictum of Kelvin, Righi referred to Volta's stern adhesion to strict scientific method and his condemnation of fanciful, speculative, or merely sensational experiments. 'What possible good', Volta wrote, 'can come out of all this, unless the observations are reduced to scale and measure. . . . What is the use of ascertaining a cause unless the quantity and intensity of the effect is determined, as well as its character or quality.'

Volta was a Fellow of the Royal Society and made his formal announcement of this newly established electric current in a letter to Sir Joseph Banks, the President, which was read on 26 June 1800.

After explaining how he was led to the construction of an apparatus which bore a great resemblance to the Leyden phial but had the singular property of recharging itself continually, he described the arrangement in the following words:[3]

'It consists of a long series of an alternate succession of three conducting substances, either copper, tin and water; or what is much preferable, silver, zinc and a solution of any neutral or alkaline salt. The mode of combining these substances consists in placing horizontally, first, a plate or disc of silver (half-a-crown, for instance) next a plate of zinc of the same dimensions; and, lastly, a similar piece of spongy matter, such as pasteboard or leather, fully impregnated with the saline solution. This set of three-fold layers is to be repeated thirty or forty times, forming thus what the author calls his 'columnar machine'. It is to be observed, that the metals must always be in the same order. That is, if the silver is the lowermost in the first pair of metallic plates, it is to be so in all the successive ones, but that the effects will be the same if this order be inverted in all the pairs. As the fluid, either water or the saline solution, and not the spongy layer impregnated with it, is the substance that contributes to the effect, it follows that as soon as these layers are dry, no effect will be produced.'

The pile when consisting of twenty pairs of plates or more would give shocks and produce a spark and he referred to a pile with sixty plates giving shocks 'as high as the shoulder'. And again, in announcing what is now so well known as his *Couronne de Tasses*, he described 'an apparatus in which the fluid is interposed between the metals without being absorbed in a spongy substance. This consists of a number of cups or goblets, of any substance except metals, placed in a row either straight or circular, about half filled with a saline solution, and communicating with each other so as to form a kind of chain, by means of a sufficient number of metallic arcs or bows, one arm of which is of silver, or copper plated with silver, and the other of zinc. The ends of these bows are plunged into the liquid in the same successive order, namely, the silver ends being all on one side, and those of zinc on the other,—a condition absolutely necessary to the success of the experiments.'

The Abstracts in *Philosophical Transactions* for 1800[3] say: 'at the close of the paper the author points out the striking analogy there is between this apparatus and the electric organs of the torpedo and electric eel, which are known to consist of membranaceous columns filled from one end to the other with a great number of laminae or pellicles, floating in some liquid which flows into and fills the cavity. These laminae cannot be supposed to be excited by friction, nor are they likely to be of an insulating nature; and hence these organs cannot be compared either to the Leyden phial, the electrophore, the condenser, or any other machine capable of being excited by friction. As yet, therefore, they can only be said to bear a resemblance to the apparatus described in this paper. The effects hitherto known of this apparatus, and those which there is every reason to expect will be discovered hereafter, are likely it is thought, to open a vast field for reflections and inquiries, not only curious but also interesting, particularly to the anatomist, the physiologist, and the physician.' Today we may add 'and to the Electrical Engineer'.

It is interesting to note in Volta's description of his observations and experiments the mental process by which he detached electricity from the static form which had held sway in man's attention for many centuries. Hesitatingly he clung to the Leyden jar, the electrophorus and the frictional machine. Shocks and sparks were the means of recognition and yet he found something remarkable in the laminated structure of the electric eel and the rebuilding up of its charge from within after a shock had been given. Suspecting that he was launching out on to a vast new world of electrical knowledge, he yet could not fully realize the extent to which his discovery had given mankind one of its greatest boons, the electric current.

REFERENCES

1. Paul F. Mottelay, *Bibliographical History of Electricity and Magnetism.*
2. Ibid., p. 210.
3. *Phil. Trans.*, Part II, 1800, p. 408.

CHAPTER II

Electro-Chemistry, the First Fruits

Although there was no continuous electric current before Volta's Pile, the rapid discharges of static electricity had produced currents which lasted for short periods, in most cases for a small fraction of a second only. The passing of the discharge through fine wire had been observed to fuse the wires and relatively robust sword blades were heated, particularly at the point where the discharge had been crowded into a small cross section. In 1772 the lightning conductor rods on St. Paul's Cathedral, consisting of iron bars four inches broad and half-an-inch thick, became red hot during a thunderstorm. Eleven years earlier, in 1761, Ebenezer Kinnersley, already referred to, had delivered what was probably the first public lecture ever given on the subject of electricity. To an audience assembled in the famous Faneuil Hall in Boston he spoke of electricity as 'a subtile fluid which does not take perceptible time to pass', while in 1769, John Cuthbertson, an English instrument-maker, in an interesting book on electricity and magnetism, gave the results of experiments in which lengths of iron wire were fused by the passage of the current from a battery of Leyden jars.

The idea of electric conduction through a liquid took form and both electrical decomposition and synthesis of water were observed. In 1797, that is three years before Volta's announcement of his pile, a remarkable set of experiments was carried out by George Pearson, an English physician. Using the discharge from Leyden jars he decomposed water into its constituent gases, hydrogen and oxygen, and then, by passing a spark through a vessel containing a mixture of the gases, caused them to reunite and become water again.

Thus the arrival of the voltaic pile coincided with a climate of scientific thought which immediately suggested important applications. Within a few months of Volta's announcement, two Englishmen, William Nicholson, a chemist who was famous for the scientific journal bearing his name, and a surgeon named Carlisle, were experimenting with a pile consisting of seventeen half-crowns alternating with copper discs and cloth separators soaked in brine when they observed that gas was set free in water in which the two wires connected to the pile were dipped. Collecting the gases they found that the negative wire produced a gas which burned in air and the positive one a gas which supported combustion. They also observed when platinum wires were used, that the volume of hydrogen was twice that of the oxygen and rightly concluded that they had decomposed water by means of this new steady current. Again, William Cruickshank working at Woolwich, made up a large pile of zinc and silver plates and showed that the hydrogen came from the wire connected to the silver end and oxygen from the zinc end.

As the news of Volta's discovery spread throughout the scientific world many experimenters took up the subject. In October 1801, Dr. van Marum, a Dutch scientist, at the request of Volta constructed a pile of one hundred and ten pairs of very large plates, alternately copper and zinc, under the auspices of the Teylerean Society and with this powerful device he fused wires, decomposed water and carried convincingly the former tentative ideas of static discharge effects into this new realm of the continuous current. Wollaston, also, about the same time, established the similarity of the chemical effects of static and voltaic electricity. There was one man, however, Humphry Davy, who above all others took Volta's contribution at this stage and welded it into a mighty foundation on which the first steps were taken to rear the great edifice of electrical engineering and electro-chemistry.

In an excellent biography written only a few years after his death by his brother, Dr. John Davy, who was twelve years his junior, we have a strikingly frank picture of a remarkable young man.[1] Born in Penzance on 17 December 1778, of an expert wood carver turned farmer, Humphry does not seem to have made much of a mark while at school, spending a good deal of his time on outdoor sports, but at sixteen, on the death of his father, his character seems to have developed rapidly. He was apprenticed to a surgeon-apothecary and, from the notebooks which he wrote at the time, we see a picture of a broad and expanding intellect. He read widely and wrote much, both in prose and verse, and it is evident that, in a quiet way, he had benefited enormously from the classical studies of his school years. Mr. Borlase, with whom he was apprenticed, was a man of culture and had considerable influence on Davy, whose studies were guided along many new channels. He devoted a year to mathematics and then turned to metaphysics, covering a wide range of philosophical works. At the age of nineteen he became interested in chemistry and read assiduously Lavoisier's *Elementary Chemistry* and Nicholson's *Dictionary of Chemistry.* It was without doubt the former, with its clear and logical exposition, which led him on to chemical experiments on his own account, and within a few months he had attracted the attention of several local notabilities as well as others who were destined in their intellectual powers to achieve national fame. A son of the famous engineer, James Watt, an educated young man, was also a personal friend.

Mr. Borlase, the surgeon-apothecary, was so impressed by the outstanding ability of his apprentice that when Davy was offered a post in the Pneumatic Institution at Bristol he agreed to cancel the indentures. The Institution, founded by voluntary subscription for the purpose of investigating the physiological effect produced by inhaling various gases, quickly gave Davy new opportunities for employing his passion for research. One gas only, nitrous oxide or laughing gas, seems to have proved a most profitable field, both in the method of preparation and in its application as an anaesthetic. Characteristic of the man, Davy carried out many of the experiments on himself, fully appreciating the dangers, but fortunately he came to no harm and, throughout the period, wrote much in the way of notebooks and essays on a wide variety of subjects, including philosophy and poetry.

In January 1801 Davy's reputation had reached Count Rumford at the Royal Institution and he was offered a post as assistant lecturer with the prospect of becoming, as he did the following year, the Professor of Chemistry at a salary of £500. In a letter to his mother at this time he told her he was asking for the specific terms of the appointment 'when I shall determine whether I shall accept it or not', and says 'I will accept no appointment except upon the sacred terms of *independence.*' At the age of twenty-three on 25 April 1801, only a few weeks after taking up the appointment, he delivered his first lecture for which he chose the subject Galvanism.

There was some controversy between the biographers regarding the way in which Davy acquitted himself on this first appearance before a fashionable London audience. Dr. Paris wrote of his 'uncouth appearance and address, of a smirk in his countenance and a pertness in his manner' while an account in the *Philosophical Magazine* describes how the audience, which included Count Rumford, Sir Joseph Banks and other distinguished philosophers were 'highly gratified and testified their satisfaction by general applause'. 'Mr. Davy', it said, 'who appears to be very young, acquitted himself admirably well. From the sparkling intelligence of his eye, his animated manner, and the *tout ensemble*, we have no doubt of his attaining distinguished excellence.' Another writer of the time said: 'Though his manners were retreating and modest, he was generally thought naturally graceful; and the upper part of his face was beautiful. I remember when he first lectured at the Royal Institution, the ladies said "Those eyes were made for something besides poring over crucibles".'

Whatever the personal impressions made on individuals it is clear that this first course of lectures created a sensation and was the beginning of a remarkable series through which Davy achieved in ten years a unique international reputation. His discoveries advanced the understanding and appreciation of the newly found steady electric current and his carefully prepared and rehearsed lectures assumed a brilliance which attracted large fashionable audiences and established the reputation of the Royal Institution for all time.

For our present purpose the subjects of particular interest in Davy's work were the development of the carbon arc and the decomposition of water and chemical compounds. He and others had produced small sparks between electrodes with the voltaic current in the first years of the century. The *Philosophical Magazine* for February 1801 contains, for instance, a description by a Dr. H. Moyes of experiments with the voltaic pile which reads as follows: 'When the column in question had reached the height of its power, its sparks were seen in daylight, even when they were made to jump with a piece of carbon held in the hand.' Similar observations were made in Germany about the same time, and in March 1802 the *Journal de Paris* contained a reference to the experiments of Citizen Robertson at the Paris Galvini Society who had 'mounted metallic piles to the number of 2,500 zinc plates and as many of rosette copper. We shall forthwith speak of his results, as well as of a new experiment that he performed yesterday with two glowing carbons. The first having been placed at the base of a column of 120 zinc and silver elements and the second communicating with the apex of the pile they gave at the moment they were united a brilliant spark of an extreme whiteness that was seen by the entire Society.'

Davy's first reference to the arc seems to be that recorded in the *Journal* of the Royal Institution for 1802 which reads: 'When instead of metals, pieces of well-calcined carbon were employed the spark was still larger and of a clear white.' It was not for some years, however, that Davy demonstrated the really spectacular effect of the carbon arc. The Royal Institution possessed a battery constructed by Cruickshank of Woolwich in which a bitumenized wooden trough had depressions filled with diluted acid into which were dipped the bimetallic elements each consisting of rectangular plates of zinc and copper soldered together. This battery became worn out by Davy's many experiments, however, and in July 1808, as the result of a special subscription organized by the Managers, Davy was able to install a new one which was the most powerful of its kind in existence. It had 2,000 double plates of zinc and copper arranged in 200 groups, each of 10 cells, the electrolyte being 108 parts of nitric acid and 25 parts of sulphuric acid in 1168 parts of water. With this new tool Davy made the first public display of the electric arc. The following is an extract from his brother's biography describing the occasion.

'When pieces of charcoal about an inch long and one-sixth of an inch in diameter were brought near each other (within the thirtieth or fortieth part of an inch), a bright spark was produced and more than half the volume of the charcoal became ignited to whiteness, and by withdrawing the points from each other, a constant discharge took place through the heated air, in a space equal at least to four inches, producing a most brilliant ascending arch of light, broad and conical in form in the middle. When any substance was introduced into this arch it instantly became ignited; platina melted as readily on it as wax in the flame of a common candle; quartz, the sapphire, magnesia, lime, all entered into fusion; fragments of diamond, and points of charcoal and plumbago, rapidly disappeared, and seemed to evaporate in it even when the connection was made in a receiver exhausted by the air pump; but there was no evidence of their having previously undergone fusion.' Thus was laid the foundations of the first application of the electric arc both to illumination and to the melting of metals, though, as we shall see later, several decades passed after Davy's demonstration, and much inventive genius displayed, before the carbon arc lamp became a practical form of illuminant. Much longer was taken to achieve the development of the electric furnace which today is such a conspicuous application of the electric current.

Davy's outstanding contribution to electrical science was in the field of electro-chemistry. Immediately on taking up his appointment at the Royal Institution he started on a series of experiments which continued for eight fruitful years, culminating in the most brilliant of his discoveries that what were known as fixed alkalies were combinations of metals and oxygen. The best account of this period is to be found in his two Bakerian Lectures delivered before the Royal Society in 1806 and 1807; in the former, entitled 'On Some Chemical Agencies of Electricity', he traced the origins to the discoveries of the electrical decomposition of water by Nicholson and Carlisle, followed by the work of Cruickshank on the electrolysis of the chlorides of magnesia, soda and ammonia with the important deduction that alkaline matter always appeared at the negative and acid at the positive pole. He described how, in 1802, he conceived the idea that all chemical decomposition might be polar and how, a few years later, to use his own words, 'I drew the conclusion "that the combinations and decompositions by electricity were referable to the law of electrical attractions and repulsions"; and advanced the hypothesis "that chemical and electrical attractions were produced by the same cause, acting in one case on particles, in the other on masses" and that the same property, under different modifications, was the cause of all the phenomena exhibited by different voltaic combinations.'

In an early notebook Davy recorded: 'If chemical union be of the nature which I have ventured to suppose, however strong the natural energies of the elements of the bodies may be, yet there is every probability of a limit to their strength; whereas the powers of our artificial instruments seem capable of indefinite increase.' Not only had he, by his remarkable mental acuity, founded electro-chemistry, a vast scientific and engineering field, but had accurately predicted the future possibilities of the electric current from the simple and inadequate batteries then available.

The report of progress in his first Bakerian Lecture was widely and authoritatively acclaimed as an epoch-making contribution to chemistry and, even during the war then raging between England and France, the Institute of France awarded him the prize founded by Napoleon for the most important discoveries in Galvinism.

Davy's most brilliant success was, however, that recorded in his second Bakerian Lecture on the decomposition of the fixed alkalies which occupied 44 pages crowded with experimental results and mature observations. Although delivered before the Royal Society within a few weeks of his making the discovery and at a time when his health was rapidly deteriorating, it shows no signs of haste or careless compilation.

Davy had found that the alkali metals, potassium and sodium, could not be separated by the electrolysis of their solutions in water; the water only was decomposed. He knew that rocks could be broken down electrically and decided to pass a current through the vegetable alkali potash in a molten state. After raising the temperature of the mass with a blow-pipe and connecting the wires from his large battery he was delighted to find that at the negative wire there appeared a curious flame which was followed, as the current was increased, by the appearance of brilliant metallic globules. His simple note, dated October 6th (1807) read as follows:

'In my first trials on potash I used strong aqueous solutions.

'Dry potash is a non-conductor: I then employed fused potash; and, in this instance, inflammable matter was developed.

Experiments

'Then a piece of potash moistened; and, to my great surprise, I found metallic matter formed.

'October 6th—This matter instantly burnt, when it *touched water* —swam on its surface, reproducing potash.

Instance

'Soda was decomposed in the same manner.'

Davy's brother biographer records:[1] 'When he saw the minute globules of potassium burst through the crust of potash, and take fire as they entered the atmosphere, he could not contain his joy—he actually danced about the room in ecstatic delight; and some little time was required for him to compose himself sufficiently to continue the experiment.'

Striking as was the discovery of the two metals potassium and sodium what was even more important was the way in which it opened up the discovery of the remaining alkalis and earths which so far had resisted decomposition. On recovering from the serious illness which followed this period of extreme mental and physical activity, Davy characteristically produced and published a long poem and then set himself to extend the field which had opened up so promisingly. In a short time he had separated and given names to barium, strontium, calcium, and magnesium in addition to potassium and sodium.

In subsequent chapters we shall consider in detail the life and work of Michael Faraday, who joined Davy at the Royal Institution and ultimately succeeded him as Professor. For continuity of treatment, however, we must anticipate at this stage one aspect of his investigation which is relevant to the subject of electro-chemistry. Among his many chemical activities Faraday took up the observation made by Wollaston in 1801 on decomposing water both by sparks from frictional electricity and by the voltaic current, so identifying for the first time the two forms. After many experiments of a quantitative nature Faraday was able to lay down what he called in one of his famous numbered notes (SS 557) the 'doctrine of definite electro-chemical action'.[2] 'Hence it results', he wrote in SS 112 and 113, 'that both in magnetic deflection and in chemical force, the current of electricity of the standard Voltaic battery for eight beats of the watch was equal to that of the machine evolved by thirty revolutions'; and again, 'It also follows that for this case of electro-chemical decomposition, and it is probable for all cases, that the chemical power, like the magnetic force is in direct proportion to the absolute quantity of electricity which passes.'

Faraday at this stage gave names to various components in the operation of electrolysis which have become standard nomenclature. He called all bodies which are decomposable by the electric current *Electrolytes* and substances into which they divide *Ions*. The part of the surface of the decomposing matter in contact with the current supply he called an *electrode* and differentiating between entrance and exit of the current he gave the name *anode* to that connected with the positive pole and *cathode* to the part in contact with the negative pole. His masterly summing up in notes 559 onwards are almost in the form of a religious creed and well worth recording here at some length.

'559. . . . I have proposed to call these bodies generally *ions*, or particularly *anions* and *cations*, according as they appear at the *anode* or *cathode*; and the numbers representing the proportions in which they are evolved electro-chemical equivalents. Thus hydrogen, oxygen, chlorine, iodine, lead, tin are *ions*; the three former are anions, the two metals are cations, and 1, 8, 36, 125, 104, 58, are the *electro-chemical equivalents* nearly.'

And again:

'562.ii. If one *ion* be combined in right proportions with another strongly opposed to it in the ordinary chemical relations, i.e. if an *anion* be combined with a *cation*, then both will travel, the one to the *anode*, the other to the *cathode*, of the decomposing body.

'563.iii. If, therefore, an *ion* pass toward one of the electrodes, another *ion* must also be passing simultaneously to the other electrode, although, from secondary action, it may not make its appearance.

'564.iv. A body decomposable directly by the electric current, i.e. an electrolyte, must consist of two *ions*, and must also render them up during the act of decomposition.

'565.v. There is but one *electrolyte* composed of the same two elementary *ions*; at least such appears to be the fact dependent upon a law, that *only single electro-chemical equivalents* of elementary *ions* can go to the *electrodes, and not multiples*.'

In subsequent paragraphs Faraday continued his thoughtful analyses, and as Tyndall said in his *Faraday as a Discoverer*, written nearly a century ago: 'From all these difficulties emerged the golden truth, that under every variety of circumstances the decompositions of the Voltaic current are as definite in their character as those chemical combinations which gave birth to the atomic theory. This law of Electro-chemical Decomposition ranks in point of importance with that of Definite Combining Proportions in Chemistry.'

For clarity of analysis, precision of observation and exposition, Faraday's conduct of this series of experiments must rank for all time among the highest in the history of scientific research.

REFERENCES

1. Dr. John Davy, *Memoirs of the Life of Sir Humphry Davy*. London, 1836.
2. M. Faraday, *Experimental Researches in Electricity*. 3 vols. Reprinted London 1839–55.

See also: James Kendall, *Michael Faraday, Man of Simplicity*. Faber, 1955.

CHAPTER III

The Primary Cell

As we have seen in discussing Davy's experiments, Volta's discovery was quickly followed by the construction of simple copper-zinc-acid cells. William Cruickshank, the Woolwich experimenter, employed an improved pile in 1800 and a few years later Pepys, the son of an English surgical instrument-maker, constructed the strongest one so far made. It had sixty pairs of zinc and copper plates, each six feet square. Following on with various modifications he finished in 1808 the enormous battery of 2,000 plates for the Royal Institution which had a total area of 128,000 square inches. Improvements were effected in the form of container and devices constructed for raising the plates out of the acid when the battery was not in use. Wollaston devised what, at the time, was considered a greatly improved arrangement of the two plates. To reduce the wastage of the zinc and increase the output he doubled up the copper, so exposing it to both sides of the zinc, securing electrical separation by using strips of cork or wood. As a careful workman Wollaston took pride in constructing miniature apparatus and made thimble-sized cells. With these he showed how it was possible to fuse some of the very fine platinum wires for which he also became famous.

About 1822 Pepys introduced a further major modification. He constructed a large spiral cell which, because by its use he was able to produce great heat, he called a calorimotor. It consisted of two strips of copper and zinc, two feet wide and fifty feet long, which were laid together with horsehair and rope separators and then coiled around a wooden cylinder. This assembly was suspended by a pulley by which it could be immersed in a tub containing 55 gallons of 40 to 1 strong nitrous acid in water.

The two decades following Volta's epoch-making contribution at the end of the eighteenth century had resulted in great experimental activity throughout the scientific world but interest was waning.

Davy's researches seemed to provide a culmination to the enquiry and a writer of the time, John Bostock,[1] an English physician, summed up the position as follows:

'We have carried the power of the instrument (the steady current) to the utmost extent of which it admits, and it does not appear that we are at present in the way of making any important additions to our knowledge of its effects or of obtaining new light upon the theory of its action.'

But Bostock was wrong. In the first place the weaknesses of the voltaic cell as the practical source of electric current were to be removed by a number of ingenious devices which were destined to make it a reliable engineering tool for an ever-widening range of application. Moreover, within almost a matter of months of Bostock's doleful prognostication, the first steps were to be taken in the development of an alternative source of current which would open the way for even more revolutionary large-scale applications. This part of the story must be postponed to our next chapter. In the meantime the rapidly increasing use of the simple primary cell, as the only available source of supply of the electric current, soon brought to light two important weaknesses which became known as *local action* and *polarization*. The former of these resulted in deterioration of the cell, even while standing out of use, through disintegration of the zinc plate, and the latter produced a gradual reduction in the output during the use of the battery in supplying current. The provision of remedies to overcome these defects occupied many years and for some time the sole practical remedy was intermittent operation, for which mechanical devices were introduced for lifting the plates out of the acid except for the actual time current was required.

Local action was soon traced to impurities in the zinc plate which caused local currents to pass in the electrolyte between one part of the surface and another. This current was not, of course, available for use in the outside circuit but it did consume energy and was wasteful. A small inclusion of impurity, for instance iron, would be electronegative to the main body of the zinc plate and a current would flow locally through the electrolyte from the zinc to the particle of iron impurity so carrying zinc into solution and eroding the electrode.

It was a quarter of a century before a practical remedy was found to combat local action and when it came it proved to be extremely simple and effective, so much so that it has remained in commercial use up to the present day. In 1828 Kemp and Sturgeon[2] found that amalgamating the zinc plate with mercury overcame the difficulty.

They rubbed the mercury on to the zinc plate with an acid-impregnated rag, so producing a clean bright surface which was not attacked by the electrolyte even when the underlying zinc contained particles of impurity.

While local action was so quickly and simply disposed of, polarization was a much more complicated problem. Users of the simple zinc-copper cells had noticed that bubbles of hydrogen gas collected on the surface of the copper plate and that brushing off the bubbles restored the action of the cell after it had fallen off in use. The reduction in current was, of course, due to two reasons, the interposition in the circuit of the high electrical resistance of the gas film, and the back E.M.F. which resulted from the hydrogen facing the zinc in the electrolyte.

The first attempt at overcoming polarization was made in 1829 by Becquerel,[3] but no practical form of cell appeared until 1836, when Professor Daniell[4] described the first self-depolarizing cell which has made his name famous.

Daniell adopted the ingenious two-fluid principle which was later to be followed by other inventors in different form. He immersed the zinc plate in dilute sulphuric acid and the copper electrode in a separate solution of copper sulphate. The two solutions were brought into contact with one another through a porous separator which in practice was an unglazed earthenware pot containing the copper plate and copper sulphate solution. With this construction the current in its passage from the zinc to the copper, instead of releasing the objectionable free hydrogen gas, caused it to combine at the separator with the copper sulphate so releasing metallic copper which was deposited from the copper sulphate, harmlessly onto the copper plate. The zinc plate was in the form of a cylinder standing in dilute sulphuric acid contained in an outer vessel and surrounding the porous pot.

Owing to defects in the Daniell cell various investigators continued their search for a better construction and during the following year many types of cell appeared with different names associated with them. In one of these, the Grove cell,[5] put forward in 1839, the outstanding feature was the use of a platinum electrode immersed in strong nitric acid contained in a porous pot which separated it from the zinc element standing in its own weak sulphuric acid. In this form of double liquid cell the hydrogen when produced was not replaced by a metal, which could be deposited harmlessly, but was allowed to form at the platinum electrode where it was immediately oxidized by the strong acid to form water. At a later period the platinum of the

Grove cell was replaced, for economic reasons, by carbon and became the equally well-known Bunsen cell.[6]

The simplest of all depolarizing methods put forward at this period was that adopted in the Smee cell. In 1840 Smee[7] constructed a single liquid cell in which he employed a plate having a roughened surface so that the bubbles of hydrogen would not adhere. To obtain the form of surface necessary platinum was deposited on silver in a finely divided state. Owing to its simple construction the Smee cell remained in practical use to a restricted extent for many years but the bichromate cell, another single liquid cell, achieved a wider popularity. The electrolyte employed was bichromate of potash in sulphuric acid or, alternatively, chromic acid. The usual and well-known characteristic form of the bichromate cell was a cylindrical glass bottle with a parallel upper portion and a bulbous lower end. An ebonite cap supported two long flat carbon electrodes between which a zinc plate carried by a metal rod could be raised or lowered at will. In this cell the offending hydrogen is eliminated by reduction of the chromic acid with the characteristic change in colour of the solution.

With the availability of modern supplies of electric current from mechanically driven generators it is difficult to assess the important position occupied by the primary cell about the middle of the nineteenth century. The uses for which the current was required extended beyond the development of the cell as a source of current and a good illustration of the position appears in a statement by an American Professor published in 1891.[8] 'Before the introduction of dynamo-electric machines and the storage battery, forty Grove cells, requiring only seven or eight pounds of nitric acid, served the writer for many years whenever a brilliant arc light was needed or projection experiments in spectrum analysis were performed.'

But the extended use of the primary cell has not been as a source of heavy current; that duty has been taken over by the outcome of electro-magnetic induction. Primary cells have continued rather to be a source of interest to the inventor for their invaluable applications in the field of light-current engineering. Among the many types proposed the one which in its various forms has most successfully held the field since 1868 is the Leclanché cell, the most characteristic feature of which is the employment of a solid depolarizer, manganese dioxide. The active plates are zinc and carbon and the electrolyte a solution of sal-ammoniac (ammonium chloride).

The Leclanché cell has appeared in many forms, the most common of which is the popular well-known bell battery. In this type the carbon plate is packed around with the depolarizer in a cylindrical porous pot which stands in a square glass container containing the sal-ammoniac solution. The zinc electrode is in the form of a rod which is housed in a suitably shaped corner of the glass vessel. Another characteristic feature in the appearance is a band of black compound around the upper edge of the cell to prevent creeping of the electrolyte.

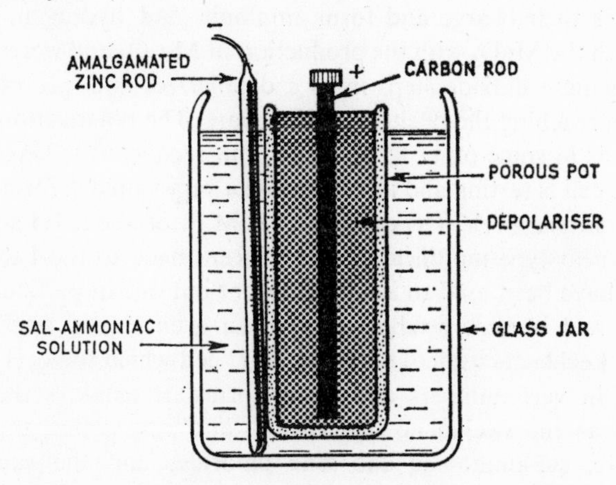

Fig. 1. Leclanché Cell. This cell employs a solid depolarizer of manganese dioxide. Invented in 1868, it has been the most popular of the many primary cells originating from that period.

In relating the whole history of electrical engineering not too much emphasis can be placed on the importance of the Leclanché cell as a factor in its progress. Invented ninety years ago, it has been, and still is, employed in an extensive variety of forms not only as a liquid cell but in the form known as dry cell, though 'unspillable' would be a more exact description.

As the Leclanché cell is a single liquid cell the porous pot is provided only for the purpose of holding the solid depolarizer and keeping it in contact with the carbon electrode. To improve the contact the manganese dioxide is mixed with grains of carbon and in an alternative form of construction a canvas container or pack is employed. Much research has been carried out on the Leclanché cell since its first introduction on such matters as the shape, size and source of the manganese dioxide and carbon grains. The purity of the amalgamated zinc rod, method of formation to increase life by reducing local action and excessive corrosion at the surface of the electrolyte and consequent loss of unused zinc are problems which have had much attention.

The components of the Leclanché cell are chemically very simple and the operation easily defined. The chlorine in the ammonium chloride combines with the zinc forming zinc chloride while the ammonium (NH_4) passes through the porous pot as charged ions. These lose their charge and form ammonia and hydrogen, which reacts with the MnO_2 with the production of Mn_2O_3 and water. Then the manganese dioxide steps in as a depolarizer and prevents the hydrogen reaching the positive carbon plate. The polarization is not so rapid as in some other forms of cell but recuperation takes place when the cell is resting and it is therefore a very valuable form of cell for intermittent work. The cell has an E.M.F. of about 1·5 volts.

Many near-type modifications have been made to the Leclanché cell and have been used to a limited extent but one in particular has emerged and become established with an even greater application than the Leclanché itself. This is the 'dry' cell which today is manufactured in vast numbers and many forms. It employs the same elements as the Leclanché cell, carbon and zinc plates, the same electrolyte, sal-ammoniac and zinc chloride, and the same depolarizing agent, manganese dioxide. Instead, however, of a liquid electrolyte in an open container the dry cell has a stiff paste which is unspillable. In the usual form, as employed for torch batteries and high tension radio batteries, the zinc plate forms the container and the negative terminal is attached to it, or is in rubbing contact with it. The positive terminal, usually protruding in the form of a brass post through the sealing compound at the upper end of the cell, is connected to a carbon rod packed in the polarizing element, which is a paste containing manganese dioxide, powdered sal-ammoniac, carbon, zinc, chloride and water. The exciting electrolyte between this and the zinc enclosing vessel is usually a white paste consisting principally of sal-ammoniac, zinc chloride and a porous matrix such as flour. Many minor modifications of detail appear in dry cells made by different manufacturers such as the use of sawdust, cardboard pads and the provision of ventilating tubes.

Historically the first gropings for a dry cell go back to within a few years of Volta's discovery. As early as 1803 Professor Hachette, of the Ecole Polytechnic in Paris, gave before the Institut Nationale[9] a description of the dry pile which he had constructed in conjunction with Desormes, a French scientist and manufacturer of chemicals.

Stöhrer's Magneto-electric Machine, 1843
ıe of several machines invented about this time in which coils were rotated near to the poles permanent magnets. Stöhrer, of Leipzig, introduced the multipolar principle with three rse-shoe magnets and six coils.
hoto: Science Museum)

(*b*) Wilde's Dynamo
This machine was separately excited by a small subsidiary dynamo.
(*Photo: Science Museum*)

Holme's Magneto-electric Machine, 1867
Holmes had made several machines of this type during the previous ten years. This one, whic was installed in the Souter Point Lighthouse in 1871 and ran for many years, is now in t Science Museum at South Kensington. It has 56 permanent magnets carried by the iron fran and 96 rotating coils. It ran at 400 revs. per minute, taking 3·2 horse power.
(*Photo: Science Museum*)

They made up what was virtually a voltaic pile but substituted, for the wet discs of porous material between the copper and zinc plates, a starch paste with various salts, gums, etc. Many constructions were proposed and tried during the following years[10] but it was not until after the advent of the Leclanché cell that any real progress was made and even then it was some twenty years before Hellesen, in 1887, produced the first really successful dry cell.[11] He used a carbon rod, and filled the space between the rod and the zinc containing case with a mixture in plaster of zinc oxide, sal-ammoniac, chloride of zinc and water. The object of the zinc oxide was to make the composition loose and porous to facilitate the movement of gases. Modifications of ingredients continued until gradually the form of the present-day dry cell evolved.

Although to do so means jumping ahead of our main story, it is logical here to trace at this stage the development of the standard voltaic cell which today is such a valuable instrument of precision in providing a reproducible standard of electrical voltage for measurement purposes. Quite early the need for such a standard became evident and the characteristics of the different types of cell as they developed were examined to determine the extent to which they could be relied upon to give a reproducible steady voltage.

In 1863 Raoult carried out a series of researches on the Daniell cell from the standpoint of adapting it as a standard for measurement purposes. His conclusions were no doubt as reliable as the means at his disposal permitted, but would scarcely stand up to examination by modern methods. In 1882 Kittler gave the name 'normal element' to a Daniell cell consisting of chemically pure copper and amalgamated zinc plates in a solution of copper sulphate and sulphuric acid of specified densities and at stated temperatures. Other electrical engineers tried alternatives and about the same time the British Post Office, to whom the question of a standard voltage was assuming importance, laid down a specification for a standard Daniell cell built up in three separate vessels. In this and in an improved two-compartment model the principal feature seems to have been the provision of steady standing conditions for the porous cell and quick assembly when tests were required. The E.M.F. was 1·07 volts. Other attempts at standard cells were made by Sir William Thomson, Lodge, Beetz, and others, but the accuracy of temperature coefficient attained seems to have been only of the order of 0·02 to 0·05 per cent per degree centigrade.

The first standard cell worthy of the name was that described for the first time by its author, Latimer Clark, in a paper to the Royal

Society on 19 June 1873.[12] It consisted originally of a tubular glass cell in the bottom of which was a pool of mercury forming one electrode. The mercury was covered with a paste formed by mixing mercurous sulphate in a saturated solution of zinc sulphate and the zinc plate in the form of a rod of the pure metal dipped into the paste.

In 1894 an Order in Council rendered legal certain electrical standards, those of resistance related to the dimensions and weight of a column of mercury and those of current by deposition of silver, were adopted internationally in 1908. At the earlier date legal status was given to the Clark cell as a standard of voltage. The volt was defined as that electrical pressure which, if steadily applied to a conductor whose resistance is one ohm will produce a current of one ampere. It is represented by $\frac{1000}{1434}$ of the electrical pressure at a temperature of 15°C. between the poles of the voltaic cell known as Clark's cell set up in accordance with a formal official specification.

In the Board of Trade Standard Clark cell contact was made with the positive mercury electrode by means of a platinum wire sealed through the lower end of a narrow glass tube which protects the rest of the wire as it is brought up through the cell to form the positive terminal.

It was discovered that the original Clark cell had certain irregularities in performance which made it unsuitable as a legal standard and Lord Rayleigh undertook a thorough investigation.[13] He found that acidity in the paste and under-saturation of the zinc sulphate solution would result in the E.M.F. of the cell being too high. Dryness of the cell and supersaturation of the solution brought the E.M.F. down, as did impurities in the mercury, so he suggested improved sealing by the use of marine glue. Modifications in methods of preparation of the ingredients also improved the temperature coefficient.

A major change in general form of the cell was introduced by Lord Rayleigh with a view to mitigating the effects of uncertainty in the contact between the zinc and the paste. The zinc could dip either into clear solution, into a layer of crystals on the surface of the paste or well into the paste itself. To obviate this difficulty he adopted the H form of cell now used so extensively. Two small test tubes are connected by a cross tube and a platinum wire is sealed into the bottom of each. One tube contains mercury and the other the zinc amalgam. Above the mercury is the mercurous sulphate and a saturated solution of zinc sulphate fills the double tube up to and above the cross tube.

With all the care possible it was found by Kelvin and others that the Clark cell had a high temperature coefficient but in 1892 Mr. E. Weston[14] invented a standard cell which has a temperature coefficient of about 0·004 per cent per degree C. The Rayleigh H form construction is adopted and the following is a description of the cell as made by a modern manufacturer.[15]

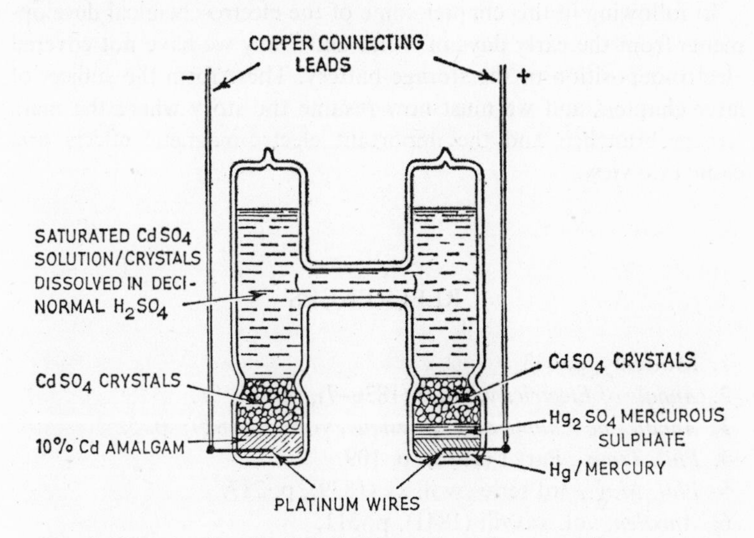

Fig. 2. Weston Standard Cell. Now adopted universally as a standard of electromotive force. The 'H' form is designed to prevent movement of crystals.

In one of the lower limbs is a layer of mercury covered by a paste of mercurous sulphate and a layer of cadmium sulphate crystals. The opposite limb contains cadmium amalgam covered by cadmium sulphate crystals and the cell is filled above the cross tube with a saturated solution of cadmium sulphate. In each of the upright tubes a constriction is formed level with the top of the layer of cadmium sulphate crystals. These crystals become loosely cemented together, and owing to the presence of the constriction, form taper plugs which hold the contents of the cell in their proper places and greatly increase its portability.

Two such cells are often assembled together as a check on one another in a case provided with a thermometer, and platinum wires, sealed into the bases of the two pairs of limbs, are brought out to four terminals on the top of the case. The case can be filled with oil to reduce temperature gradients and a typical range of voltage change for such a cell on test is as follows:

15° C.	..	1·01877
20° C.	..	1·01860
25° C.	..	1·01838

In following in this chapter some of the electro-chemical developments from the early days of Volta and Davy we have not covered electro-deposition or the storage battery. These form the subject of later chapters and we must now resume the story where the main stream branched and the important electro-magnetic effects first came into view.

REFERENCES

1. Mottelay, p. 443.
2. *Annals of Electricity*, vol. i (1836–7), pp. 81, 88.
3. *Annales de Chimie et de Physique*, vol. xli (1829), p. 5.
4. *Phil. Trans.*, Part I (1836), p. 109.
5. *Phil. Mag.*, 3rd series, vol. xv (1839), p. 287.
6. *Annalen*, vol. xxxviii (1841), p. 311.
7. *Phil. Mag.*, 3rd series, vol. xvi (1840), p. 315.
8. Henry S. Carbart, *Primary Batteries*, Boston, 1891.
9. Mottelay, p. 375, and *Annales de Chimie et de Physique*, pp. 76 etc., May 1816.
10. *The Electrician*, vol. xlii (1889), p. 185.
11. Park Benjamin, *The Voltaic Cell*, p. 202.
12. *Phil. Trans.*, 1874.
13. *Phil. Trans.*, Part II, 1885.
14. *The Electrician*, vol. xxx (1893), p. 741.
15. Cambridge Instrument Co., London.

CHAPTER IV

Electro-Magnetism Emerges

Although the first two decades of the nineteenth century saw remarkable developments in the evolution of the steady current, in its production and in its chemical and heating effects, there was at first no hint of the major part it was to play as the main basis for electrical engineering through electro-magnetic phenomena.

There had been a suspicion for many years that electricity and magnetism were in some way related. Magnetism had two polarities as had static electricity and the law of inverse squares operated in both spheres. Navigators knew that their compass needles were weakened and could indeed be reversed in polarity by the effects of lightning. In some of his experiments Benjamin Franklin had controlled the action and had magnetized sewing needles by the discharge from a battery of large Leyden jars. Steel knives in a box struck by lightning were found to have become magnetized.

As a result of all this interest in the subject, Volta's discovery of the steady current was born into a climate which turned many thoughts in the same direction—what would be the magnetic effects of this new steady continuous current? Before the real answer was found several experimenters had made claims, one of the earliest being Romagnosi, an Italian, who communicated a paper to the *Gazetta di Trento* in 1802[1] which was interpreted as 'establishing the directive influence of the Galvanic current upon a magnetic needle.' The results seem to have been uncertain, however, and the general view is that any small movement of the needle which Romagnosi, and others who tried similar experiments observed, was due entirely to electrostatic attraction or repulsion and not to the electromagnetic effect of a steady current. In fact there is evidence that their observations of deflections of a magnet by the use of the voltaic cell were obtained with the cell on open circuit! The needle was acting simply as an electrometer—Mottelay, who made a careful analysis of all the evidence, concluded[2] that 'No satisfactory results were in fact obtained until Oersted made his famous discovery which forms the basis of electro-magnetism.' This was in the year 1820, as epoch-making a date as had been 1800, for Oersted's discovery triggered off a spate of electrical progress which has never since diminished.

For the immediate advance of electrical knowledge at this time we are indebted to three great men, Oersted the observer and originator, Ampère the thinker and analyser, and Arago, with his famous disc and other devices which posed further problems. These three laid a great foundation but it was the further brilliant experiments of Michael Faraday which stimulated major progress towards modern electrical engineering.

Hans Christian Oersted (1777–1851) was born at Rudköbing on the island of Langeland in the Baltic. His father was an apothecary and Hans, with his younger brother, who later became a famous jurist and statesman, both possessed remarkable natural gifts and a thirst for knowledge, so that it is not surprising that after instruction in Copenhagen in the classical languages, Hans Christian entered the University there and commenced a career of outstanding brilliance. Like many men of science of that period Oersted occupied himself not only with the study of natural science and medicine but with aesthetics and philosophy. He distinguished himself in these subjects and at the age of twenty-two was created Doctor of Philosophy for his thesis 'Dissertatis de forma metaphysices elementaris naturae externae'. Going to Berlin later he published his *Recherches sur l'identité des forces electriques et chimiques*, in which he laid the foundations of the electro-chemical system. He was appointed to the Chair of Physics at Copenhagen in 1806, a post which he held up to his death in 1851.

Oersted, in spite of his great work for science, never lost his love for the philosophical and in the last work he published, *The Soul of Nature*, he expressed that view of the world which had always sustained him.[3] Nature was a manifestation of the Deity's combined wisdom and creative power. The laws of nature were to him reason's laws and the true and beautiful were but different views of what is rational. His lectures at the University, warm and animated, enraptured his audience and spread widely an interest in nature study among his compatriots. With the same intent he took a lively interest in the founding of the Danish Society for the Advancement of Natural Science and of the Copenhagen Polytechnic Institute, for which he worked untiringly for many years. Madsen summed up his character thus: 'Eminent as a scholar, equally great was he as a man, modest and lenient in his judgment of others, strict with regard to himself, benevolent, always ready to help others with advice and deed; himself truthful in the highest degree, he demanded truthfulness from others as their first duty.'

Oersted visited England several times and was on friendly terms with Davy, Wollaston, Faraday, Wheatstone and others. In 1824 Davy visited him in Copenhagen, and, in his biography, Dr. John Davy[4] quotes some notes which he made on various people he had met on the tour. Of Oersted he wrote: 'Chiefly distinguished by his discovery of electro-magnetism, he was a man of simple manners, of no pretensions, and not of extensive resource; but ingenious, and a little of a German metaphysician.'

This then was the manner of man who finally produced the key to unlock the mystery of electro-magnetism. In a generous tribute paid to Oersted by J. J. Fahie[5] on the occasion of the Faraday Centenary he said *à propos* of the many tentative approaches already made to the question, 'In trying to elucidate these hazy notions, Oersted, by a happy impulse or by chance, closed the battery circuit which hitherto had always been left open, and to his delight he saw the needle move from its position of rest. This occurred at a private lecture to advanced students in the winter of 1819–20. With kindling eyes, it is said, he searched the faces around him, and while announcing the result in a few hurried words he pointed to the apparatus with trembling hands and invited the students to make the experiment for themselves. Thus, at last, was discovered the long sought secret of the connection between electricity and magnetism.'

During the next few months Oersted continued his experiments and in the spring of 1820 he wrote an account of his discovery which was published on 21 July and many copies sent to scientists, societies and publications in Denmark and abroad. The treatise was published in Latin with the title *Expirimenta circa effectum conflictus electrici in acum magneticam*, and an English translation was made for the Society of Telegraph Engineers by Rev. J. E. Kempe Rector of St. James's, Piccadilly.[6]

In this document Oersted first of all describes how he discovered the importance of closing the circuit to obtain any effect—a point not noticed by previous experimenters—and then, as the movement of the needle was feeble, he solicited help from his friend Esmarck, Minister of Justice, to construct a larger battery which he describes. Speaking of what takes place in the conductor and in the space surrounding it as the *conflict of electricity* and with the conductor disposed above and parallel with the needle, the end of the needle next to the negative pole of the battery moves towards the west. At a distance of three-quarters of an inch he obtained a deflection of 45°. With the needle *under* the wire the needle moved in the opposite direction. The report continued to give the results and observations on many different dispositions of the current-carrying wire relative to the needle and contained some shrewd observations which lead to the conception of lines of force in air.

'Electric conflict can only act upon magnetic particles of matter. All non-magnetic bodies seem to be penetrable through electric conflict; but magnetic bodies, or rather their magnetic particles, seem to resist the passage of this conflict, whence it is that they can be moved by the impulse of contending forces. That electric conflict is not inclosed in the conductor, but as we have already said is at the same time dispersed in the surrounding space, and that somewhat widely is clear enough from the observations already set forth.

'In like manner it is allowable to gather from what has been observed that this conflict performs gyrations, for this seems to be a condition without which it is impossible that the same part of the joining wire, which, when placed beneath the magnetic pole, carries it eastward, drives it westward when placed above; for this is the nature of a gyration that motions in opposite parts have an opposite direction.'

Thus was described, in what appears to us rather quaint language, a discovery epoch-making in the history of mankind and one destined to affect the whole of civilized life through the contributions of electrical engineering. The publication produced widespread interest and many experimenters turned at once to this new field of exploration so dramatically offered. Honours were showered on Oersted in all directions including a fellowship of the Royal Society. The frontispiece of the *Philosophical Transactions* for 1821 carried the following paragraph conspicuously displayed: 'The President and Council of the Royal Society adjudged the Medal of Sir Godfrey Copley's Donation, for the year 1820, to Professor John Christian Oersted, of Copenhagen, for his Electro-Magnetic Discoveries', while the same issue[7] carried a communication made by Davy to Wollaston, who was then President, in which he described an interesting series of experiments repeating those of Oersted. After a generous tribute which reads: '. . . the discovery of the fact of the true connection between electricity and magnetism seems to have been received for

M. Oersted, and for the present year', he explains how, using his battery of 100 plates of 4 inches, the south pole of a common magnetic needle placed under the communicating wire (the positive end of the apparatus being on the right hand) was strongly attracted by the wire overcoming the magnetism of the earth.

'I threw some iron filings on a paper, and brought them near the communicating wire, when immediately they were attracted by the wire, and adhered to it in considerable quantities, forming a mass round it ten or twelve times the thickness of the wire; on breaking the communication, they instantly fell off, proving that the magnetic effect depended entirely on the passage of the electricity through the wire.'

He then fastened several steel needles, in different directions, by fine silver wire to a wire of the same metal and placed them in the electrical circuit of a battery; they all became magnetic, '. . . those under the wire (the positive end of the battery being east) had their north poles on the south side of the wire and their south poles on the north side, and thus those placed over, had their south poles turned to the south, and their north poles turned to the north'. On breaking the circuit only those needles carried transversely retained their magnetism.

Davy also anticipated lines of magnetic force for he goes on to say: 'I placed some silver wire of one-twentieth of an inch, some of one-fiftieth in different parts of the voltaic circuit when it was completed and shook some steel filings on a glass plate above them; the steel filings arranged themselves in right lines always at right angles to the axis of the wire.'

Davy's paper contains much further discussion of detail including the application to terrestrial magnetism and in a footnote he says: 'I find by the *Annales de Chimie et de Physique*, for September, which arrived in London November 24, that M. Arago has anticipated me in the discovery of the attraction and magnetising powers of the wires in the Voltaic circuit.—Since I have perused M. Ampère's elaborate treatise on the electro-magnetic phenomena, I have passed the electric shock along a spiral wire twisted round a glass tube containing a bar of steel and I found that the bar was rendered powerfully magnetic by the process.'

Of all those who followed up Oersted's discovery the most active, and the one whose contributions have left the most permanent and extensive mark on the course of electrical engineering, was André Marie Ampère (1775–1836). Ampère was a professor at the College of

France and at the Ecole Polytechnique in Paris. He was also Inspecteur General of the University of Paris for some years, a member of the Académie des Sciences, and, as a scientist of great distinction, frequently referred to as possessing the qualities of a genius.

Ampère was born at Lyons during the troublous times of Louis XVI and Marie Antoinette and was fourteen years of age at the storming of the Bastille. During the *terreur* which accompanied the accusation, condemnation and execution of Louis, Ampère's father went to the guillotine and at eighteen the mind of the young man was seriously deranged through grief at the loss. His father had been a great companion and virtually his only teacher. Finding that he had a bent for mathematics at an early age his father had encouraged him in every way so that at the age of twelve André was studying the works of Euler, Bernouilli and Lagrange with great profit. He had a remarkable memory and astonished all who associated with him by his outstanding intelligence.

At Polemieux, a few miles north of Lyons, there may still be seen the house where Ampère regained his mental composure. The little white house and garden where he had lived as a boy stand in a peaceful countryside, while, in an enclosure nearby, is the beautiful stone monument erected to his memory by friends some years after his death. Here he married in 1799 and moved to Lyons where he obtained a teaching appointment in the Ecole Centrale, a foundation with a distinguished history extending over many centuries.

In 1804 Ampère lost his wife after a few brief happy years and moved to Paris. Here his life's work took shape and for many years he published important papers on mathematical subjects. But it was in 1820 that Ampère made his greatest contribution. On September 11 of this year a member of the Academy of Sciences returning to Paris from Geneva announced the results of Oersted's experiments which at once intrigued Ampère. He commenced that remarkable series of experiments for which he is so justly famous and within seven days—that is within two months of Oersted's original announcement—he communicated what was to be the first of a most important series of memoirs to the Academy of Sciences, Paris. In these he enunciated principles and laws which are today the basis of our everyday electrical engineering in many branches. His little *bon homme* swimming in the wire with the current still gives us the clue for determining the direction of the resulting magnetic field.

In his first memoir Ampère enunciated the law for determining the position taken up by the needle in relation to the position of the conductor carrying the current and its direction of flow. He also showed that the current in its course through the battery had the same effect on the magnetic needle as had the current on the external circuit, thus establishing the conception of a complete circuit. Without actually demonstrating them, for the time was too short, Ampère described a number of pieces of apparatus which he was later able to construct. He predicted that coils of wire carrying current would act as magnets. In analysing the forces on a magnetic needle due to current in a nearby wire he introduced the idea of the astatic needle. In one of his memoirs he states 'When a magnetic needle is withdrawn from the directive action of the earth it sets itself under the influence of the voltaic current in a direction which makes a right angle with the conducting wire and has its south pole to the left of it. If M. Oersted obtained deviations of less than a right angle it was solely because the needle was not withdrawn from the influence of the earth's magnetism, and consequently took up an intermediate position—the resultant of the two forces acting upon it. There are several ways of withdrawing a magnetic needle from the earth's action. A simple one consists in attaching to a stout brass wire—two needles of equal strength in such a manner that their poles are in opposite directions, so that the earth's directive force upon one is neutralized by the action in the opposite direction which it exercises on the other. The needles are so arranged that the lower one is just below the coil of wire and the upper one close above it.'

Ampère had concluded that not only should the electric current have an influence on a magnetic needle but also that conductors carrying current would have an effect on each other.[8]

'Two electric currents attract one another when they move parallel and in the same direction, and they repel one another when they move parallel and in opposite directions.'

He also made it quite clear that these effects were absolutely different from the attractions and repulsions of ordinary electricity. The many detailed conclusions arrived at are too numerous to discuss here but in a treatise entitled 'Exposé des nouvelles découvertes sur l'electricité et le Magnetisme de M. M. Oersted, Arago, Ampère, H. Davy, Biot Erman Schweizer, De la Reve, etc.', which he published in 1822, they are all discussed at length.

Maxwell, later commenting on Ampère's early communications on Oersted's discovery, said it was one of the most brilliant achievements in science, for the whole theory and experiment 'has leaped fully grown and fully armed from the brain of the Newton of electricity, and summed up in fundamental formulae the basis of all electro dynamics.'

Another French physicist who took an active part in elucidating the problems opened up by Oersted's discovery was Arago (1786–1853), professor of mathematics at the Polytechnic School in Paris. He had been a pupil of Hachette at the Polytechnique and been associated with him in his construction of the dry pile.[9] Well known as joint founder with Gay Lussac in 1816 of the famous *Annales de Chimie et de Physique* Arago communicated to the French Institute his discovery that an electric current passing through a wire in the form of a coil developed magnetism in a needle placed in the coil.[10] He also noticed that iron filings which were attracted by a wire carrying a current rose before touching the wire and concluded therefore that they became magnetized individually. Arago was thus the discoverer of magnetic induction by the electric current.

The experiment for which Arago is most famous is that frequently demonstrated in juvenile lectures and known as Arago's Disc.[11] After observing that when a magnetic needle was oscillating near to a plate of metal its oscillations became damped as if it were standing in a viscous medium, he constructed the device which has taken his name. A circular copper plate is mounted on a vertical spindle and rotated by a belt and pulley in a glass-covered box beneath a pivoted magnetic needle. The effect is to cause the needle to deviate from its true stationary position and if the copper plate is rotated fast enough the needle will follow it in its rotations. When the direction of rotation of the plate is reversed, so does the needle slow down, stop and accelerate again in the opposite direction.

Although Arago did not understand that eddy currents in the disc were the cause of the action in his famous experiment and he seems to have made a greater reputation in other scientific fields, e.g., on the polarization of light, it must always stand to his credit that he inspired Faraday at a later date in his epoch-making discoveries. Arago seems to have possessed exceptional personal qualities. Humboldt spoke of him as 'one gifted with the noblest of natures, equally distinguished for intellectual power and for moral excellence'. For almost half a century they were intimate friends and their ever-increasing intimacy became such as to lead to a perfect unity of thought on scientific subjects. 'Pray remember me', he wrote, 'to . . . but especially to him I hold dearest in this life, to M. Arago.'[12]

Another step in the development of electro-magnetism was made independently by Schweigger, a chemist of Halle (1779–1857). In July

1820, Schweigger announced his *multiplier* which was 'a conducting wire twisted upon itself and forming one hundred turns' which produced a magnetic effect, one hundred times greater than a wire with a single turn![13]

By this time the name galvanometer was being applied probably due to Comming of Cambridge[14] to the instrument in which a magnetic needle was deflected by the current in a surrounding coil of wire. The device soon attracted attention both for purposes of measurement and for communication.

Fig. 3. Schweigger's Multiplier. Schweigger, a German chemist, found in 1821 that by making a coil of 100 turns he obtained a hundred-fold increase in Oersted's influence of a current on a pivoted magnetic needle. Thus was the galvanometer born.

The observation made by Arago that a current surrounding a piece of iron induced magnetic properties, bore fruit in a very practical direction. William Sturgeon (1783–1850), of Woolwich, who—as we have seen in the previous chapter—was well known for his interest in electrical devices and overcame local action in batteries by amalgamation of the zinc plate, taking Arago's result, constructed the first electro-magnet. Taking a bar of soft iron a foot long and half-an-inch in diameter, he bent it into the form of a horse-shoe, gave it a coating of insulating varnish and wrapped round it bare copper wire carefully keeping the turns separate from one another. There were 16 turns and Sturgeon found that the current from a single voltaic cell with plates 130 square inches in area would enable this electro-magnet, the first to be constructed, to carry a weight of nine pounds. This was a remarkable performance and far beyond anything achieved previously, and again aroused great interest in scientific circles. Sturgeon exhibited both the horse-shoe electro-magnet and straight ones at a meeting of the Royal Society of Arts in London in 1825. His biographer, Joule[15], describes how this amateur electrician began to send announcements of his discoveries and inventions to the leading scientific periodicals of the day. In 1824 no fewer than four papers from his pen appeared in the *Philosophical Magazine* and in 1825 the *Transactions of the Society of Arts* carried a description of his electro-magnetic apparatus. For this work the Society awarded Sturgeon their Silver Medal and a purse of thirty guineas. In all he produced over fifty separate papers and after his death in 1850 a tablet of unusual interest was erected in Kirby Lonsdale Parish Church, Lancashire, which pays tribute to his discovery of the electro-magnet, the amalgamated zinc battery and other outstanding contributions to electrical engineering, including, as we shall later see, the commutator and the electro-magnetic engine.

The improvement of Sturgeon's electro-magnet was taken up by Professor Joseph Henry (1797–1878), of New York, who during the next few years made valuable suggestions which brought this important fundamental component of all electrical engineering to a point of great practical value. His first idea was to bring the turns of wire closer together so that instead of lying spirally they would be almost circumferential. From Ampère's analysis this would appear to approach the theoretically required conditions more nearly. In one of his published statements Henry said: 'In the arrangement of Arago and Sturgeon the several turns of wire were not precisely at right angles to the axis of the rod, as they should be to produce the effect required by the theory, but were slightly oblique, and so each tended to develop a separate magnetism. But in winding the wire over itself the obliquity of the several turns compensated each other and the effect was at the required right angle.'

Thus was the progress of electrical engineering for the first time made dependent on the adoption of an *insulated* conductor to keep the turns electrically separated from one another. Henry insulated his wire with a silk covering and in June 1828 exhibited an electro-magnet in which the winding consisted of many turns of wire one-thirtieth of an inch diameter with a resulting greatly increased lifting power. He continued to experiment with still greater numbers of turns until he encountered what was to him a new problem though today one which to us it would seem to be fundamentally obvious —the limitations of current imposed by Ohm's Law. Turns added beyond a certain point were found to be less effective in increasing the lifting power. Two alternative solutions presented themselves, the first to add a second coil from the same battery to the same magnet core, and the second to add to the 'propulsive' force of the current to enable it to pass through a greater number of turns of wire

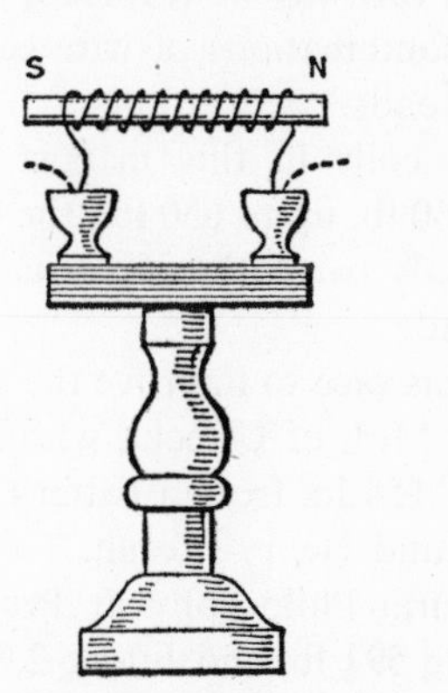

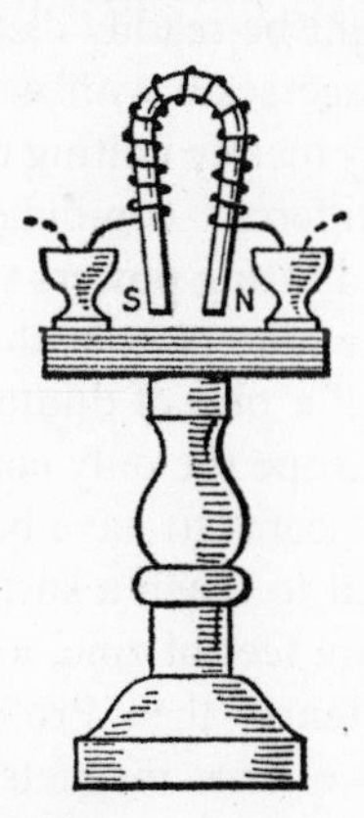

Fig. 4. Sturgeon's Electro-magnet. William Sturgeon (1783–1850) of Woolwich was an early and enthusiastic electrical experimenter. He produced electro-magnets about 1821. By winding 16 turns of copper wire around a half-inch bar of iron 12 inches long and bending it into a horseshoe shape, he was able to lift a weight of nine pounds. He made contacts in cups of mercury.

by adding to the number of the battery elements. With a horse-shoe core half-an-inch diameter and 10 inches long and 30 feet of fine copper wire, Henry was able to hold a weight of 14 lb. with a current obtained from a battery with only 2½ square inches of zinc. Adding a second winding of 30 feet raised the lifting power of the magnet to 28 lb., over three times that of Sturgeon's wonder magnet.

In 1830 Henry went still further and produced an achievement which is best described in his own words: 'a bar of soft iron 2 inches square and 20 inches long was bent into the form of a horse-shoe. A piece of iron of the same bar and weighing 7 lb. was filed perfectly flat on one surface to act as armature or lifter. The flat ends of the horse-shoe were also truly ground to the surface of the armature. Around this horse-shoe 540 feet of copper bell-wire were wound in nine coils of 60 feet each. These coils were not continued round the whole length of the bar, but each coil, according to the principle above mentioned, occupied about 2 inches and was wound several times backward and forward over itself. The several ends of the wires were left projecting and all were numbered so that the ends of each coil might be readily distinguished. In this way we formed a magnet on a large scale with which several combinations of wire could be made by merely uniting the projecting ends.'

By different combinations of the coils in this magnet Henry obtained lifting powers varying from 60 lb. up to 650 lb., the battery having a zinc plate with an area of only two-fifths of a square foot with half a pint of dilute sulphuric acid.

In Europe the only competitor in this race to improve the electro-magnet seems to have been Professor Moll of Utrecht, who in 1830 managed to reach a sustained load of 154 lb. from a battery having 11 square feet of zinc, a long way behind Henry's result. Two years later, Henry, then Professor of Natural Philosophy at Princeton, made two more magnets, one weighing 59½ lb. and lifting 2,064 lb., the other weighing 100 lb. and holding 3,500 lb. In Britain as late as the 1851 Exhibition no such results had been achieved, for on that occasion claims were made for a mammoth magnet weighing one hundredweight with 50 yards of wire in its winding which would lift a load of one ton.

It is very interesting today to see how intelligent men like Henry groped in the dark over such simple ideas as the Ohm's Law relationship in a circuit. Using the sustaining force of his magnets as the criterion he was the first to work out practically the essential differences between what he first called the *quantity magnet* and the *intensity magnet*. The former had a short low-resistance coil and the latter a long high resistance coil. He found that the best result was obtained on a *quantity magnet* where the current was provided by a *quantity battery*, i.e. a single cell, while the *intensity magnet* required an *intensity battery*, that is one with many cells in series. An *intensity battery* was useless on a *quantity magnet* and a *quantity battery* useless on an *intensity magnet*. Henry did not appreciate, as we do through the outcome of Ohm's Law, that the situation in the coil was being largely controlled by the drop of potential in the battery itself.

(*a*) Werner von Siemens' Dynamo, 1867
One of the first self-exciting machines. (*Photo: Science Museum and Deutsches Museum*)

(*b*) Gramme's Dynamo, about 1876
One of several machines tested in a competition by Trinity House to determine their suitability for lighthouse supply.
(*Photo: Science Museum*)

Plate XII

(*a*) Edison's Dynamo, Vertical Type, 1870–1880
Typical of Edison's early dynamos the field magnets were unduly long. Association with Hopkinson later resulted in a shortening of the cores to what became usual practice.
(*Photo: Science Museum*)

(*b*) Brush Dynamo, 1878 (*Photo: Science Museum*)

REFERENCES

1. Fahie, *History of Electric Telegraphy*, p. 259.
2. Mottelay, p. 383.
3. Memoir. *Journal Soc. Telegraph Engineers*, vol. v (1877), p. 469.
4. John Davy, *Memoirs of Sir H. Davy*, vol. ii, p. 216.
5. *Journal Institution of Electrical Engineers*, vol. 69 (Nov. 1931), p. 1351.
6. *Journal Soc. Tel. Engineers*, vol. v (1877), pp. 459–73.
7. Sir H. Davy, *Phil. Trans*, 1821, pp. 7–19.
8. William Sturgeon, *Annals of Electricity*, vol. i (1836), p. 121.
9. H. Boissier, *Memoire*, Paris, 1801.
10. *Ann. de Chimie et de Phys.*, vol. xv, p. 94.
11. *Phil. Trans.*, 1825, p. 467.
12. Mottelay, p. 481.
13. *Cornhill Magazine*, vol. ii (1860), p. 61.
14. Silvanus Thompson, *Electricity and Magnetism*, 1900, p. 185.
15. *Memoirs Lit. Phil. Soc. Manchester*, 1857.

CHAPTER V

Birth of the Electric Telegraph

In his classic history of the telegraph[1] Fahie records how the editor of *The Electrician* 75 years ago, declined to publish articles on the subject unless they omitted all the non-electric part dealing with 'fire-, flag- and semaphore-signalling, acoustic, pneumatic and hydraulic telegraph, etc., etc.' Fascinating as is the story of the many attempts over the centuries to transmit intelligence from place to place by these simple devices, they cannot be considered as a factor in the history of electrical engineering although some of them achieved remarkable results in their time. The semaphore system of Chappé, for instance, set up between Paris and Lille in 1794, conveyed messages between the two cities, a distance of nearly 150 miles, in a matter of two minutes.

The employment of magnetic and electrical effects for the transmission of intelligence had been predicted long before the advent of the steady current had provided a sound foundation for practical application.

For centuries there has existed in literature a myth known as that of the sympathetic needles. This was no doubt based originally on the ancient observation that one magnetic needle floating on water would respond to the presence of another and, without contact, could be made to point to different letters on a surrounding ring. The fiction arose that even at great distances the same transmission of intelligence could be effected so long as the two magnets, one at the sending end and one at the receiving end, were made sympathetic with one another. In 1558 Porta, a Neopolitan philosopher, wrote a book on Natural Magic which went into many editions and in one of these he said:[2] 'And to a friend that is at a far distance from us and fast shut up in prison, we may relate our minds; which I doubt not may be done by two mariners' compasses, having the alphabet writ about them.'

During the early half of the eighteenth century a good deal of experimenting was carried out by Stephen Gray,[2] a Charterhouse pensioner, while Dr. Watson, a member of a Royal Society Committee, brought to light the essential difference between conductors and insulators. It was discovered that a damp hempen cord suspended by silk loops could be made to transmit a static charge from a Leyden jar. With a metallic wire so supported the results were still more effective and a circuit was set up in July 1747 across the Thames over old Westminster Bridge. The circuit was completed through the body of an assistant, who held the far end of the wire in one hand and with the other touched the water with a metal rod. His reaction completely vindicated the result which had been anticipated. Other experiments

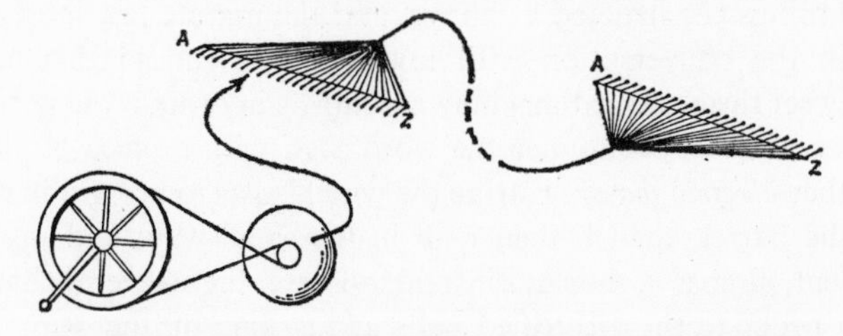

Fig. 5. Le Sage's Telegraph. In this early telegraph the frictional charge from a rotating sulphur ball was conducted by a portable conductor to one of twenty-six lines representing the alphabet. At the receiving end the operator noted which wire sparked to earth.

followed in various parts of London and the length of circuit increased to several thousand feet. Franklin in 1748 in Philadelphia and De Luc in 1749 in Switzerland extended the range but remarkable as it seems today, there was no suggestion of the discoveries being applied to telegraphy.

Five years later, on 17 February 1753, a most outstanding event took place in the form of an anonymous letter in the *Scots Magazine*, Edinburgh, which described in considerable detail a quite feasible telegraph to be operated by a frictional electric machine. The author proposed a series of wires between two places, each line representing one letter of the alphabet. By means of simple spring switches, the near ends could be brought into contact one by one with the 'gun-barrel' of the frictional machine while the far ends were terminated in metal balls suspended over an earthed metal plate. Beneath these balls lay pieces of paper, one to each, with the letters of the alphabet written on them. When a particular wire became charged by the pressing of its transmitting contact it attracted the paper below, thus indicating which switch had been operated at the sending end.

The letter, signed C.M., is reproduced in full on pp. 68–71 of Fahie's book[1] and contains many quaint phrases such as 'Let, then, a set of wires equal in number to the letters of the alphabet, be extended horizontally between two given places, parallel to one another, and each of them about an inch distant from that next to it. At every twenty yards end let them be fixed in glass, or jeweller's cement, to some firm body, both to prevent them from touching the earth, or any other non-electric, and from breaking by their own gravity. . . .'

'All things constructed as above, and the minute previously fixed, I begin the conversation with my distant friend in this manner. Having set the electrical machine a-going as in ordinary experiments, suppose I am to pronounce the word *SIR*; with a piece of glass, or any other *electric per se*, I strike the wire S so as to bring it in contact with the barrel, then i, then r all in the same way; and my correspondent, almost in the same instant, observes these several characters rise in order to the electrified balls at his end of the wires. Thus I spell away as long as I think fit; and my correspondent, for the sake of memory, writes the characters as they rise, and may join and read them afterwards as often as he inclines.'

As a variation 'C.M.' proposed that the wires be connected to a series of bells of different tones. 'And thus, by some practice, they may come to understand the language of the chimes in whole words, without being put to the trouble of noting down every letter.'

In the final paragraph of C.M.'s letter we have what must be one of the earliest, if not the earliest, references to insulating a conductor. 'Some may, perhaps, think that although the electric fire has not been observed to diminish sensibly in its progress through any length of wire that has been tried hitherto, yet as that has never exceeded some thirty, or forty yards, it may be reasonably supposed that on a far greater length it would be remarkably diminished, and probably would be entirely drained off in a few miles by the surrounding air. To prevent the objection, and save longer argument, lay over the wires from one end to the other with a thin coat of jeweller's cement. This may be done for a trifle of additional expense, and as it is an *electric per se* will effectually secure any part of the fire from mixing with the atmosphere.'

The significance of the C.M. letter written in 1753 was not fully realized for a whole century, but in 1853 the question of the identity of the writer was raised in *Notes and Queries*[3] and other journals. Various searches were made by different investigators, including a diligent schoolmaster in Renfrew—the address given at the head of the original letter—but without much success.

Some attributed the letter to Charles Marshall, who could make 'lichtnin' write and speak' and who could 'licht a room wi' coal-reek'.[4]

Among those interested was Sir David Brewster, who considered he had solved the mystery[5] by giving the credit to a 'gentleman in Renfrew of the name of Charles Morrison, who transmitted messages along wires by means of electricity, and who was a native of Greenock and bred a surgeon . . . he was connected with the tobacco trade in Glasgow, was regarded by the people in Renfrew as a sort of wizard, and was obliged, or found it convenient, to leave Renfrew and settle in Virginia where he died.' There is also a suggestion in the correspondence that as he was likely to be ridiculed by many of his acquaintances he would publish his papers only with his initials. Even this story, picturesque and attractive as it is, cannot entirely establish the identity of 'C.M.', which must perhaps remain one of the unsolved mysteries of electrical history.

During the next few years following the appearance of the famous C.M. letter, many suggestions were made for constructing telegraph systems based on the transmission of a static charge. In 1767 Joseph Bozolus, a natural philosopher at the College of Rome, suggested a telegraph which depended on visual observation of the received sparks. He seemed to be more adept, however, at writing Latin verse descriptive of the telegraph than of explaining how the sparks were to be translated into letters and words.

The first serious attempt at producing a static electricity telegraph was made in 1774 by Lesage, a Frenchman living in Geneva. He employed twenty-four wires communicating with twenty-four simple electrometers with pith balls each identified as one of the letters of the alphabet. At each end the wires were arranged horizontally as the keys on a harpsichord. For practical distances he passed the wires along inside an underground tube of glazed earthenware fitted at intervals with partitions perforated with holes to keep the wires apart.

Lomond, 1787, a Frenchman, is claimed to have reduced the static electric telegraph to a single wire and a report of the time states that 'you write two or three words on a paper; he takes it with him into a room and turns a machine enclosed in a cylindrical case, at the top of which is an electrometer, a small fine pith ball; a wire connects with a similar cylinder and electrometer in a distant compartment, and his wife by remarking the corresponding motions of the ball, writes down the words they indicate'.[6] No details of operation are given but possibly synchronized clocks as used later by Ronalds were employed. A rather fantastic suggestion was made in 1795 by a Spaniard, Salva, who employed twenty-two pairs of wires for a twenty-two letter alphabet. At the receiving end each pair was held by a man who called out his letter when he felt the shock. In due course Salva resorted to simpler detectors in the form of tinfoil plates and reduced the number of wires.

The early history of the electric telegraph is at times a confused mixture of over-optimistic prediction with no experimental support and real practical achievement which led to useful progress. A good example of the former is to be found in a letter written by a Louis Odier, a distinguished physician of Geneva, to a lady of his acquaintance in 1773. 'I shall amuse you, perhaps, in telling you that I have in my head certain experiments by which to enter into conversation with the emperor of Mogol, or of China, the English, the French, or any other people of Europe, in a way that, without inconveniencing yourself, you may intercommunicate all that you wish, at a distance of four or five thousand leagues, in less than half an hour! Will that suffice you for glory? There is nothing more real. Whatever be the course of those experiments, they must necessarily lead to some grand discovery, but I have not the courage to undertake them this winter.' Nor did he ever do so, although that did not prevent the great Fahie heading a paragraph 'Odier's Telegraph'.

A very different story is that of the Ronalds Telegraph (1816). Although still following on the use of static electricity at a time when the steady current had become available, Ronalds' device showed considerable originality and possessed some ingenious features. It held possibilities of practical application far in advance of anything previously suggested. This particular pioneer is of more than usual interest to telegraph engineers as The Institution of Electrical Engineers—formerly the Society of Telegraph Engineers and Electricians—possesses the fine library of 2,000 books and hundreds of interesting pamphlets which Ronalds collected during his long life.

Francis Ronalds (1788–1873) was born in London. His father was a City merchant and Francis attended schools at Walthamstow and Cheshunt but seems to have terminated his academic career for an office desk at fifteen without having created much impression. Four years later his father died and at this early age he became responsible for the conduct of a business with a turnover of about £150,000 per annum. This was not his *metier*, however; he was much more interested in carrying out chemical experiments than in studying trading accounts and by the time he was twenty-five he was creating some interest among his neighbours by his loud detonations and their fear of being killed by the 'lightning which he brought into the place'. He says in his writings that, in fact, two or three of his neighbours were killed, but these were only unprincipled rats who devoured his pony's corn.

In the year 1814 Ronalds became acquainted with De Luc, a celebrated natural philosopher of Swiss extraction living in London and employed as reader to Queen Charlotte, the wife of George III. De Luc was busily engaged on his dry pile of gilt paper and laminated zinc which kept small bells ringing for several years and these experiments must have confirmed Ronalds' interest, for he constructed a dry column containing 1,000 pairs of elements to which device he added a ratchet and pawl arrangement by which the pile produced rotation of a pointer round a dial.[7]

By this time Ronalds had moved to the house now known as Kelmscott House in Upper Mall, Hammersmith, and it was here that he constructed the telegraph system through which his name became so famous. The best description of his achievement is contained in a small book which he published in 1823[8] and reprinted in 1871, from which it is clear that his first concern was whether the electric fluid in its static form could be made to travel over long distances without undue delay. For this purpose, in 1816 he set up two wooden structures in his garden twenty yards apart between which he strung iron wire backwards and forwards forming a continuous length of more than eight miles. The wire was insulated at the 37 hooks on each of the 19 bars at both ends by silken loops and the two ends brought out to two pith-ball electrometers.

It has been suggested by certain writers that this was Ronalds' telegraph line, but a careful reading of his original account shows that the object of this original structure was entirely to settle the preliminary question—Is time lost in transmitting the charge? Having set up the aerial circuit he carefully made the following observations. 'When the line was charged by a Leyden jar the electrometers at near end and far end diverged at exactly the same moment, and on discharge, by being touched by the hand, collapsed as simultaneously.

When a person placed in series with two insulated inflammable air pistols received a shock, the pistols discharged at exactly the same moment. The pistols, too, exploded together. Thus was Ronalds convinced that the transmission of electric signs was instantaneous and he could proceed with the completion of his ideas.

After considering various ideas he decided on an underground cable. He dug a trench 4 feet deep, 525 feet long, and in it laid a trough of wood 2 inches square coated with pitch. He placed lengths of thick glass tube in the trough—the joints between lengths being made with soft wax surrounded by a protecting tube—and finally drew

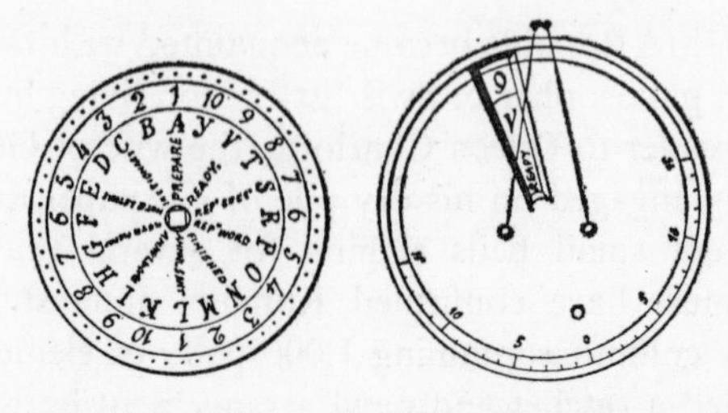

Fig. 6. Ronalds' Telegraph. The illustration shows the ingenious basis of Ronalds' single-wire telegraph operated by frictional electricity which, although quite practical, was declined by the Admiralty in 1816. A disc carrying the letters of the alphabet was rotated by watch mechanism so that they appeared in the aperture on a cover plate. The sent letter appeared as the simple electrometer indicated the arrival of the charge.

in the wire. After a final application of pitch to the box the trench was filled in with earth. Anyone with a knowledge of moisture problems in the laying and maintenance of underground cables for the voltages which Ronalds was employing and the serious effect of even a small leak must applaud his courage and the skill with which he ensured success in this pioneer effort.

In the apparatus used at both ends of his line for transmitting and receiving signals, Ronalds again showed great originality. For the first time he adopted the device of synchronizing two instruments and using this arrangement to secure the transmission of the necessary number of letters and figures over one single line. To the seconds arbor of a clock he fixed a light circular plate divided into twenty equal sections. Each division carried a letter and a figure; and a second plate was placed in front of this in such a manner that it could be rotated by hand when required. This plate had a narrow segment cut away so that it showed one figure and one letter of the under plate. Thus the clock movement brought different letters and figures in view in turn and the timing of the appearance of any particular letter could be adjusted by moving the front plate. In front of the two plates a pith ball electrometer with two balls was suspended, so allowing the charged or discharged state of the line to be observed at the same instant as the letter showing at the time. Thus by keeping the clocks in synchronism and charging or discharging the line at the sending end as a particular letter was showing the operator at the receiving end simply noted the same letter in view at the time indicated by the electrometer.

By the simple convention of applying a higher charge than usual the receiving operator read certain service instructions instead of letters when he noticed the electrometer giving an unduly high deflection. Thus, if one of the clocks has gained or lost, the repetition of the word 'prepare' enabled the operator to bring A in view so synchronizing them once more. To obviate the need for constant watching a gas pistol was fired by a contact connected with the signal 'prepare'.

With this ingenious scheme developed at such an early date and so thoroughly worked out in every detail, it is difficult to understand why it was never tried out on a commercial scale. The story of its rejection by the Admiralty is now well known, but for completeness, Ronalds' letter bringing the invention to the notice of the Government and the laconic reply which he received from the Secretary of the Admiralty are given.

'Upper Mall, Hammersmith.

'July 11. 1816.

'Mr. Ronalds presents his respectful compliments to Lord Melville and takes the liberty of soliciting his lordship's attention to a mode of conveying telegraphic intelligence with great rapidity, accuracy and certainty, in all states of the atmosphere, either at night or in the day, and at small expense, which has occurred to him whilst pursuing some electrical experiments. Having been at some pains to ascertain the *practicability* of the scheme, it appears to Mr. Ronalds, and to a few gentlemen by whom it has been examined, to possess several important advantages over any species of telegraph hitherto invented, and he would be much gratified by an opportunity of demonstrating those advantages to Lord Melville by an experiment which he has no doubt would be deemed decisive, if it should be perfectly agreeable and consistent with his lordship's engagements to honour Mr. Ronalds with a call, or he would be very happy to explain more particularly the nature of the contrivance if Lord Melville could conveniently oblige him by appointing an interview.'

'Admiralty Office
5 August.

'Mr. Barrow presents his compliments to Mr. Ronalds, and acquaints him, with reference to his note of the 3rd inst., that telegraphs of any kind are now wholly unnecessary, and that no other than the one now in use will be adopted.'

In following through the story of the application of static electricity to telegraphy we have jumped ahead, for soon after Volta's discovery announced in 1800, many minds turned to the application of the steady current to signalling purposes. The first practical scheme was that put forward by Sömmering, a German physiologist.[9] Between 9 July and 6 August 1809, Sömmering constructed an apparatus in which direct current caused decomposition of water at the receiving end. There were 35 wires each designating a letter or figure and at the transmitting end the wires terminated in a series of copper cups indicating the different characters. A voltaic cell could be connected to any one of them at will. At the receiving end 35 glass tubes collected the bubbles of gas released by the current, so indicating by the identifying labels on them by which line the current was being received, and which letter or figure was intended. Sömmering's invention was shown first to the Bavarian Academy of Sciences and then to the French Academy who appointed a committee of scientists to report on its possibilities. Apart from the main principles of Sömmering's proposal it is interesting to note this early attempt to produce an electric cable for underground use. He was able to transmit intelligence through a 1,000-foot length of cable, a remarkable achievement for the time. In the construction of the cable the copper wires were given a coating of gum lac, then covered with silk thread and laid together. The whole assembly was then coated with more hot gum lac and finished with a ribbon impregnated with the same material. Wide interest was taken in the idea in different continental countries and in March 1812 a record transmission distance of 10,000 feet was attained.

Following on Sömmering's announcement, Schweigger, whose multiplier we have already noticed in the previous chapter, suggested various methods of reducing the number of conductors necessary in the electro-chemical telegraph.[10] With two batteries of different strengths, introducing changes in the duration of the signals, the use of a code and the changing of the interval, he was able to transmit messages effectively through two wires only. But Schweigger had himself laid, perhaps unwittingly, the foundation of the practical electric telegraph, for it was by the application of his multiplier that the next outstanding steps were taken.

Baron Schilling, a German attached to the Russian Embassy in Munich, is generally credited with the invention of the first electro-magnetic telegraph. He had been associated with Sömmering in developing the electro-chemical telegraph and for many years made experiments during which time Oersted's discovery was announced. As we have seen, this was followed within a few months by Schweigger's multiplier and Ampère's analysis. Schilling in 1832[11] set up a magnetic needle and a multiplier, to some extent in accordance with ideas suggested by Ampère and others previously, but original in that it employed a single needle and a code constituting various combinations of movements. The needle, suspended in a horizontal position by a silk thread, carried a disc of paper, black on one side, and white on the other. The code followed the following pattern:

A Black followed by White.
B Three Blacks in succession.
C Black. White. White.
D Black. Black. White.
and so on.

To reduce the time required to complete the swing of the needle and prevent useless oscillation, Schilling anticipated the now well-known principle of damping: he allowed the lower end of the needle suspension to dip into a small bath of mercury. The calling signal was a simple trip which released a clockwork alarm bell.

Schilling carried on his experiments for several years and showed them in many countries. Only a short time before his death in 1837, through the interest displayed by the Emperor Nicholas and the recommendations of a Russian Commission of Inquiry, arrangements were made for a cable to be laid in the Gulf of Finland between St. Petersburg and Cronstadt. But the honour of first establishing a working telegraph passed to the German physicists Gauss and Weber, who operated a circuit daily with needle instruments at Göttingen from 1833 to 1838. To secure increased sensitivity these two pioneers employed a small mirror on the suspension system to increase the sensitivity of the instrument, a device adopted by Poggendorff[12] a few years earlier and, later, applied with such great success by Lord Kelvin and many others.

The year 1837, the year of accession of Queen Victoria, is renowned also for its significance in the history of the electric telegraph for it was in that year that two Englishmen, Sir Charles Wheatstone and William Fothergill Cooke combined to bring the telegraph into daily use. Another Englishman, though forgotten today, Edward Davy, had made important contributions and a few years later an American, Samuel Morse, established the telegraph in the Western Hemisphere, having added not only the code associated with his name, but many technical improvements.

Fahie publishes[13] many pages of manuscript in connection with the pioneer work of Davy, a chemist in the Strand, along with an interesting memoir by his nephew from which it certainly appears that his contributions were unduly overshadowed by those of Cooke and Wheatstone. He exhibited a working needle telegraph in London and was in close negotiation with the railway companies. Unfortunately at a time when his system was in a less advanced state than that of Cooke and Wheatstone he had to emigrate to Australia for private reasons, leaving the field clear for his rivals.

Sir Charles Wheatstone was born near Gloucester in 1802 and within a few years his parents moved to London, where at 128 Pall Mall his father taught the flute and made musical instruments. As a boy Wheatstone was shy and sensitive and at fourteen was apprenticed to a musical-instrument maker at 436 Strand. He displayed no interest in the business, however, but developed a great taste for reading and spent all available pocket money on books. After reading Volta's account of his electrical experiments in French Charles repeated many of them and soon extended his interest in scientific pursuits, particularly in the fields of acoustics and optics. In 1834, at the age of thirty-two, he was appointed to the Chair of Experimental Physics at King's College, London, where, though a poor lecturer, he proved to be a brilliant experimenter. He had become interested in the various attempts at home and abroad to develop the electric telegraph when in February 1837 he received a visit from Cooke.

William Fothergill Cooke was a medical officer in the Indian Army who, during a leave period in Europe, had been attending lectures at Heidelberg. Here he had heard of Schilling's experiments and became so interested that he abandoned medicine and decided to develop the telegraph to a commercial stage. In 1837 he exhibited a three-needle system in London, but feeling that his scientific knowledge was inadequate he sought Faraday's advice at the Royal Institution. Faraday and a contemporary, Dr. Roger, recommended him to go and see Wheatstone, who was interested in the same field. After exchanging views the two enthusiasts formed a formal partnership, Cooke assuming the business responsibility for launching the scheme and Wheatstone looking after the scientific and technical matters.

Davy is said to have failed because he offered his backers only 20 per cent of the proceeds: we do not know how much Cooke and Wheatstone offered, possibly nothing, as Cooke seems to have been capable of selling the idea to the authorities and very soon the railway companies became interested. Among their first patents was one for the five-needle telegraph and a calling device which made a mercury contact start a clockwork alarm. The needles were assembled on a diamond-shaped board in such a way that when the signal was received the deflected needles pointed to the desired letter on the board.

The first full-scale experiment was carried out between Euston and Camden Town stations on The London and North Western Railway Co., a distance of less than two miles, but the results were so epoch-making that the sensations of Cooke and Wheatstone on hearing the message come through on that memorable 28 July 1837 have often been recalled. Mr. Cooke was at Camden Town and Professor Wheatstone in a dingy little office at Euston lit by a tallow candle. After Wheatstone had sent a message and was receiving a reply he was thrilled with the import of what they had done. 'Never did I feel such a tumultuous sensation before, as when, all alone in the still room I heard the needles click, and as I spelled the words, I felt all the magnitude of the invention pronounced to be practicable beyond cavil or dispute.'

Unfortunately, in spite of the success of the experiment the reaction of the railway company directors was cool. They saw no future for the new-fangled device and requested its removal, and for two years no further progress was made. In July 1839, however, the Great Western Railway Company agreed to instal a line between Paddington and West Drayton, a distance of thirteen miles, and two years later extended it to Slough. In addition to its daily use an exhibition was staged at Paddington for the public who paid an admission fee of one shilling to see this 'marvel of science, which could transmit fifty signals a distance of 280,000 miles in a second'.

The event which, more than any other, brought the telegraph to the notice of the public, was the arrest, in 1842, of the murderer Tawell at Paddington Station. Mr. John Tawell, a respectable resident of Berkhamsted, was known by the police to be in the habit of visiting a woman, Sarah Hart, at her home near Slough. Early one morning she was found dead and a man had been seen leaving her home in rather suspicious circumstances some time before. On enquiry at the Great Western station they learned that a man answering to their description had just gone off by the slow train to London. 'But', said the inspector, 'why not try this new telegraph?' At once the message was sent indicating that the man wanted was dressed as a Quaker with a brown coat reaching almost to his feet. There seems to have been some delay owing to the fact that the five-needle instrument had no signal for Q and KWAKER was not at once understood. When Tawell stepped out of the train at Paddington, however, he was shadowed to a New Road omnibus and ultimately arrested. In due course he was tried for the murder and executed. The excitement, first over the manner of his arrest, and secondly through startling revelations of his past career, seems to have done as much, if not more, for establishing the telegraph of Cooke and Wheatstone than all their technical and business ability. Joint stock companies were set up to exploit the invention for commercial purposes—The Electric Telegraph Company which paid Cooke and Wheatstone £33,000 for their invention; the Magnetic Telegraph Company, and others—so that by 1868 over 16,000 miles of telegraph line had been erected. By 1870 the systems had been incorporated into one national undertaking and taken over by the Post Office.

The early lines were crude in the extreme and the five wires for the Euston–Chalk Farm circuit were carried in longitudinal grooves cut in the top and sloping sides of wooden bearers. In consequence electrical failures were common. One of these, on the Fenchurch Street–Blackwall line proved, according to Fleming,[14] the direct cause of a great improvement in the needle system. Three out of the five wires broke down and the telegraph clerks devised a code which enabled them to continue working with only two needles in use. Finally it was found that one needle was all that was required. When the movement of the needle was restricted by stops on the dial a convention of 'dots' to the left, and 'dashes' to the right was adopted, so that various combinations of dots and dashes gave all the letters and figures required.

From the needle telegraph to the relay, or repeater, was only a step. The movement of the needle was made to complete a contact with a cup of mercury and this in turn controlled a current much stronger than that which was necessary to operate the needle. The important principle of polarizing the needle was also introduced; by disposing a fixed permanent magnet close to the needle the strength of the magnetism was increased, so making the needle much more sensitive to the incoming currents in the operating coil.

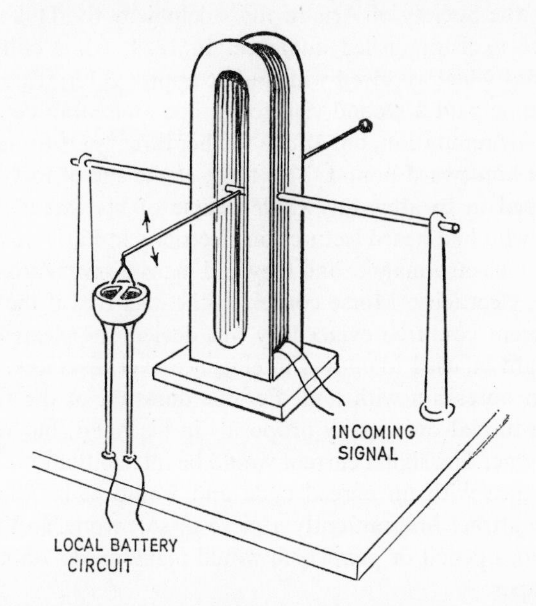

Fig. 7. Wheatstone's original relay. The local circuit is closed by a double contact fixed to the moving needle dipping into two mercury cups.

Wheatstone also introduced the first direct reading telegraph. At the sending end a pointer, operated by hand, could be moved around a vertical circular scale carrying the letters of the alphabet. The arbor of the moving part carried contacts so that intermittent currents were sent out, the number of pulses being decided by the movement of the pointer from one letter to another. At the receiving end a similar disc had a pointer operated by an escapement which was controlled magnetically by the incoming pulses, and the letters sent out were then received in turn and available for reading.

While Cooke and Wheatstone were enjoying their success in

England an American whose name was to become world famous was making important contributions to telegraphy in the United States. As in the case of Cooke, Samuel Morse was neither scientist nor engineer. Born of cultured and talented parents in Charlestown, Massachusetts, in the year 1791, he graduated at Yale in 1810 and thenceforth devoted himself to art. Both in the United States and in Europe he painted industriously and also achieved considerable success in sculpture. He was presented with a prize and gold medal in 1812 at the Society of Arts in the Adelphi by the Duke of Norfolk before a distinguished audience. In 1829, while still producing paintings which received wide public approval but little financial return, Morse paid a second visit to Europe and continued to build up his artistic reputation, but after a further three years found himself once more homeward bound. This time, at the age of forty-one, the fates stepped in to alter the whole course of his career. A fellow passenger who had heard lectures on electricity in Paris was carrying with him an electro-magnet and they had many conversations on this new thing, electricity. Morse conceived the idea that if the presence of the current could be evinced by this device, the electro-magnet, then it might be used to convey intelligence over distances. The idea became an obsession with him. For the duration of the six-weeks' voyage he turned over many proposals in his mind, but ultimately settled on one. The signal current would be intermittent on a pattern in accordance with an agreed code and would cause the electro-magnet to attract intermittently a piece of soft iron. To this would be attached a pencil or pen which would make marks on a moving strip of paper.

On arrival back home in New York, Morse gradually deserted the brush and canvas, leaving his national picture 'The Signing of the First Compact on Board the *Mayflower*' unfinished, and returning the 300 dollars which had been subscribed for it. Instead he turned to his ideas of the telegraph and was fortunate at the same time in being appointed as the first Professor of the Arts of Design in the new University of New York. This allowed him to escape from the low financial ebb which had again overtaken him, and his experiments went ahead. We do not hear how the Arts of Design progressed, but by 1836 he had devised the simple relay so that, as he said, 'If I can succeed in working a magnet ten miles I can go round the globe.'

Morse was fortunate in enlisting the co-operation of a Professor Gale, a chemist, who helped him to make improvements in his apparatus and to set up a more powerful battery, and of a young

Plate XIII

(*a*) Thomson Houston's Globe Dynamo, 1885
This machine originated in the United States for supplying constant current to arc lamps in series. Three armature coils connected together at one end are wound at different angles on a spherical core and the other ends are brought to a 3-bar commutator. Thus two coils are always in series between the brushes which are automatically adjusted circumferentially as required by the load. The machine shown is sectioned to indicate the construction.
(*Photo: Science Museum*)

(*b*) Edison Hopkinson Dynamo, 1883–6
This machine shows the early influence of Hopkinson in improving the magnetic circuit, which brought up the efficiency to about 95 per cent, a great step forward. (*Photo: Science Museum*)

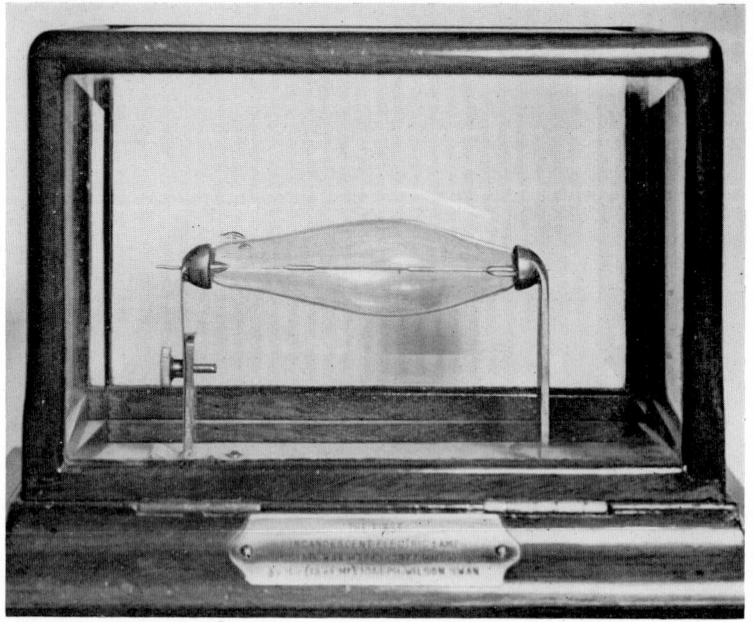

(*a*) Swan's First Incandescent Filament Lamp, 1878
This historic item was the first practical incandescent lamp exhibited to the public. It consists of a straight thin carbon conductor one twenty-fifth of an inch diameter sealed into an evacuated glass container.
(*Photo: Associated Electrical Industries Ltd.*)

(*b*) Single Phase Alternators, Grosvenor Gallery, about 1883
This early installation in Bond Street, London, was started as a local private plant to provide lighting for shop-keepers and householders in the neighbourhood, but rapidly developed into an important public supply system. The photograph shows two separately excited Siemens single-phase alternators each of 250 kW, belt-driven by portable steam engines.
(*Photo: London Electricity Board*)

man, Alfred Vail, of a mechanical turn of mind, who not only co-operated in the design and in making equipment, but, what was even more vital at this time, secured financial backing. On 3 October 1837 Morse obtained his patent and Vail improved the printer so that it produced clear dots and dashes. The Morse Code, so well known by the name of the inventor, seems really to have been the work of Vail, for we read[15] 'Vail tried to compute the relative frequency of all the letters in order to arrange his alphabet; but a happy idea enabled him to save his time. He went to the office of the local newspaper and found the result he wanted in the type-cases of the compositors.' Thus was established the Morse Code which has survived for so long and in so many forms of signalling.

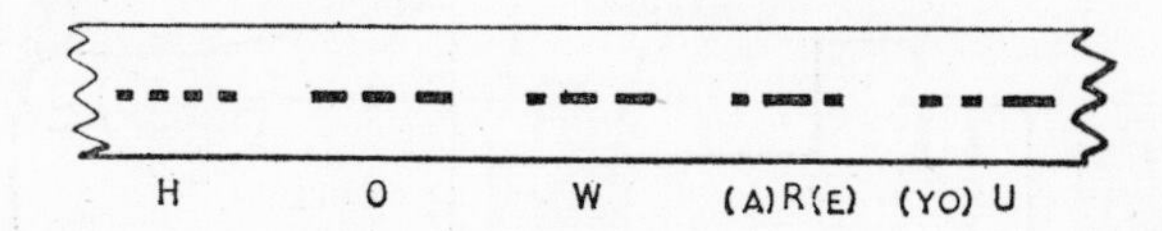

Fig. 8. Morse Telegraph Code. The well-known code consisting of dots, dashes and spaces, which bears Morse's name, was the joint product of himself and an assistant, Vail, working at Menlo Park, the laboratory of the famous American inventor.

After four years of political string-pulling, frustration and living on the verge of starvation, Morse had his system approved by Congress and he received an appointment with a salary of 2,500 dollars a month to superintend the erection of a line connecting Baltimore and Washington. Serious difficulties arose with underground conductors but a conductor on poles was carried through and on 23 May 1843 messages were transmitted by Morse from the Capitol and received by Vail at Baltimore. The line was opened on 1 April 1845 as a public service and after the Postmaster-General of the day had declined to purchase the invention for 100,000 dollars, Morse proceeded to secure the support of private enterprise. The Western Union amalgamated some of the earlier participants in the venture and went ahead in stretching a network of circuits over the United States, the most dramatic of which was the New York to San Francisco line completed in 1861.

Morse's original receiver recorded the signals by indentation on the paper strip but an inked wheel was introduced later and so led to the Morse Inker. The simple Morse sounder ultimately used in such large numbers was not, as might be supposed, developed *per se*, but arose out of the more complicated inker in a curious manner. When the recording instrument was introduced the clerks became accustomed to the sound of the clicks and could soon read the message entirely by ear without reading the dots and dashes on the tape. This annoyed Morse, who regarded the printed record as the fundamental principle of his invention. He therefore threatened with dismissal any who adopted the unorthodox procedure, but the situation was

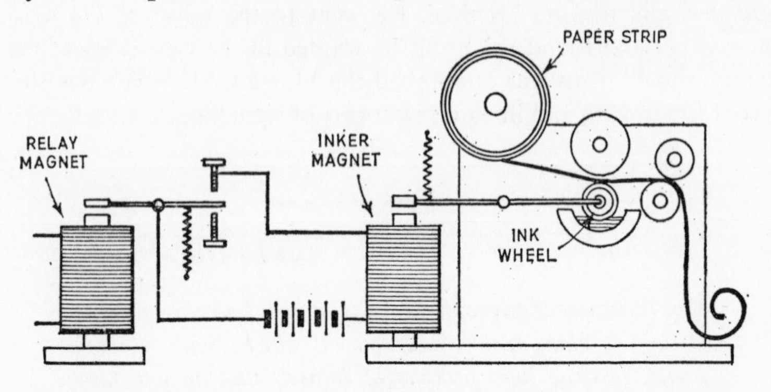

Fig. 9. Morse Inker. As the paper strip was moved along by continuously running rollers, the incoming signal currents operated the inker magnet by means of the relay and the ink wheel marked the strip.

too strong for him—the electro-magnet with its clicking armature was all that was necessary to receive the message—and thus the sounder was born.

It is interesting today to recall the size and weight of the early telegraph equipment used by Morse. He thought at first that his receiver coils should be wound with wire the same size as that used for the line circuit, and used No. 16 gauge copper insulated with a winding of cotton thread. The bobbins were thus 18 inches in diameter and 3½ inches long. The electro-magnet weighed 158 lb. and we read that two men were required to lift it.[16] Today, of course, the coils of telegraph instruments are wound with wire only a few thousandths of an inch in diameter, and their weight in comparison is measured in ounces.

This period in telegraphic history cannot be brought to an end without reference to two more important inventions of Wheatstone: the A.B.C. instrument and the automatic transmitter. The former arose out of his 'Letter Showing Telegraph' referred to earlier. It brought into use the discovery of Faraday regarding electro-magnetic generation of current and to that extent is out of chronological order at this stage of our story. The A.B.C. system was so successful, was

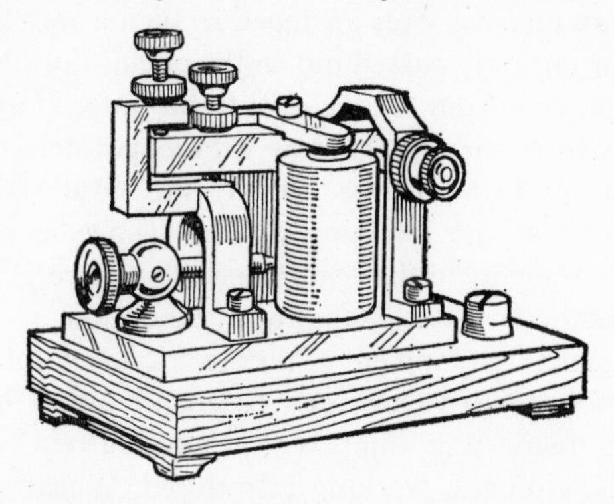

Fig. 10. Morse Sounder. The bar carrying the iron strip was pulled down by the two magnet coils carrying the received current and struck the left-hand stirrup fitting. The lower adjusting screw pulled, through a spiral spring, the lower end of a right-angled extension from the bar. The signals were read from the clicking sounds as the bar struck the upper and lower points of contact.

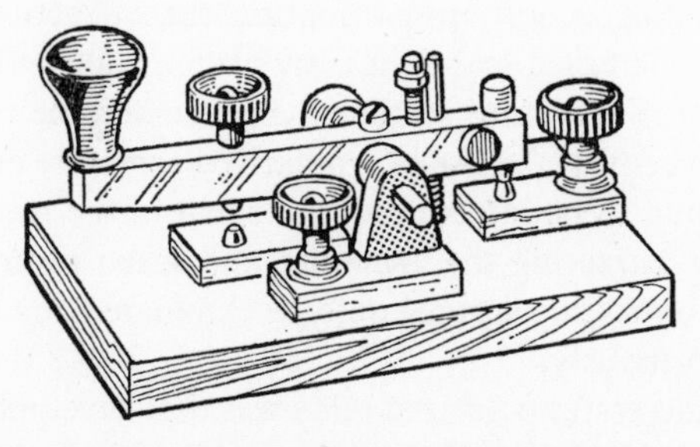

Fig. 11. Morse Key. In its simplest form indicated, the key was held out of contact with the line terminal by a spring. Depressing the key completed the circuit.

employed so widely, and is such an essential part of Wheatstone's contribution that it must be included. Although invented in 1840 it was widely used in British Post Offices right up to the first World War.

In the Wheatstone A.B.C. system the transmitter consisted of a small electro-magnetic alternator driven by a crank handle. This was geared to a pointer turning around a horizontal circle on which were engraved the letters of the alphabet. Outside each letter was a press button with an internal lever arranged in such a way that the alternations of current only passed out to the circuit during the time the pointer was moving from '12 o'clock' to the letter in question. Thus supposing the instrument is standing with its pointer at zero, button G is pressed and the crank turned, the needle moves to G. Similarly a dial at the receiving end has a needle which follows exactly the movements of the sending needle and, receiving the same number of electric impulses, stops at the same letter.

By this time the development of the electric telegraph was following three closely related channels. To enable longer circuits to be operated the relay was improved, double-current working was adopted and automatic working introduced. Wheatstone invented a simple form of relay in which a contact on the needle closed a local circuit by mercury cups and brought into use a local battery to operate an electro-magnet. Morse, in the United States, possibly owing to the longer distances, quickly developed an electro-magnetic relay with a simple double-arm electro-magnet having an armature carrying the local contact. The great stride forward in efficiency of operation was in the introduction of the polarizing principle in the relay. Instead of the coils which carried the receiving current attracting a soft iron armature the attractive force made available was increased many times by placing a powerful permanent magnet near the armature so that the latter became magnetized by induction. Moreover, the use of a polarized armature enabled double current working to be introduced; a current in the receiving coils in one direction attracting the armature and when reversed, repelling it. Thus the controlling spring could be eliminated and the sensitivity increased enormously.

The most successful polarized relay was that developed by Siemens which has since had extended use in the British Post Office and in many other administrations throughout the world. This relay takes the form of a vertical brass cylinder about 6 inches high and 3 inches in diameter. The upper end is closed by a hinged glass cover. The receiving coils surround the two vertical limbs of an iron core which terminate at their upper ends in shaped pole-pieces in close proximity to one another. A light soft-iron armature is carried on a vertical spindle to one side of the magnetic axis so that one end of it can

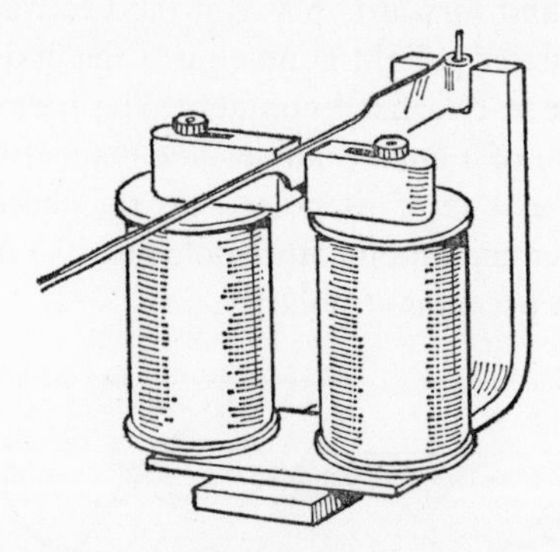

Fig. 12. Siemens' Original Telegraph Polarized Relay, c. 1858. In the simplest early relays the contacts were kept apart by a controlling spring or by gravity. Through the introduction of a polarizing magnet, a much more sensitive control became possible by employing magnetic bias.

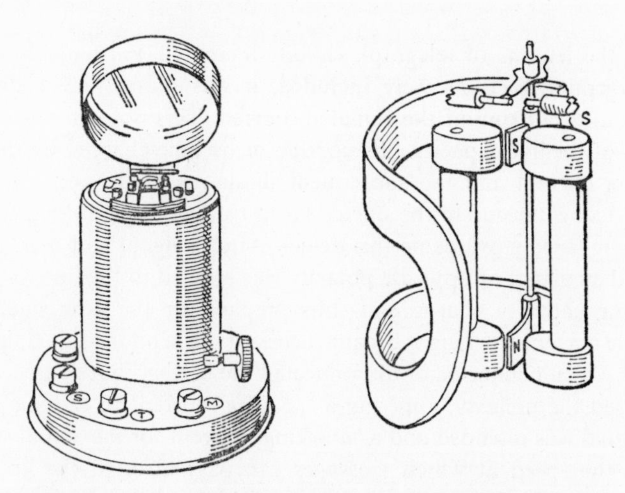

Fig. 13. Siemens' Polarized Relay. External appearance and diagrammatic sketch. By the rearrangement of the polarizing magnet, Siemens produced a sound, sensitive and reliable instrument which was employed in large numbers throughout the world.

move backwards and forwards between the two main poles. On the same spindle is carried a light contact arm the end of which moves horizontally between two fixed contacts. The permanent polarizing magnet in the form of a curved horse-shoe magnet housed inside the cylindrical case, has one of its poles near the centre of the electro-magnet at the lower end and its other pole near the moving armature, so inducing in it a permanent polarity.

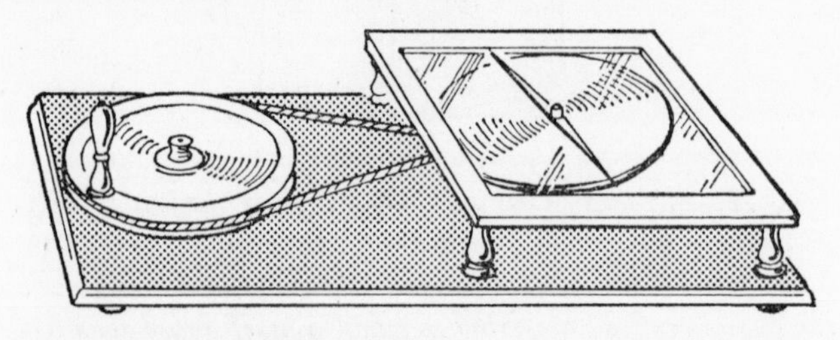

Fig. 14. Arago's Disc. In this famous experiment which puzzled Faraday for some time, a rotating copper disc dragged an independently suspended magnet around with it.

As the lengths of telegraph circuits increased, particularly where underground sections were included, it was found that a slowing down and blurring of the signals occurred. This was, of course, the effect of the line capacity; the conductor became charged by the signalling current and the subsequent discharge at the receiving end caused an extension of the signal. The introduction of double current working largely overcame this defect. After a signal had been transmitted a current of opposite polarity was applied to the line to sweep out the capacity charge and thus prepare for the next signal. In double current working it became necessary to send the signal, not by a key which connected or disconnected the battery, but by one which reversed the polarity, applying a 'spacing' current to the line when no signal was intended and a 'marking' current for the actual signal.

As the speed at which messages can pass through the line and equipment of a telegraph circuit is much higher than can be achieved by manual operation, attention was early directed to the possibility of automatic sending and receiving. The first to propose a practicable scheme was Bain in the year 1846. He prepared strips of paper with long and short perforations which were drawn over a metal roller under a spring contact. The circuit was completed whenever a gap allowed the stylus to fall on to the cylinder and was otherwise broken. By arranging the gaps in the paper as dots and dashes, the effect was the same as with key signalling but the speed could be very considerably increased; it became possible to dispose of four hundred messages an hour on a single circuit. Bain punched the tapes separately by hand and so was able to feed the work of many clerks into one line, an important matter on an extensive and costly system.

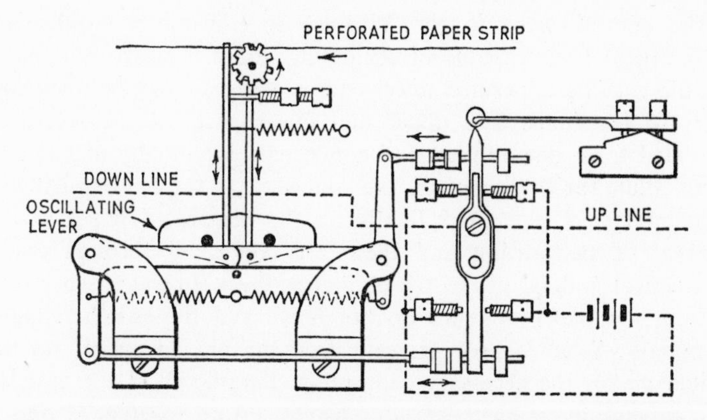

Fig. 15. Wheatstone's Automatic Transmitter. This remarkable invention, made by Professor Wheatstone, employed punched paper tape in the transmitter. The engagement of steel pins with the holes in the paper as it was drawn through the instrument caused the vertical contact rod to oscillate in accordance with the signals punched in the tape. By employing several operators punching the tape, the one machine was capable of transmitting at high speed, so increasing the economy of expensive lines.

The greatest of Wheatstone's triumphs was his invention of the automatic telegraph, not only for the outstanding contribution which it made to long-distance communication at the time, but also because, by his remarkable ingenuity and foresight, he produced apparatus which continued to be used practically unaltered in its fundamentals for well over half a century. In the Wheatstone system the message was first translated into a code of punched holes in a paper tape, the tapes were then passed through the high-speed transmitter which operated as a telegraph key employing the Morse code. It sent dots and dashes to the line by means of a marking current and intervals by a spacing current. The basis of the receiver was a polarized relay actuating an inking wheel which marked a paper tape.

The perforations on the punched tape in the Wheatstone system are not distinguishable as dots and dashes. A centre line of small holes is provided for driving the paper forward and the message perforations occur along each side of this centre row. A Morse 'dot' is registered by two holes on opposite sides of the centre in line with one another, a 'dash' by two holes staggered, one a space ahead of the other.

The central piece in the Wheatstone transmitter mechanism is a vertical lever capable of oscillating about a centre horizontal spindle with its upper and lower ends making contact between two pairs of terminals. The upper and lower parts of the lever are insulated from one another and connected individually to line and earth, while the fixed contacts are connected to the sending battery. Two light steel pins press lightly on the perforated paper at the interval of perforation and these are connected through crank mechanism and springs to the oscillating lever. In the absence of a hole in the paper the pins do not move and the switch remains stationary. When a hole presents itself the pin advances and the switch makes the necessary connection, long or short for dash or dot, depending on whether both pins go forward together or one at a time.

REFERENCES

1. J. J. Fahie, *History of Electric Telegraphy*, 1884.
2. Mottelay, p. 153.
3. *Notes and Queries*, 15 Oct. 1853.
4. *Cornhill Magazine*, vol. ii (1860), pp. 65–6.
5. *The Home Life of Sir David Brewster* (Edinburgh, 1869), p. 207.
6. Fahie, *History of Electric Telegraphy*, p. 92.
7. *Phil. Mag.*, vol. xlv, 1815.
8. Francis Ronalds, *Descriptions of an Electric Telegraph*, London, 1823.
9. Hamel, *Journal Soc. Arts*. London, vol. vii, 1859, pp. 595–605.
10. *Journal fur die Chemie und Physik*, vol. ii, p. 240.
11. Mottelay, p. 422.
12. *Ann. der Phys. und Chem.* vol. vii (1826), pp. 121–30.
13. J. J. Fahie, *A History of Electric Telegraphy*.
14. Ambrose Fleming, *Fifty Years of Electricity*, 1921.
15. J. Munro, *Heroes of the Telegraph*, London, 1891, p. 59.
16. Sabine, *The Electric Telegraph*, London, 1867.

CHAPTER VI

Faraday's Great Contribution

We have seen in the previous chapters how electrical engineering sprang from the ancient lore of magnetism and static electricity, extending back over a thousand years and culminating at the end of the eighteenth century in the production of the steady voltaic current. The immediate practical applications were in the field of electro-chemistry and the carbon arc. During a period of twenty years many investigators produced new ideas which resulted in the perfected voltaic cell as a reliable source of steady current.

In 1820 the magnetic effects obtainable from the electric current had emerged and this had opened up a new electrical era in which the principal practical application was the electric telegraph launched commercially, after many attempts, in 1837. Up to this point electrical engineering depended entirely on the voltaic current and the stream continued, unabated and apparently self-sufficient, for another quarter of a century. During this period, however, another tributary commenced to flow and presently entered the main stream, making electrical engineering a vastly greater thing than it had been before. A new source of current was discovered which had possibilities of application previously undreamed of and to one man, Faraday, are we indebted for the brilliant series of experiments which led to this result.

Michael Faraday was born at Newington in the Surrey outskirts of London on 22 September 1791, his father an impecunious blacksmith; when he was five his parents moved to rooms over a coach-house in Jacobs Well Mews, Charles Street, Manchester Square. At thirteen, Michael entered the employment of a bookseller and newsagent, Mr. Riebau, at No. 2 Blandford Street, off Marylebone High Street, a shop which still exists, and five years later the family moved to 18 Weymouth Street, where his father died soon afterwards.

During his apprenticeship as a bookbinder, Faraday soon developed a taste for reading the books which, in this way, came to his notice, particularly those bearing on scientific subjects. He kept notebooks of his reading in which the subject of electricity shows prominence over all others. The Secretary of the Athenaeum was a visitor to Riebau's shop and one day, noticing that one of the young employees was engrossed in the contents of a book which he was in the process of binding, engaged him in conversation. To his surprise he found the young man not only curious on the contents of an article on Electricity in the old *Encyclopaedia Britannica*, but very fully informed on the subject, or as he records 'a self-taught chemist of no slender pretensions'. Not only did reading and note-taking occupy the attention of this young bookbinder but, with every available moment and to the limit of his resources, he carried out scientific experiments. He constructed an electrical machine and the following description of a home-made voltaic pile is taken from a letter he wrote to a friend at the time:

'I, Sir, I my own self, cut out seven discs of the size of halfpennies each! I, Sir, covered them with seven half pence, and I interposed between seven, or rather six, pieces of paper soaked in a solution of muriate of soda! But laugh no longer, dear A.; rather wonder at the effects this trivial power produced. It was sufficient to produce the decomposition of sulphate of magnesia—an effect which extremely surprised me; for I did not, could not, have any idea that the agent was competent to the purpose.'

There is evidence in the personal notes which Faraday made in his late 'teens that he had a considerable knowledge of chemistry, due entirely to his own efforts in reading and attendance at one or two sporadic lectures. We know that one of his favourite books was the then popular *Conversations on Chemistry* by Mrs. Marcet, the versatile wife of a Swiss doctor, who lived in London. Many years afterwards, when he had become famous, he was delighted to have her consult him on his own discoveries and so to give him the opportunity of knowing personally one who had had such an influence on his career.

One day Faraday accidentally heard of a series of lectures by a Mr. Tatum at a shilling a time and, to his eternal credit, his brother Robert, now established as a blacksmith, found the ten or twelve shillings necessary. But the outstanding event in Faraday's early life was the introduction which he obtained to the Royal Institution. Sir Humphry Davy was delivering a course of four public lectures in

1812 and one of the shop's customers, a Mr. Dance, noticing Faraday's insatiable thirst for scientific knowledge, obtained tickets for him. Many writers have referred to the excitement with which the young enthusiast sat in the gallery of that beautiful lecture hall in Albemarle Street—just over the clock which still today strikes the clear signal for the precise entry of the lecturer through the same swing doors. He took full notes of the lectures and at the end copied them out in his clear handwriting; he bound them with his own hands, complete with a title page in which he dedicated the book to Davy. The book is preserved as a treasured possession in the archives of the Royal Institution and on the occasion of the Christmas Lectures for Juveniles in 1949 it was the author's privilege to show it, by means of television, to an audience of several million people and to read over with them word by word one of the more interesting pages.

Davy, struck by the originality of the gift and no doubt greatly impressed by Faraday's enthusiastic account of the lectures, wrote a kindly acknowledgment and arranged to see him. In the Minutes of the Managers' Meeting dated 1 March 1813, Sir Humphry Davy is recorded as recommending Michael Faraday, aged 22 years, whose 'habits seems good, disposition active and cheerful, and manner intelligent', to fill the vacancy of assistant in the laboratory at twenty-five shillings a week together with the use of two rooms at the top of the house. Thus commenced an association which was to last for nearly fifty years, and one in which this modest man laid a sure foundation for immortality.

Throughout this long period Faraday was engaged on a wide variety of scientific activity, mostly in the chemical field, including major investigations resulting in the discovery of new compounds, on alloy steels and optical glass. While the subject of electro-magnetism was ever present in his mind, the periods during which he produced major results for the advancement of electrical engineering were, firstly, during the year following Oersted's discovery, 1820–21, and again ten years later. The former period covered the development of what became known as electro-magnetic rotation and the latter, of still greater significance, of electro-magnetic induction. The most complete source of information on this work is the *Diary* published in 1932–6.[1] Faraday's notes on these two fields of electro-magnetism are contained in the first volume.

As we have seen earlier, several investigators, following Oersted's discovery, spoke of electro-magnetic rotation, particularly prominent among them being Ampère. Their ideas were far from clear, however

Wollaston, for instance, seemed to think that a wire should rotate on its own axis in the presence of a magnetic field although Davy, in Faraday's presence, showed that a magnet would deflect the electric arc.[2] On repeating Oersted's experiment in a slightly different form Faraday saw the significance of the *mutual* effect of the current and the magnet and came to the conclusion that the force was applied to the pole of the needle tending to make it travel round the wire carrying the current. Moreover a North pole should travel one way and the South pole in the opposite direction. We read in the *Diary*[3] how he tried in various ways to secure practical evidence of this rotation. He set up a vertical wire, described in his own words as follows: 'To establish the motion of the wire a connecting piece was placed upright in a cork on water; its lower end dipped into a little basin of mercury in the water and its upper centred into a little inverted silver cup containing a globule of mercury.' The mercury was suspended in this way upside down through amalgamation with the silver cup and so provided a frictionless moving contact at the top of the wire. But it was only when he cranked the wire that he began to make progress, for in these conditions the wire, instead of being thrust away, began to move around the pole of the magnet. These various stages in the discovery were all passed through in one day, 3 September 1821, and in conclusion Faraday wrote in his diary: 'Very satisfactory, but make more sensible apparatus.' The next day he fixed a magnet vertically in wax at the bottom of a basin of mercury with its upper pole projecting above the surface of the mercury. The wire, making contact with the mercury at its lower end and the upper inverted cup was found to travel in a circular track around the magnet as soon as the current was switched on.

From this experiment it was only a step to making the magnet rotate around the wire. This was done the same day by loading one end of a bar magnet so that it hung vertically in the mercury with its upper end projecting from the surface and bringing the wire down in a rigid form to the mercury bath. When the current was switched on the magnet rotated round the wire.

For the next few days Faraday investigated many matters of detail including the position of the poles in a magnet and the effect of the earth's field on a floating solenoid and then returned to his chemical experiments.

To crystallize these two effects Faraday had made for him by an instrument-maker a device in which the same current passed in turn through two tubular mercury baths. In one of these the magnet among them being Ampère. Their ideas were far from clear, however, rotated around a fixed conductor and in the other the conductor rotated around a fixed magnet.[4] This experimental success was of great significance in the history of electrical engineering for, after all, it demonstrated the possibility of converting the electric current into continuous mechanical motion. The first electric motor had been constructed.

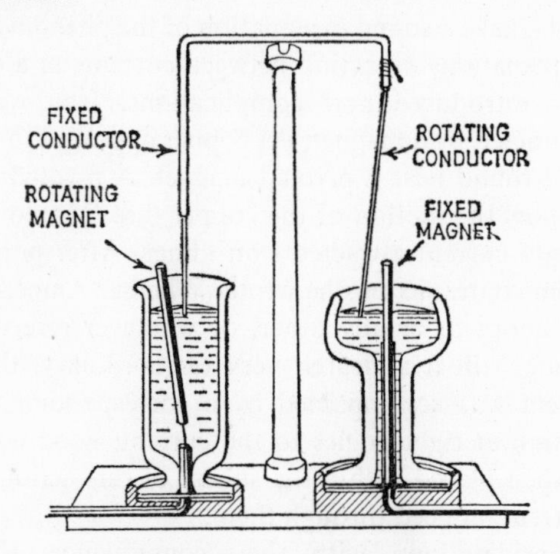

Fig. 16. Faraday's Electro-magnetic Rotations. In this experiment, Faraday produced the basis of the rotating electric motor as well as demonstrating the interrelation between the electric current and magnetic fields. The effect produced both rotation of a magnetic pole around a conductor (left-hand side) and the rotation of a conductor around a magnetic pole (right-hand side).

Three months later Faraday returned to the subject, and on Christmas morning, 1821, he demonstrated to his brother-in-law, George Barnard, how the rotation did not even require the presence of a magnet. By setting up the rotatable conductor in the correct position relative to the angle of magnetic dip, continuous rotation could be obtained due entirely to the earth's field. In a letter written later, Barnard described the excitement of the scene in which Mrs. Faraday had been called down to participate. He wrote 'All at once he [Faraday] exclaimed "Do you see, do you see, do you see, George?" as the wire began to revolve. One end I recollect was in the cup of quicksilver, the other attached above to the centre. I shall never forget the enthusiasm expressed in his face and the sparkling in his eyes.'

For the next ten years Faraday was chiefly occupied with his chemical interests, although from time to time he returned to considering electro-magnetic effects. He studied everything which had been published since Oersted's epoch-making observation, and in particular was interested in Ampère's theories, which he predicted, quite correctly, gave a sound explanation of the phenomena. Ampère based his explanations on action between currents at a distance but Arago's disc introduced new complications. Iron was the only substance attracted by a magnet, yet copper appeared to be magnetic when whirled round near a pivoted magnet. A pivoted magnet was caused to follow the motion of the copper disc. Moreover a copper wire carrying a current attracted iron filings. After pondering over these apparent contradictions he wrote 'Whether Ampère's beautiful theory were adopted, or any other, or whatever reservations were mentally made, still it appeared very extraordinary, that as every electric current was accompanied by a corresponding intensity of magnetic action at right angles to the current, good conductors of electricity, when placed within the sphere of this action should not have any current induced through them.

'These considerations, with their consequence, the hope of obtaining electricity from ordinary magnetism, have stimulated me at various times to investigate experimentally the inductive effect of electric currents. I lately arrived at positive results; and not only had my hopes fulfilled, but obtained a key which appeared to me to open out a full explanation of Arago's magnetic phenomena.'

Faraday's next and most outstanding discoveries in electromagnetism were made from 29 August to 16 December 1831.[5] In their application to electrical engineering they constitute clearly the origins of (i) the transformer, (ii) the alternator, (iii) the continuous generation of direct current.

As Professor Cramp pointed out in his Faraday Lecture given on the occasion of the Faraday Centenary in 1931,[6] the only tests available in 1831 for determining whether a body was electrified or not were the gold-leaf electroscope, the sensation on the tongue, the kicking of a dead frog or evidence through a spark, while the presence of a current in a wire was settled by a rise in its temperature, decomposition of a weak acid by a pair of metal plates forming part of the circuit or the effect on a nearby compass needle. Faraday soon adopted the latter, with the refinement of Schweigger's Multiplier, to which Ampère gave the name 'galvanometer'.

In 1831 Davy had been dead two years and Faraday, having expressed to the Royal Society Committee a wish that he should relinquish his investigations on glass which had occupied so much of his spare time, so that he could devote more attention to 'such philosophical enquiries as suggested themselves to my own mind', he turned with great energy to the experiment which quickly proved his crowning triumph. He had a burning desire to produce electricity from magnetism, and on 29 August 1831 he achieved success with the well-known ring experiment. His diary reads as follows:

'Have had an iron ring made (soft iron), iron round and $\frac{7}{8}$ inch thick and ring 6 inches in external diameter. Wound many coils of copper wire round one half, the coils being separated by twine and calico—there were three lengths of wire each about 24 feet long and they could be connected as one length or used as separate lengths. By trial with a trough each was insulated from the other. Will call this side of the ring A. On the other side but separated by an interval was wound wire in two pieces together amounting to about 60 feet in length, the direction being as with the former coils; this side call B.'

This famous coil, the first transformer, is preserved in the Royal Institution and still carries parts of the original paper labels A and B. His description continues:

'Charged a battery of 10 pr. plates 4 inches square. Made the coil or B side one coil and connected its extremities by a copper wire passing to a distance and just over a magnetic needle (3 feet from iron ring). Then connected the ends of one of the pieces on A side with battery; immediately a sensible effect on needle. It oscillated and settled at last in original position. On *breaking* connection of A side with battery again a disturbance of the needle.

'Made all the wires on A side one coil and sent current from battery through the whole. Effect on needle much stronger than before.'

It is not clear what Faraday expected to happen to the galvanometer—some feet away from the coil—when he switched on the current but, apparently, he looked for a steady deflection: magnetism converted into electricity. He appears to have been very puzzled by the intermittent nature of the current in the secondary winding and for a week he carried out many other experiments using coils and magnets. Some failed through the insensitivity of his apparatus and misled him. But he carried on and when moving a flat wire helix between opposite magnetic poles the effect on his galvanometer was so pronounced that he recorded with ardent enthusiasm: 'Hence here distinct conversion of Magnetism into Electricity.' He then found that effects were still obtainable when the iron ring core was replaced by a wooden one though these were much more feeble.

The second major discovery in this series occurred on 17 October 1831. He covered a hollow paper cylinder ¾ inch diameter and about 8 inches long with a helix consisting of eight coils of copper wire each 22 to 32 feet in length and connected them in parallel to the galvanometer. Into this coil he inserted the end of a cylindrical bar magnet, then thrust it in quickly the whole length of the cylinder. There was a sudden deflection of the needle, but this died out at once and with the magnet lying in the coil no current was produced. On withdrawal of the magnet he observed a second kick and these intermittent effects led him to speak of 'a wave of electricity'.

About this time Faraday realized the inadequacy of his equipment at the Royal Institution and arranged to repeat some of the experiments on a large magnet, belonging to the Royal Society, which was kept at the house of his friend Mr. Christie. On 28 October 1831 he paid a visit with his coils and galvanometer. First of all he inserted a soft-iron cylinder ¾ inch diameter and 13 inches long in his now famous coil 'O' and connected it to the galvanometer at a distance of 10 feet. To quote the *Diary*, p. 380:

'By connecting the two poles (magnetic) by the soft-iron cylinder, when connection between the galvanometer and wires was not made, the galvanometer was very slightly affected, so little as to be barely sensible. But when wires were connected, then on making or breaking the magnetic contact with the iron cylinder—a powerful pull *whirling the Galvanometer needle* round many times was given.

'As the helix or cylinder were moved to or from the magnet, *not touching*, corresponding effects were exhibited by the galvanometer.'

And again:

'Brought helix O up suddenly between the large poles of the magnet; it having no iron bar in its axis. The needle was strongly affected; and also upon its removal as in former cases. This of course a mere effect of approximation and that not very near—not subject to any objection founded on motion of the iron exerting a momentary peculiar action at time of becoming a magnet—is directly connected with Arago's experiment.'

Arago's experiment had been bothering him during the whole

Maiden Lane Power Station, London, 1889
llans vertical high-speed engines coupled to Edison Hopkinson dynamos, two of 84 kW acity and two of 50 kW, supplied the famous Gatti's restaurant in the Strand and the Adelphi eatre. An extensive supply system, the Charing Cross Electricity Supply Co., grew out of pioneer effort. (*Photo: The Engineer*)

Regent's Park Station, St. Pancras, 1891
op 6-pole machines driven by triple-expansion Willans engines, an outstanding example of n dynamo design and station layout which preceded the turbine era.
oto: Central Electricity Generating Board)

Plate XVI

Eccleston Place Station, 1892

course of these electro-magnetic experiments of his own. On the very day of his great 'transformer' discovery he wrote:[7] 'May not these transient effects be connected with causes of differences between powers of metals in rest and in motion in Arago's experiment.' Again, after his series of successful induction experiments in early October 1831 he remarked: 'Still doubt that pure electromagnets will produce Arago's effect.'[8] On October 28 he experimented with a revolving copper plate 12 inches in diameter and about one-fifth of an inch thick, mounted on a horizontal brass axle. By means of small magnets six or seven inches long fixed to the pole pieces of Mr. Christie's powerful magnet he produced a short magnetic gap in which the copper plate could be revolved. To make contact with the rotating plate without introducing an undue amount of friction he amalgamated the edge with mercury and then applied two metal brushes connected to the galvanometer at different points on the disc. On November 4 he arrived at the arrangement in which one contact is made on the periphery of the disc and the second one on the axle and produced *powerful* currents which lasted as long as the plate revolved.

Faraday summarized all these results in his famous paper read before the Royal Society on 24 November 1831.[9] This covered all three of his recent remarkable experiments, the original ring experiment of 29 August, the magnet plunged into a coil on 17 October, and the generation of the first continuous current by induction. A most astonishing feature of this series is the fact that the actual discoveries were made in only ten days of experimentation, but the popular idea that they were sudden, unpremeditated results is quite wrong. The pages of the diary and the scores of carefully numbered notes show that Faraday gave lengthy and profound consideration to the many observations which he and others before him had made. Faraday directed his experiments with an intelligent appreciation of the underlying principles and the capacity to formulate a theory which led him to plan his experiments in such a manner as to produce results quickly. If there were fortunate accidents in his rich mine of discoveries he certainly deserved them all through the careful mental and practical preparation he made before carrying them out.

REFERENCES

1. *Faraday's Diary*, ed. by Thomas Martin. 7 vols. G. Bell and Sons, London.
2. *Faraday's Diary*, p. 45.
3. *Faraday's Diary*, p. 49.
4. *Quarterly Journal of Science*, vol. xii, pp. 186, 283.
5. *Faraday's Diary*, pp. 367–97.
6. *Journal I.E.E.*, vol. 69 (1931), p. 1358.
7. *Faraday's Diary*, p. 369.
8. *Faraday's Diary*, p. 375.
9. *Phil. Trans*, 1832, pp. 125–62.

CHAPTER VII

Development of the Dynamo

Faraday's achievements in 1831 quickly led to the production of practical machines for converting mechanical into electrical energy—dynamos, magnetos and alternators—and electrical energy into mechanical energy through the electric motor. The two forms of energy were found to be completely convertible and the one device would operate in either direction at will. Faraday himself carried the matter a stage further beyond his electro-magnetic rotations, his coil and magnet experiment, and his disc generator, through a number of simple though illuminating extensions of his original ideas—including his earth inductor. It was, however, through the news of his discoveries reaching the attention of many inventive minds that the vast movement was started which from that day to this has not once abated.

Within a few months of Faraday's announcement, an Italian, Signor Salvatore dal Negro, of Padua, made the first tentative steps by setting up four coils on a table and four small magnets on a carriage which could be advanced up to the coils and withdrawn again mechanically. This oscillating engine seems to have been rather a crude piece of apparatus, but the results satisfied the inventor, who wrote:[1] 'From the little I have done and from what I have said it follows that being able by this method to sum up the simultaneous action of an indefinite number of electric currents, this my battery may become fulminating.' While dal Negro was possibly overstating his own claims, there is no doubt that the outcome of Faraday's fundamental discovery was explosive in its influence on the progress of mankind. At the time of writing this chapter (1960) the Central Electricity Generating Board announce that they have just met an instantaneous demand of over 20,000 million watts. Signor dal Negro's generator, push and pull as fast as he could, would not have been able to produce more than a fraction of a watt, yet today many single generators have individual outputs of over one hundred million watts without, of course, the expenditure of any human muscle whatever.

But, to be practical, machinery must rotate and the first electromagnetic generator with this characteristic was invented by M. Hypolite Pixii, of Paris, about the same time as dal Negro's reciprocating machine appeared. Pixii fixed a soft-iron horse-shoe provided with two bobbins pole downwards from a wooden gantry and rotated a steel horse-shoe magnet below about a vertical axis with

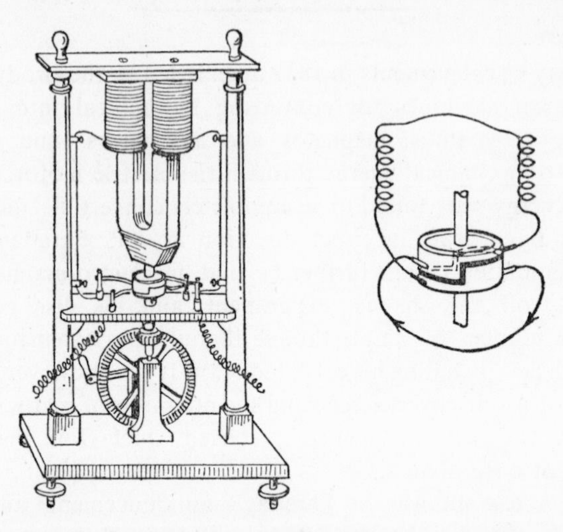

Fig. 17. Pixii's Magneto-electric Machine. At the end of 1831, Monsieur Hypolite Pixii of Paris rotated a permanent horse-shoe magnet about a vertical axis so that its poles came close to the poles of a wire-wound soft-iron core fixed in a suspending frame. He carried away the induced current through an early form of commutator.

its poles close to the bobbins. Rotation of the magnet by means of a hand crank and gearing produced an alternating current or a rapid succession of sparks. The machine was first made public at the meeting of the Academie des Sciences in Paris on 3 September 1832.[2]

There was a curious feature in Pixii's machine. He seems to have wanted more than was provided by the change of magnetic flux through his armature coils and deliberately inserted an interruptor to raise the output voltage and produce more sparks. This he did by suspending a wire point near the surface of a pool of mercury so that the vibration of the machine as the magnet was rotated caused the point to make intermittent contact with the mercury. On this machine Pixii used 3,000 feet of wire in his coils and gave very powerful shocks. In 1832, on the suggestion of Ampère, Pixii made a second machine to which he fitted a commutator, so converting the alternating current into direct current, at least unidirectional though obviously undulating.

In June 1833, the meeting of the British Association for the Advancement of Science was held in Cambridge and among the objects of scientific interest exhibited was a magneto-electric machine made by Saxton. This had a horizontal magnet made up of twelve horse-shoe plates and the armature, rotating on a horizontal spindle, was disposed opposite the ends of the magnet. The armature consisted of a four-armed soft-iron cross, to each of the arms being attached soft-iron cores carrying four coils. A feature of Saxton's machine which is of great interest today is one which showed how ideas of Ampère-turns, and dependence of output voltage on number of turns, were beginning to take shape. He wound opposite coils A and B with a double wire of one-sixtieth of an inch diameter, 400 yards long, for giving shocks, and the opposite pair of coils C and D with 22 lengths each 75 feet long of one-thirtieth of an inch diameter for generating sparks. The spindle which carried the 4-pole armature ran along the centre of the horse-shoe magnet and was driven by a belt from a vertical pulley and hand crank. Here, again, point collectors were used over a mercury bath, but Saxton added a device on the rotating spindle so that either coils A, B or C, D could be employed at will. A contact could be moved around through ninety degrees to ensure the point of breaking with the mercury being timed to the instant when the coil passed through the appropriate position relative to the magnet pole. Saxton's magneto-electric machine was on public view at the Adelaide Street Gallery, London, from August 1833 and received the approval of Professors Faraday, Daniell, and Wheatstone. In his statement Saxton recounts how in November 1833 Count di Predevalli brought over from Paris a Pixii machine which was operated alongside the Saxton machine in the Adelaide Gallery and found to give more powerful shocks than his own. He drew the conclusion that the difference in performance was due to the lengths of wire employed.

A London instrument-maker, Edward M. Clarke, of 39 Charles Street, Parliament Street, and later of 9 Agar Street, West Strand, became so active in the production of magneto-electric machines as to style himself in a paper which appeared in the *Philosophical Magazine* of October 1836 as 'Magnetician'.[3] In an earlier contribution published in March 1835 and dated 13 January 1835 he had described how he had been engaged for some time past in the manufacture of magnetic electrical machines and how he had been led to an important conclusion on the nature of their output. Holding in his hands the terminating wires he received a slight continuous thrilling sensation but on bringing them together and again separating them he received a powerful instantaneous shock passing through the arms. Without knowing it, this pioneer was of course using the phenomenon of self-induction, which became such a feature many years afterwards in magnetos used for ignition purposes.

In his 1836 communication Clarke came into public conflict with Saxton. With the aid of a detailed illustration he described a construction in which a powerful horse-shoe magnet was fixed to a wooden stand with its poles downwards. A soft-iron horse-shoe armature was carried on a horizontal spindle and rotated in close proximity to the sides of the magnet pole pieces. He employed alternative pairs of coils, one which he called the *quantity* armature consisting of 40 yards of thick copper bell wire[4] and the *intensity* armature in which he used 1,500 yards of fine insulated copper wire. Clarke summed up the performance of his machine as follows:

With the quantity coil:	Large and brilliant sparks.
	Induces magnetism.
	Ignites gunpowder.
	Produces scintillation from a small steel file.
	Produces rotary motion of delicately suspended wire frames round poles of vertical horse-shoe magnet.
With the intensity coil:	No person out of the hundreds who have tried it could possibly endure the intense agony it is capable of producing.
	It deflects a gold-leaf electroscope.
	Charges a Leyden Jar.

The words used by Clarke in 1836 which roused Saxton were these: 'From the time Dr. Faraday first discovered magnetic electricity, to the present my attention, as a philosophical instrument maker, has been entirely devoted to that important branch of science, more especially to the construction of an efficacious magnetical electrical machine which after much anxious thought labour and expense I now submit to your notice.'

Unfortunately Saxton had not published a detailed description of his machine, but in reply to Clarke he now did so[5] and concluded: 'I think it will be evident from the preceding statement that the magneto-electrical machine which Mr. Clarke has brought forward . . . is a piracy of mine.' Mr. Clarke returned with a long explanation of the improvements he had made. He had stopped vibration and had done other vital things and had had Saxton machines to repair!

While this dispute had been proceeding the English maker of magnets, Mr. Sturgeon, then Lecturer in Experimental Philosophy at the Honourable East India Company's Academy, Addiscombe, had introduced one more variation into the construction. He made a magneto with a shuttle type of coil which revolved *between* the poles of the magnet, 'so as that the wire may strike, at right angles, the most formidable group of magnetic lines'. Sturgeon also produced the first metallic—as distinct from mercury—commutator. In the same communication he says: 'My unio-directive discharger . . . is by far the most happy contrivance I have yet hit upon in this class of apparatus. It consists principally of four or more semi-cylindrical pieces, properly attached to a revolving spindle.' He goes on to show that as a result he has entirely dismissed the mercury discharger as a complete nuisance to the operator.

A few years later Stöhrer of Leipzig introduced the multipolar idea into magneto design in his attempts to reduce the vibration. Stöhrer made a six-pole machine by setting up three horse-shoe magnets with their poles uppermost forming a circle. Above these he rotated a frame carrying six coils and a commutator. From the designs available and scrutiny of the model on exhibition at the South Kensington Science Museum this attempt of Stöhrer clearly carried the tentative efforts of his predecessors far along the way to a reliable commercial machine although it was still hand driven.

At this period in development the 1851 Exhibition gave a great impetus to the progress of the new electrical engineering. No better assessment of the position can be obtained than by reference to the reports of the juries of experts which were set up by the Royal Commission to grant awards for inventions announced at the Exhibition. This book with its 800 or more pages makes fascinating reading today. Makers of electro-magnets were competing with one another to produce the greatest lifting power and one mammoth—for those days—with 50 yards of winding in the coils weighed a hundredweight and could lift a load of one ton. Crude attempts at making a practical commutator were on display and the judges' reference to one of these, 'the nicety of construction consisting in the adjustment of the moment of union and disruption so as to obtain an effective difference of action always in the direction between the two opposite poles of the permanent magnet', is interesting. Great ingenuity appears to have been displayed in the Exhibition in the production of a wide range of 'electro-magnetic engines' and considerable interest was evinced in their operation. Both the rotating and reciprocating arrangement of coil and magnet were adopted and, from the dimensions referred to, it is clear that the toy stage had been left far behind. We read, for instance, that in one machine for which a prize medal was awarded the horizontal movement of the piston with a section 6 inches square was 16 inches and that pulls of 160 lb. were recorded on test. The mechanical generation of the electric current, at any rate on a small scale, was well established at the time of the Exhibition.

Permanent magnets were employed and the rotating coils moved in front of the magnets. By the addition of the commutator a limited number of machines were actually being used for electro-plating which, by then, was becoming an industrial process. Woolrich, some years before, had produced a continuous as distinct from an intermittent, unidirectional current by employing on a magneto machine more coils than magnets and taking the current only from the coils active at the moment of contact. These machines were already in use in Birmingham for electro-plating and Wheatstone had carried the principle further by employing five separate armatures with individual commutators in series so picking off the current in a continuous form. The Woolrich machine in the Science Museum is arranged to supply two plating baths simultaneously.

About the time of the 1851 Exhibition there was great interest in applying electricity as the source of illumination for lighthouses, and this gave a considerable impetus to the design of magneto-electric machines. Professor Nollet of Brussels, who was engaged on a scheme for producing oxygen and hydrogen from water by electrolysis for limelight, took out a British patent in 1850 for an improved Stöhrer machine to be driven by steam power. A few months later William Millward of Birmingham obtained a patent for a machine with quite new features. He arranged a series of eight horse-shoe magnets in a circle with their poles pointing inwards and rotated a wheel carrying sixteen bobbins which passed between the poles of all the magnets in turn. The following year, 1852, Nollet's English agent filed a further application in which Millward's ring was duplicated with the magnets turned through ninety degrees so that the coils instead of passing between the poles ran past them. An Anglo-French company, 'La Compagnie de l'Alliance', was formed to develop these ideas but Nollet died and the project fell through. An Englishman, Professor Frederick Hale Holmes, however, who had worked with the French Company, returned to England convinced that he could design a machine which would meet the lighthouse requirements. As it turned out, the progress of the development of dynamo electric machines for the next fifteen years became channelled through lighthouse illumination and the greatest contributor was Holmes, whose design held the field for many years.

The story of Holmes's success is told in a comprehensive paper read by Sir James Nicholas Douglass at an International Exhibition held in Melbourne and printed in the *Proceedings of the Institution of Civil Engineers* in 1880.[6] On receiving a request from Holmes to be allowed to supply a machine and arc lamps for use in lighthouses the Elder Brethren of Trinity House sought the advice of Faraday. A preliminary trial took place at Blackwall in 1857 under the direction and 'to the great delight of Faraday'. The machine, which was 5 feet long, 5 feet wide and 4¼ feet high, weighed 2 tons. The frame was of wood and carried 120 coils carried by 5 rings with 24 coils in each. The rotating component had 36 compound permanent magnets, each weighing 50 lb., on 6 wheels each having 6 magnets. A D.C. output was obtained through a large commutator fitted with rollers and the power was 2½ horse power.

The result of this trial was approved by Faraday, while Holmes received an order for two machines to be installed in the South Foreland lighthouse. It was stipulated that the speed must not exceed 90 revolutions a minute and Holmes produced a considerably modified design. This time the magnets were stationary, were 60 in number, each weighing 48 lb., and carried in three vertical planes. The two coil-carrying wheels each with 80 coils rotated at 90 r.p.m. and the frame, 9 ft. 3 in. long, 5 ft. 6 in. wide and 9 ft. 6 in. high, was made of iron and weighed 5¼ tons. Each machine absorbed 2¾ h.p. and was driven by a non-condensing steam engine through a belt drive.

On 8 December 1858 at South Foreland high lighthouse, the electric light was thrown on the sea for the first time and the romance of the event caused Douglass to write: 'Thus were magnets serving, not only in the compass to direct the mariner in his course, but also in producing a most intense light to warn him of danger and guide him on his path.' In his report Faraday said: 'I beg to state that in my opinion Professor Holmes has practically established the fitness and sufficiency of the magneto-electric light for lighthouse purposes so far as its nature and management are concerned. The light produced is powerful beyond any other that I have yet seen so applied and in principle may be accumulated to any degree; its regularity in the lantern is great, its management easy, and its care can therefore be confided to attentive keepers of the ordinary degree of intellect and knowledge.'

Encouraged by Faraday, Trinity House decided to continue with electric light for their lighthouses and, after some teething troubles, a further Holmes machine was taken into use in June 1862 which ran for thirteen years, and five years later two more were ordered for Souter Point, between Sunderland and Shields. This 1867 design had fixed magnets and rotating coils and, before installation, the machine was sent to the Paris Exhibition. It ran at 400 r.p.m., had no commutator and took off the A.C. with wire brushes. These two machines remained in service until 1900 and one has been preserved in the Science Museum at South Kensington.

The Alliance Company had come back in France with an improved machine. At this same period a French design of magneto-machine with new features appeared, devised by A. de Meriteus, in which coils were replaced by a distributed winding to give a more uniform current. These machines became popular both in France and Britain and many were installed here and abroad; but other influences were coming to bear on the design of machines for generating the electric current and the days of the simple magneto-electric machine were numbered. It is interesting to note, however, that during their popularity for lighthouse lighting, figures indicating remarkable improvements in efficiency were obtained. In the Holmes machine installed in 1862 at Dungeness lighthouse, which had a beam strength of 19,000 candle power, the costs compared with oil were 29·12 pence per hour for oil and 86·7 pence for electricity, a ratio of 100 to 298, but when corrected for the greater strength of the electric beam the cost per candle power per hour was 0·1165 pence for oil and 0·1294 pence for electricity, a ratio of 100 for oil to 111 for electricity. To run ahead of our chronological sequence for a moment, similar tests taken on the Lizard light in 1873 supplied by a Siemens dynamo gave figures of 0·1047 pence per hour per candle power for oil and 0·0147 for electricity, a ratio of 100 for oil to 14 for electricity. In spite of all this interest, however, Douglass tells us that in 1880 there were only ten electric lighthouses in the world—five in Britain, three in France, one at Odessa, and one at Port Said.

As long ago as 1838 the principle of using electro-magnets in dynamo-electric machines instead of permanent magnets was being explored. The Abbés Moigno and Raillard laid the foundation for replacing the permanent magnet in a magneto-electric machine by an electro-magnet[7] through an observation on the result of energizing a large lifting magnet by the current from a magneto machine, the magnet of which would only lift a relatively small weight. Moigno says that in July 1838 he and the Abbé Raillard were able to carry on M. Pouillet's great magnet a load of 600 kilogrammes by the small machine of M. Billand, which would carry only a few grammes. It was only a step to converting the big magnet with its enormous (for those days) lifting power into a magneto-electric machine and so continuing the multiplication of the effect which was looked upon as a conversion of magnetism into electricity.

About the same time Professor Page of Washington was interested in the same problem and found that by adding a current-carrying coil around the permanent magnet he 'increased the effects of the original magnet greatly'.[8] In Germany another inventor, Sinsteden, took up the same idea in 1851. In his scheme four-coiled armatures with cylindrical commutators were used; the permanent magnet of the first machine, capable of lifting 200 lb., gave a current to the electro-magnet of the second machine which would lift 500 lb. This in turn fed a third machine the magnet of which would lift 1,000 lb. Thus was there a vista of enormous currents opened up far beyond the capacity of the most gigantic batteries known. At this stage, in 1856, Dr. Werner von Siemens made the fundamental invention of the shuttle-wound longitudinal armature rotating in a closely encircling tunnel in the magnet pole pieces. The wire was wound in deep longitudinal slots and the ends brought out to the two curved contact plates of a simple commutator.

Throughout its history electrical engineering has been under a debt to many German engineers and among these no name stands higher than that of the family Siemens. Of three brothers, Werner, William and Karl, the reputation of William (1823–83), particularly in this country, places him among the pioneers, as founder of the great firm of Siemens Brothers.

Carl Wilhelm Siemens, as he was christened, was the seventh son of a Hanover farmer of liberal education and studious habits. His mother is recorded as being a woman of refined tastes and fond of poetry. At an early age, however, William lost both parents and his brother, Werner, seven years his senior, assumed the position of trusted adviser. While still a young army officer, Werner saw the possibilities in the development of the electro-telegraph and soon afterwards decided to devote all his time to electrical pursuits, which he carried to great success all through his life, honours being showered upon him. The other brothers were successful in business pursuits and a significant harmony existed between them for many years which emerged in a remarkable *Siemens Stift* founded for the promotion of good feeling in the family. Once every five years, all members, rich and poor, came together at a pleasant locality in the Harz Mountains to pass a day or two in social intercourse and help was provided for the less fortunate members from a special fund.[9]

This family relationship was no doubt a vital factor in the subsequent life of William Siemens, for when Werner had a patent for a system of electric plating to dispose of, he commissioned William to bring it to England.

He arrived in London in the year 1843 to visit the firm of Elkingtons, who were already established in the plating business in Birmingham. The young representative—he was under twenty years of age—had all the disadvantages of a foreigner for carrying out his assignment, but he was a hard worker with a technical college education embracing mathematics, chemistry and physics and with, moreover, ambition and a skill in advocacy. His brother had agreed terms in which William, if successful, was to receive £800 for his part in the transaction. In this he was successful and with a sum, so considerable in those days, in his pocket, he forthwith adopted England as his future home, with high hopes of success in electrical and other fields of engineering.

After some relatively unproductive years during which he worked on ideas for a regenerative furnace, he turned to the manufacture of electric cables. In the meantime his brother Werner in Germany had become associated with Halske. When the German company of Siemens and Halske was formed, William became the London agent, and on 1 October 1858 a London company was established with William Siemens taking one-third of the profits. Thus was the basis laid for Siemens Bros., though not until 1864 did the firm assume its present name.

William Siemens soon became a recognized expert in submarine cables and published papers on the subject from both the theoretical and practical angles. At the historic Commission of Inquiry in 1859–60 he was an important witness and in 1874 he designed and launched the first cable ship, *Faraday*.

When the dynamo became a practical proposition, he had turned his attention to its design and in February 1867 had communicated a paper to the Royal Society in which he demonstrated that permanent magnets were not necessary to convert mechanical into electrical energy. In this paper he said 'the result obtained by this experiment

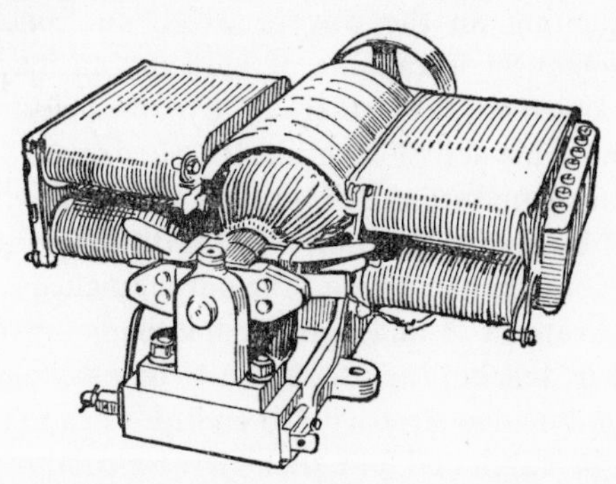

Fig. 18. Siemens' Dynamo, c. 1877. The field magnets were composed of fourteen bars of wrought iron, seven above and seven below the armature, and bent to produce the armature tunnel. The strips were connected by iron yokes at the two ends. The field coils were arranged to give consequent poles so producing a vertical field through the armature. This was formed by winding iron wire on a wooden core. Twelve groups of longitudinal windings held by wooden pegs were connected to a 24-segment commutator. The machine gave an output of 20 amperes at 50 volts.

is remarkable, not only because it demonstrates this hitherto unrecognized fact, but also because it provides a simple means of producing very powerful electrical effects'. The first patent was taken out on 31 January 1867 and within a few years he had developed the practical machine so well known by his name in which the field magnet was in two parts, the poles facing one another on opposite sides of the armature. This actual machine was probably used later by Hopkinson at King's College.

Almost simultaneously with Siemens, Wheatstone and Varley published the results of similar experiments which they had carried out, resulting in a good deal of confusion on priority of ideas. Moreover, a further claimant appeared in Wilde, who in 1866 brought the various experimental suggestions into the form of a commercial machine. He had found that by starting with a permanent magnet which would sustain a load of 40 lb., using this as a magneto machine, taking the current from this to field magnets of a second and then repeating the process he could at the third stage obtain a lifting power of 25 tons. A description of Wilde's invention was read by Faraday before the Royal Society in March 1866.[10]

After commenting on the way in which an 'indefinitely small amount of magnetism is capable of inducing an indefinitely large amount of magnetism or further, of dynamic electricity', Wilde describes in detail his actual experiments. He drove a magneto with a $1\frac{5}{8}$ in. diameter armature at 2,500 r.p.m. and found it made a 3-in. length of 0·04 iron wire red hot. Passing the current to the field magnet of a second machine (5 in.) the output melted 8 inches of the same wire and heated a 24-in. length red hot. By going from the $1\frac{5}{8}$ in. machine to a 5 in. machine and then to a 10 in. machine he produced an evolution of dynamic electricity so enormous as to melt pieces of iron rod 15 in. in length and $\frac{1}{4}$ in. diameter. He continued:

'The illuminating power of the electricity from the *intensity* armature is of the most splendid description. When an electric lamp furnished with rods of gas carbon half an inch square was placed at the top of a lofty building the light evolved from it was sufficient to cast the shadows from the flames of the street lamps a quarter of a mile distant upon the neighbouring walls. When viewed from that distance the rays proceeding from the reflector have all the rich effulgence of sunshine.'

An unfortunate controversy occurred about the year 1900 over Wilde's claims to have invented the dynamo. He refused to accept the terms of the citation adopted by the Royal Society of Arts in awarding him their Albert Medal as he considered they had not accorded him adequate credit for what he had done. Later on, in Silvanus P. Thompson's famous book, *Dynamo Electric Machinery*, the author stated that Siemens coined the term *dynamo electric machine* to describe his (Siemens') self-exciting machine. Wilde again came to the attack and the case went to the Courts, but was dismissed. Further correspondence in the columns of *The Electrician*[11] suggested some confusion between the recollection of what Siemens had claimed in his Berlin discourse and what was actually printed in the proceedings. Whatever the facts, the seventh edition of

Dynamo Electric Machinery contained modified wording on this point to the effect that the actual record of the Berlin Academy does not support the contention that Siemens was responsible for the term Dynamo Electric Machine. Thompson stuck to his point in the correspondence stating that both Werner and Charles Siemens were using the term dynamo-electric machine in 1866–7 to denote a self-exciting machine. In the confusion some help was given by Mr. Brooke, who had suggested[12] that the first section of a compound term signifies the acting cause and the second the resulting effect so that a dynamo-electric machine was one in which energy is applied to produce an output of electricity.

In his early experiments Wilde had come very near to discovering the possibility of starting up a dynamo from the residual magnetism in its core and subsequently depending entirely on the armature current for the excitation of the field magnets. In March 1867 he went the whole way and effected self-excitation by diverting some of the armature current through the field winding on some machines which he constructed for the electro-deposition of metals. He also supplied machines for the first use of electric searchlights on battleships. Many minds were studying the problem at this time, Farmer in America, Wheatstone and Varley in England and Werner Siemens in Germany, with the result that in the course of a few months the principles of both 'shunt' and 'series' windings were recognized and adopted. Siemens connected his field coils in series with the supply to the load while Wheatstone adopted the shunt principle from the start.

Wilde seems to have been singularly unfortunate in the overlapping of his own inventions with those of others and, to anticipate our narrative by some twenty years, an interesting example of this occurred later in connection with the running of alternators in parallel. In 1883 Dr. John Hopkinson read a paper before the Society of Telegraph Engineers on 'Some Points on Electric Lighting', in which he failed to give credit to Wilde's earlier work. In the following year, however, he made amends for, in his paper on 'Theory of Alternating Currents, particularly in reference to Two Alternate Current Machines connected to the same circuit'[13] appeared the following:

'My attention has only today been called to a paper by Mr. Wilde published by The Literary and Philosophical Society of Manchester, December 15, 1868, also *Philosophical Magazine*, January 1869. Mr. Wilde fully describes observations on the synchronising control between two or more alternating-current machines connected together. I am sorry I did not know of his observations when I lectured before the Institution of Civil Engineers, that I might have given him the honour which was his due. If his paper had been known to those who have lately been working to produce large alternate-current machines, it would have saved them both labour and money.'

The various machines being constructed about 1867–8 all suffered from the defect that the strength of the current fluctuated to an inconvenient extent, but a solution was found in the invention of the ring wound armature. This device, which was suggested about 1860 by an Italian, Pacinotti, was virtually unnoticed until the same idea was re-invented by Gramme, a Belgian, in 1870. Pacinotti's machine had a circular magnet inside which rotated on a vertical spindle a toothed iron wheel. On this wheel were wound sixteen coils between the teeth. The coils were joined in series and each junction connected to one of sixteen commutator 'bars'. Thus very much steadier current was obtained than previously. When Gramme, attracted by the same idea, constructed his machine he adopted an entirely different form of field magnet and rotated the ring armature on a horizontal spindle between field magnet poles above and below. The armature differed from Pacinotti's in having an increased thickness and it had many more sections, thus leading very clearly to the modern many-bar commutator and a still more uniform current. Gramme-type machines were produced in large numbers and some of them remained in use for many years. The Gramme machine opened up a new era in electrical engineering: the electric current was no longer experimental but could now be made available on a practical scale from power-driven dynamos of reliable construction. In its already extensive applications to electro-plating and arc lighting the electric current was established as a vital factor in civilized life and applications were being suggested in many directions.

Now that practical dynamos were being produced on a commercial scale the question of efficiency and cost became important and modifications were introduced with these factors in mind. Hefner Alteneck in 1873, for instance, modified the 'ring' and 'shuttle' designs of armature by adopting the drum principle, in which copper conductors were wound over the external surface of a drum in a symmetrical fashion. This construction, in comparison with the ring design, eliminated idle copper and enabled coils to be preformed and applied to the armature complete, in place of the irksome threading

Plate XVII

(*a*) Original Parsons Turbine, 1884
Within a few years of completing his engineering apprenticeship, Charles Parsons produced a number of inventions, and before the age of 30 became interested in the fundamental problem of generating electricity—should it be low-speed engines driving high-speed dynamos by belt or direct coupled to multipolar dynamos? He adopted as an alternative solution the idea of direct drive from a turbine operated by steam and the photograph shows his first model, which opened up a new era in power supply. Its output was five horse-power and it ran at 1800 r.p.m.
(*Photo: Science Museum*)

(*b*) First Condensing Turbo-generator, Forth Banks Station, 1891–2
After constructing a dozen or more turbo generators of increasing size he carried out the first turbine installation in this public supply station with units of 200 horse-power and employing condensers. They ran at 4800 r.p.m. and generated current at 1000 volts.
(*Photo: C. A. Parsons Ltd.*)

(*a*) First Tandem Turbogenerator, 1900
One of two single-phase sets to run at 1500 r.p.m. for the city of Elberfeld. They were the largest yet built anywhere in the world and achieved outstanding results on test.
(*Photo: C. A. Parsons Ltd.*)

(*b*) Fulham Power Station, 1901
An example of the last phase of slow-speed engines—Corliss in this case—driving large-diameter multipolar machines. The near machine has the upper half of the stator removed. Generators: 300 kW, 3000 volts, 50 cycle, 2 phase.
(*Photo: General Electric Co. Ltd.*)

of the conductors through the ring. At first the coils were laid on a smooth cylindrical surface and retained in position by pegs and binding wire, but very soon this gave place to present-day practice in which they are laid in longitudinal slots on the periphery of the drum.

In 1877 a Government Commission on the various electric generators available was appointed in England which, among other experts, contained the names of Tyndall, Douglass and Sabine. This resulted in the appointment of a similar commission in America and a valuable summary of their findings was published by a French author, Th. Du Moncel.[14]

In some of the early dynamos made under a joint patent by Werner von Siemens and F. von Hefner Alteneck, wooden cores were used with iron wire wound over the surface before the conductors were applied. One of these machines supplied current for the lighting of the Science Museum at South Kensington from 1878 to 1899, and is now on display, partially sectioned, in the museum. The field magnet was formed of seven horizontal strips of wrought iron laid side by side and curved up in the centre of their length to lap over the armature while another seven bars curved down in the middle of their length to pass under the armature. The fourteen bars were connected together at both ends by iron spacers and on each of the four grouped sections were disposed one of the four field coils. The coils were so connected as to give the upper centre curved portion one polarity and the lower portion the opposite polarity so as to produce a vertical field in which the armature rotated.

The firm of Siemens Halske in Germany became very active in the manufacture of improved dynamos and the horizontal design was soon replaced by a vertical and then by the 'over' type. In 1882 compound winding of the field coils was introduced, the series winding being added to the usual shunt coils in order to compensate for voltage drop on load. In 1885 many vertical machines were exhibited including three having outputs of over 100 kilowatts and by 1890 dynamos weighing a ton and a half producing currents of 1,500 amperes with armatures 2 feet in diameter and 3 feet long were in production. Ten years later, the firm Siemens Brothers showed at the Paris Exhibition a machine for 1,570 kilowatts running at 200 revolutions a minutes.

In 1880 there came a great impulse to electrical progress through a paper read by Swan on the new filament lamp which he had invented and started to manufacture at Newcastle.[15] The announcement was a historic landmark and aroused national interest in electrical engineering. It was seen as a promising field for young men both in its industrial opportunities and its educational requirements. Important manufacturing enterprises sprang into being and University courses were taken aimed at a career in electrical engineering. In the development of the direct current dynamo at this time two names stand out prominently—Colonel Crompton and Dr. John Hopkinson. This does not mean that they alone represented electrical engineering; Crompton, for instance, employed Gisbert Kapp and James Swinburne. The young man Ferranti, then a student at University College, during time off from his studies took an amateur but active interest in the electrical installations which Crompton's firm was carrying out at Alexandra Palace and Sir William Thomson, later Lord Kelvin, was in close touch as a consultant.

Crompton set up a factory at Chelmsford to construct arc lamps and imported Gramme dynamos from the Continent. These proved unsatisfactory, however, for series arc-lamp circuits and he changed over to a machine introduced by Bürgin, a Swiss, which was soon being made in large numbers. The technical difficulties at this time were a serious handicap. As Crompton has pointed out,[16] they had at the time no satisfactory names for the electrical phenomena with which they had to deal—such as current, electro-motive force, or pressure, and the resistance of the circuit, names which were only introduced at the Paris Exhibition in 1881. Moreover, they had no formulae on which to design their machines. When they found they could produce a dynamo which would work six or eight arc lamps in series it was a noteworthy achievement, and they proceeded to take orders for railway yard and dock lighting. By his enthusiasm, foresight and business flair, Crompton soon secured for this country preeminence both for the manufacture of dynamos and the carrying out of important lighting installations.

The other great name from this period was that of Professor John Hopkinson (1849–98), closely associated with Cambridge and King's College, London. As a mathematician of a very high order, Hopkinson contributed effectively to the fundamental development of dynamo design, power transmission and alternating current working but, before considering his position in electrical engineering, let us take a look at the environment in which as a boy and a youth he developed those sterling qualities so strongly indentified in his character during later years.

Hopkinson came of that hard-working, middle-class, non-conformist stock, with its deep religious convictions and high intellectual ambitions, which was such a striking social feature of industrial England during the latter part of the nineteenth century. His father was a Lancashire millwright who by dint of hard work and ability had risen to be partner in a long-established and important Manchester firm of engineers. As an indication of his strength of character, we read in the autobiography of Alfred Hopkinson[17] how, after an exacting week controlling a large staff in the factory, he would start off early on Sunday morning to conduct the Sunday School for men, which was such an important feature of the times. Here some four or five hundred would gather for religious and social instruction, men of all ages, including children, parents and grandparents.

John Hopkinson's mother must have been an equally remarkable person for she produced a family of thirteen children, several of whom achieved outstanding careers: John, the eldest, became the electrical engineer, while Alfred, the second son, rose to a position of eminence as a K.C., Member of Parliament and Vice-Chancellor of a University. Both father and mother played a vital part in the intellectual achievements of their brilliant children. The father, always interested in the developments of electricity going on at the time, set up simple experiments in an attic and so secured John's early interest.

Electric traction was not yet known and there was no electric power. The telephone was scarcely thought of and radio was of course a remote dream, but with glass cylinder and rubber, a Leyden jar, a stool on glass legs, and a plating bath, young John's appetite was stimulated. The mother's influence, while as stimulating, took another turn. She encouraged the boys to read widely and a copy of Locke's *On the Human Understanding* took its place alongside the Bible.

John early developed a critical faculty, for, at the age of twelve, he wrote from his boarding-school: 'I have got a better opinion of Mr. Hill (a master) than I had: he is much improved!' Intelligent and industrious, he was already making satisfactory progress in his studies and within the next few years he developed an extremely independent—even a perverse—spirit. His approach to religious matters was more reticent than his parents would have wished and his sprawling untidy writing was in strong contrast to their exquisite neatness. But in spirit he was sound and he was soon singled out for the best possible educational facilities.

In 1865, when John was sixteen, his father, with considerable vision, sent him to the recently founded Owen's College to take a three-year course in theoretical engineering. Here he came under the influence of good teachers and was inspired by success in mathematics to look towards a career at Cambridge. In 1867 he obtained an open scholarship of £100 and entered Trinity College.

With a determination to reach the top, Hopkinson applied himself under the great tutor Routh and, although he also played hard, he kept on to his goal. During the course he took the London B.Sc. as an external student and in 1871 became Senior Wrangler as the result of his Mathematical Tripos. A little later he took first place in the Smith's Prize and was elected to a Fellowship at Trinity. This was much more than a personal triumph for Hopkinson: it marked the beginning of an era in the admission of nonconformists to Cambridge. After much agitation, subscription to the Thirty-nine Articles of Religion had only recently been removed from the requirements for the M.A. and nonconformists were still excluded from fellowships. It was largely due to the position taken by the few, including Hopkinson, that religious tests in the Universities were abolished. Hopkinson frequently said in later life, 'They made a lion of me and made me roar against the disabilities of the nonconformist.'

On going down from Cambridge, Hopkinson spent a few months in his father's works gaining practical experience and then took up a post with the famous firm of Chance, the optical glass manufacturers of Birmingham. About the same time he married Evelyn Oldenbourg, a young woman of German extraction and high intellectual attainments. In later life, she wrote a charming autobiography which contains much of historical interest on the Victorian scene as well as personal reminiscences of Hopkinson.[18]

Hopkinson remained with the Chance organization for about five years and then in 1877 set up in consulting practice in London, his original interest being lighthouse illumination in which he retained his contact with the firm. During this period, important papers on different phases of electrical engineering began to appear from his pen and some will be referred to later. To complete this biographical note, reference must be made to his forensic skill which produced an outstanding court success in his famous 3-wire case in which he was defending his own patent against infringement.

As his practice extended, Hopkinson was offered by King's College, in the University of London, the post of Professor of Electrical Engineering, which in 1889 he accepted. In this position he attained great distinction and was consulted on a wide range of electrical subjects at the highest levels. Unfortunately, this brilliant career was brought to an untimely and tragic end by an accident in the Alps. Hopkinson was an experienced mountaineer and took every opportunity of following his beloved recreation in Switzerland. In the early hours of 27 August 1898, at the age of forty-nine, with his eighteen-year-old son Jack and two daughters, twenty-three and nineteen, he said goodbye to his wife in an Arolla hotel as they set out to climb on the Petite Dent. They never returned and their bodies are interred in the little cemetery at Territet.

This brief summary cannot convey an adequate estimate of the outstanding contribution made by Hopkinson. As both mathematician and engineer, he reached the pinnacle of achievement in his day, a fact which can only be appreciated by a study of the many papers he wrote. A valuable analysis of our indebtedness to this remarkable man is made in the Centenary Lecture which was delivered on 14 November 1948 by a later holder of the Chair of Electrical Engineering at King's College, Professor James Greig.[19]

The contributions of Hopkinson to electrical engineering were unique. What the practical dynamo designer was mostly in need of was information about the magnetic circuit. No other evidence on this fact is necessary than a comparison of the appearance of different machines as they were being made—long slender field magnets as exhibited by Edison, short stout ones by Siemens and Crompton. Hopkinson, after his distinguished career at Cambridge and his experience with Chance Brothers, produced a paper 'On Electric Lighting' before the Institution of Mechanical Engineers, covering the research work which he had already carried out. This included the results of tests on a Siemens dynamo which is believed to be the one now preserved at King's College in the Strand. From then on there appeared many papers under his name before the Royal Society and the Engineering Institutions. In 1885 a comprehensive paper on 'Magnetization of Steel, Cast Iron and Soft Iron' was read before the Royal Society[20] by J. W. Gemwell, and the following year Hopkinson contributed a paper on 'Dynamo Electric Machines',[21] in which he gave the results of a series of careful tests carried out in the works of Mather and Platt Ltd. He defined the investigation as one to give 'an approximately complete construction of the characteristic curve of a dynamo of given form from the ordinary laws of electro-magnetism, and the known properties of the iron, and to compare the results of such construction with the actual characteristics of the machine'. Hopkinson had been greatly influenced by Maxwell's treatise on *Electricity and Magnetism* published in 1872 and he used this as a basis for his dynamo design. Thus was established the concept of the magnetic circuit with its magneto-motive-force magnetic flux and reluctance. No doubt Crompton was influenced in the major changes which he introduced in the design of his machines to obtain more powerful fields—large sections, wrought instead of cast iron—by this published work of Hopkinson's and the close friendship which developed between the two men.

Other features introduced by Hopkinson were the well-known 'Hopkinson' or 'Back-to-Back' test for determining the characteristics of two similar machines with only a small driving power, and the 3-wire system of distribution over which there was an important court case and in which he was completely vindicated. He also designed the famous 'Manchester' type dynamo as well as introducing considerable improvement in the Edison machines by eliminating the multiple magnet cores. Later we shall have occasion to refer to his contributions in the alternating current field.

A type of dynamo designed specially for series arclighting circuits which had a great vogue about 1880 was known as the 'open coil' type. In both the ring and drum type of armature, already referred to, the winding is continuous in itself and is independent of the commutator, which only makes contact with different parts of the continuous circuit in turn. In the open-coil type of machine the individual ends are normally disconnected from the commutator and the external circuit and are only connected at parts of the revolution when, owing to their position in the magnetic field at the moment, they are generating a suitable e.m.f.

Two quite different forms of 'open coil' dynamos achieved success in the arclighting era, both designed by Americans. In 1878 Mr. C. F. Brush patented a machine with two U-shaped field magnets with their poles facing one another across a gap in which the armature rotated on a horizontal axis. There were four magnet coils, one on each limb of the two magnets, and the magnet cores had cast-iron extensions on the pole pieces. The armature, of cast iron, was of the ring type and had eight coils recessed in grooves, the opposite pairs of which were connected in series and carried to a pair of segments on the special commutator. The peculiar scheme adopted in the commutator was to connect in addition to opposite pairs in series, two other pairs in parallel and leave the remaining pair—at the moment of inaction—entirely disconnected. A special device reduced the voltage should one or more of the arc lamps in the series being supplied become extinguished. The early example now displayed in the Science Museum ran at 1,200 r.p.m. and gave a current of 10 amperes at 50 volts, or 10 amperes at 90 volts when the speed was raised to 1,325 r.p.m. During the time Brush arclighting machines were in use many improvements were made to them, including substitution for the cast-iron armature one of laminated iron plates.

The second 'open coil' arclighting dynamo to make its appearance was devised by Dr. Elihu Thomson and Professor E. J. Houston, who founded the Thomson Houston Company at Lynn, Massachusetts, in 1883. In the Thomson Houston machine the field system consisted of two large-diameter coils with their common axes horizontal surrounding tubular cast-iron cores with large cup-shaped pole pieces facing each other, so forming a spherical space. In this gap the ball-shaped armature was accommodated. The core was wound over with varnished iron wire to improve its magnetic properties and the conductors were in the form of three coils spaced at an angle of 60 degrees to one another. The coils were connected together at one end and the other ends were brought out to a three-part commutator. In this way the circuit between brushes included two coils. The machine was designed to give a constant current of 10 amperes and the voltage was regulated, in accordance with the number of arc lamps connected, by an ingenious regulator in which a relay-operated magnet rotated the brush system. The field windings were series-connected and an air blast was provided to blow out the spark at the brushes as they broke contact.

The other great American name in electrical engineering at this time was, of course, that of Edison. At the age of about thirty, after many years of experiment and invention in the field of telegraphy, this remarkable man became interested in the development of the electric light and with it the design and construction of dynamos. A fellow countryman, William Wallace, had built the first American dynamo in February 1874 and in 1879 Edison made his first machine which was supplied to the ill-fated Jeannette Arctic expedition. It was intended to operate one arc lamp with some incandescent lamps in series and was later rated by him as a 15-lamp machine. It had a simple bipolar magnet with a drum armature of soft iron wire and was belt driven. Within a few months Edison and his colleagues at the New Jersey laboratories produced a second machine on a quite ambitious scale with field magnets 54 inches high and weighing 1,100 lb., which was the largest dynamo made up to that date. The cores of wrought iron were 6 inches in diameter, and what is interesting today, all the surfaces between the ends of these and the yoke and pole pieces were carefully scraped to produce the closest possible contact.

Edison was very interested in magnetism and no doubt through his experience in making telegraph instruments had formulated the idea of a magnetic circuit as there is an electric circuit. From his writings it appears that in his dynamo design studies he was always influenced by his observations, as a boy, on the enormous increase in attractive force between a magnet and its keeper as their distance apart was reduced to small dimensions. Thus he appreciated the factors which must be observed to secure the maximum field strength and, with the crudest of equipment at Menlo Park, he carried out valuable tests.

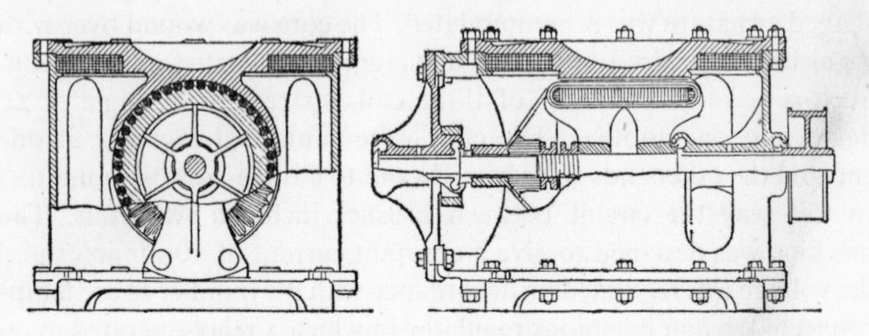

Fig. 19. ASEA Dynamo, 1883. A 30-volt 15-ampere 600 r.p.m. dynamo made in Sweden for the magnetic separation of iron ore. The machine was pot-shaped and almost completely enclosed.

Current at that time was measured in Webers and his assistant Upton established a conclusion which later was recognized as fundamental in dynamo design, that the product weber-turns, or as we now say, ampere-turns, was the fundamental criterion in the magnetic circuit. With the current measured by electrolytic deposition and the power input by a rough belt dynamometer the famous second dynamo, known as 'Long-Legged Mary Ann', gave an efficiency of 82 per cent, no mean achievement. By inserting thermometers into holes in the iron in various parts of the iron cores, he was able to locate his losses. It has been claimed with much justification that, with far less mathematical equipment than his successors, Edison paved the way in dynamo design followed by Hopkinson, and by Kapp, with whom he later collaborated. He no doubt relied to a great extent however on his assistant Upton, a graduate of Princeton and pupil of Von Helmholtz.

During the next few years Edison built a number of dynamos which were characterized by their long multi-core magnets and massive construction and, through new companies formed abroad, exported them to England and other European countries. One which became famous as the 'Jumbo', first exhibited at Paris in 1881 and subsequently installed in London at the Holborn Viaduct Station in 1882, was rated as 1,500 light 110 volt and was steam driven by a 150 h.p. Porter-Allen engine. Similar machines were installed at Milan and in the Pearl Street Station, New York.

The decade from 1880 to 1890 proved to be a period of rapid development in the dynamo. Detailed design received much attention so that efficiency and reliability were raised to high levels. Figures of 95 per cent in efficiency were attained and continuous running for long periods on load without failure or overheating became common practice. This situation led quickly to the development of public supply systems and this again resulted in a further impetus to the production of new and improved forms of generator which has continued up to the present day.

REFERENCES

1. *Phil. Mag.*, 3 series, vol. i. Pub. July 1832.
2. *Ann. Chimie. Phys.*, vol. 50 (1832), p. 322; and vol. 51 (1832), p. 76.
3. *Phil. Mag.*, 3 series, vol. 9 (1836), p. 360.
4. Bell wire meant in those days, of course, the uninsulated wire used for mechanical bells.
5. *Phil. Mag.*, 3 series, vol. 9 (1836), p. 360.
6. Douglass, *Proc. I.C.E.*, 1878–9, Part III, vol. 57, p. 77.
7. Moigno, *Telegraphie Electrique*, Paris, 1838, p. 72.
8. *American Journal of Science and Arts*, vol. 34, (1838) p. 364.
9. William Pole, *Life of Sir William Siemens*, Murray, London, 1888.
10. *Proc. Royal Soc.*, vol. 15 (1866), p. 107.
11. *The Electrician*, vol. 52 (1903), pp. 60, 177.
12. *Proc. Royal Soc.*, vol. 15 (1867), p. 409.
13. *Soc. of Telegraph Engineers*, vol. 13 (27 Nov. 1884), p. 496.
14. Le Comte Th. Du Moncel, *Electric Lighting*. English translation, Routledge, 1883.
15. *Proc. Soc. Telegraph Engineers*, vol. 9 (10 Nov. 1800), p. 339.

16. R. E. Crompton, *Reminiscences*, Constable, 1928.
17. Sir Alfred Hopkinson, *Penultimate*, London, 1930.
18. Hopkinson, Evelyn, *The Story of a Mid-Victorian Girl*.
19. Greig, John Hopkinson Centenary Lecture, 1948.
20. *Proc. Royal Soc.*, vol. 39 (1885), p. 374.
21. Hopkinson, *Dynamo Electric Machines*, Proc. Royal Soc., vol. 40, p. 326, 1886.

CHAPTER VIII

The Electric Light

The first early glimmer of the electric light appeared during those fruitful years following Volta's announcement of the continuous current in 1800. Many experimenters, attracted to this new field of investigation, observed the sparks which appeared at broken contacts carrying the current, but only a few, prominent among them being Humphry Davy, noticed the striking features of the carbon arc. In 1808 Davy, with the aid of his new powerful battery, made available by a public appeal for funds, demonstrated the amazing results of separating the tips of two horizontal charcoal rods through which the current was passing; but it was a further thirty years or more before any attempt was made to turn the discovery to use in the production of a practical illuminant. During a second thirty years the development of the arc lamp proceeded unchallenged until the incandescent lamp appeared on the scene in 1880. From then on, although the use of the arc lamp continued to extend for special uses, it was soon replaced for most purposes by the new rival which also opened up vast new spheres of usefulness in which the arc lamp had no possible application. The incandescent lamp provided, for the first time, a unit of illumination which was steady, required no attention in use, and could be constructed in sizes as small as needed for particular applications. In more recent times both arc and incandescent lamps have had to give ground increasingly to gas discharge lamps.

One step in the progress was described by Professor Silliman, founder of the *American Journal of Science*, in the *Scientific American Supplement*.[1] He wrote: 'Undoubtedly the earliest exhibitions of electric light from the voltaic battery were those made with the deflagrator [battery] of Dr. Hare and Professor Silliman at New Haven in 1822 and subsequently on a magnificent scale at Boston in 1834 when an arc of over five inches diameter was produced by the simultaneous immersion of 900 large-sized couples of Hare's deflagrator. But no means had then been devised for the regulation of the electric light to render it constant, and although Silliman, as early as 1842, used the light successfully to produce daguerreotypes, the process of invention had yet to make further use of the discoveries of science before electrical illumination was possible.'

Two years later hand-regulated arc lamps were used in a production at the Paris Opera House and inventors were beginning to study mechanism whereby carbons could be drawn apart to 'strike' the arc when the current was switched on and advanced towards one another as the points were consumed in use. In an early form due to Archereau the upper carbon was fixed while the lower one floated between a spring (or pulley with weight) which pressed it upwards and a solenoid which pulled it downwards. With the carbons cold and in contact the main current which passed through the solenoid when switched on pulled down the lower carbon and struck the arc. This reduced the current until a balance was effected and the arc then continued to function steadily.

When direct current was used it was found that the upper positive carbon burnt away twice as fast as the negative one and, in order to maintain the arc in the same position, a two-to-one rack and pinion gearing was introduced so that the upper carbon always moved up and down twice as far as the lower one and maintained the arc in the same position. A great deal of ingenuity was displayed in devising mechanisms for automatically taking up the burning away of the carbons and many names are associated with this early development—Wright, Staits, Wallace, Serrin, Linton and others.

The Foucault Dubosq lamp had two trains of clockwork, one of which brought the carbons together, while the other separated them. The selection of mechanism to operate the carbons was effected by an electromagnet connected in series with the arc, the armature of which was opposed by a spring. When the armature moved it released one or other of the gear trains by the withdrawal of a detent so that when the carbons were in contact with one another or too close together the heavy current through the electro-magnet released the withdrawing train of wheels, so increasing the length of the arc. Similarly, when the carbons had burnt away too far and the arc had become too long, the increased resistance reduced the current in the electro-magnet, the opposing spring took charge, the advancing train of gears took over and the arc was gradually shortened. The Dubosq lamp had the distinction of being selected for use in the first English lighthouse lit by electricity when the South Foreland installation went into operation on 8 December 1858.

In 1867 Holmes, in addition to designing the generating equipment for a new lighthouse installation at Souter Point, near Sunderland, improved on existing lamp regulators. The construction of the Holmes lamp was simple and not very different in principle from former designs. The chief difference was in the use of cat-gut bands working over pulleys in place of the previous rack and pinion. His carbons were $\frac{1}{4}$ in. square, the upper one 12 in. long, and the lower one 6 in.[2] Before its final installation in the lighthouse the light was erected in Paris on a high scaffold for exhibition purposes, where it attracted a great deal of attention particularly in view of the rivalry existing between the English manufacturers and the French Alliance Company. In the same year arc lamps were also displayed at the Centennial Exhibition in Philadelphia and various trial installations were carried out in London at the office of *The Times*, on the Embankment, and on Holborn Viaduct.

Within the next ten years the Brush system developed in the United States. Charles F. Brush designed a simple installation consisting of generator and single arc lamp, one of which, sold to Dr. Longworth of Cincinnati and set up outside his residence in 1877, created a great deal of interest.[3] Another development by Brush was the production of 4-light sets in which the generator had four separate armatures each supplying one lamp. Several of these were installed in Wanamaker's store in Philadelphia. Within the next few years Brush established a strong position in this field including a small public lighting system in Cleveland and lamps on high towers in Madison and Union Squares, New York.

Brush invented the clutch type of regulator for arc lamps which has possibly been more widely used than any other principle. The regulating solenoid operated from overhead on the positive carbon and lifted one side of a ring surrounding the carbon holder. The ring being loose tilted and so gripped the carbon. When the solenoid again became inactive the ring was released and the carbon fell. The solenoid had two windings, series and shunt, which gave the action greater stability. With the carbons touching and the heavy current switched on to the series coil the arc was struck, the rise of the ring beyond that necessary to form the arc being prevented by a fixed projection. The shunt coil across the arc then came into action, opposed the series coil and created a balance. The effect of burning away of the carbons was corrected by the slight release of the ring, causing the upper carbon to fall a short distance, its descent being controlled by a dash pot. To these various important features Brush added a second set of carbons with automatic change-over so as to double the life between re-carboning; the system was so successful that it was adopted in many cities both in America and in Europe.

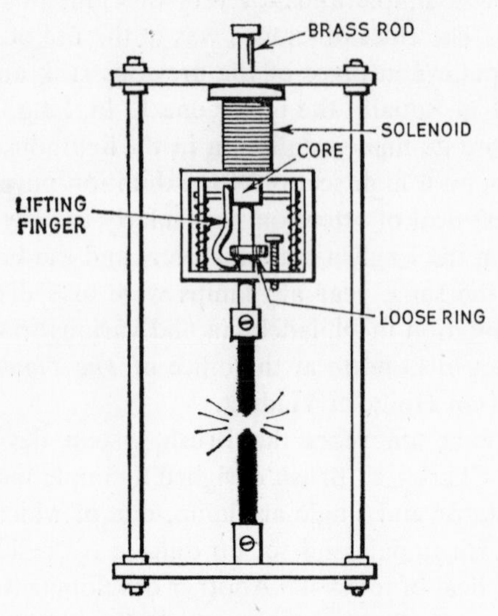

Fig. 20. Brush Arc Lamp. About 1875 Brush, in America, invented a device for adjusting the carbons of an arc lamp by means of a slipping clutch controlled by the current flowing through the arc.

In a text-book written by A. A. Campbell Swinton in 1883[4] there appeared the following statement: 'The electric candle invented by M. Paul Jablochkoff has done much to render electric illumination feasible. This candle, which has been much employed in Paris and is at present burning nightly on the Thames Embankment, is extremely simple and entirely without mechanism.'

The Paris installation consisted of sixteen of these 'candles' erected in the Avenue de l'Opera in 1877 with resulting widespread enthusiasm. Because of their brilliance they created great interest in this and many other countries. Paul Jablochkoff, a Russian, was in Paris on his way to attend the 1876 Conference in Philadelphia when he became interested in the manufacture of the Gramme dynamo in the factory of Messrs. Breguet. He naturally took up problems of electrical illumination and concerned himself with the many designs and operating problems involved in the arc lamp. He conceived the idea of constructing a source of light formed by two moulded carbon rods placed side by side and separated by a layer of insulating material composed chiefly of kaolin clay. The 'candle' was supported in a vertical position and the arc was formed at the upper end by a temporary short circuit through a thin strip of carbon fixed by the makers with a paper band. It then continued to burn for several hours, the separating substance being gradually consumed at the same rate as the carbons. To ensure that both carbons burnt away together and the tips remained opposite one another alternating current had, of course, to be employed and special generators were evolved for the purpose. As a single 'candle' 9 in. long would only last for about 1½ hours, several were mounted in one globe lantern with an automatic switch which connected a new one as each burnt out. Notwithstanding their simplicity and cheapness extended trials over several years showed that such inherent defects as the inability to restart after extinction through wind or a failure of the supply and the excessive waste of partly consumed candles made the system uneconomic and it fell into disuse.

At this time, about 1878, attention was turned to a complete revolution in electric lighting. Although the arc lamp was the only practical form of illumination using the electric current alternatives had been suggested for some years past and several experimenters were trying out their ideas. As early as 1838 a Belgian, Jobart, heated a thin rod of carbon in a vacuum, hoping to maintain incandescence, an attempt which was repeated by J. W. Starr, an American, in 1845. Starr made a lamp with a rod of plumbago assembled in the space at the top of a mercurial barometer tube. Others, including a Sunderland man, Staite, and Moleyns of Chelmsford, tried glowing platinum, while in 1859 Du Moncel obtained reasonably good results with carbon filaments made from sheepskin. In 1874 a Russian physician, Dr. M. Lodyguine, and his assistants exhibited in London an improved type of carbon rod in vacuum which was so successful that some two hundred lamps of the type were installed in the St. Petersburg Dockyard and the Russian Academy of Sciences made a considerable award to the inventor.[5]

To carry the history of the arc lamp beyond the stage when it was superseded for many purposes by the incandescent lamp, reference must be made to several fundamental changes in design. Although numerous modifications of the mechanism for controlling the carbons were made by Pilsen, Brockie-Pell, Crompton and others they all left the problems of rapid consumption of the carbon and the need for costly servicing unsolved: a 10-ampere arc lamp would lose up to two inches of carbon an hour. To meet this problem Marks in 1893 introduced the enclosed arc in which the arc takes place inside a small glass tube: this restricted the flow of oxygen and so slowed down the consumption of carbon. Although this increased the power required it reduced the maintenance to something like one-fifth.

The enclosed arc was followed by the flame arc in which flame-producing salts such as the fluorides of magnesium, barium, strontium and calcium were added in the form of cores to the carbon rods. The idea originated in Bremer's patent.[6] As early as 1899, and for several decades, the improvement in the colour of the light and reduction in consumption per candle-power by more than 50 per cent no doubt postponed the date when the threat of the incandescent lamp for general use became insistent. The latter and the more recent gas discharge lamps have resulted in the arc lamp, as a means of general illumination, becoming to all intents and purposes a historic novelty.

Soon after 1875 a problem of major importance in electrical engineering took shape, that of 'subdividing the electric light', and much confused discussion took place. Arc lighting held the field but there was wide recognition of the fact that with the arc lamp the units of lighting were too large for convenience. It was necessary to do two things: first to reduce the size of the individual light and secondly to make them more independent of one another than they were when connected in series as was the general practice. In 1875 Professor M. G. Farmer, of Newport, Rhode Island, suggested abandoning the series system and connected lamps in parallel with one another. His success in connecting 42 lamps on 42 separate parallel circuits was acclaimed as an achievement. He had overcome the difficult, indeed the well-nigh impossible, task of subdividing the light produced by a single current,[7] a point of view which today seems ludicrous. But parallel connection was only a minor part of the problem: the real task before the inventor was the production of an alternative to the arc lamp, a small light unit of simple and cheap construction capable of being used in all situations particularly in the domestic sphere but the sporadic suggestions of the years 1838 to 1875 pointed the way. The solution would be found in an incandescent filament supported in an evacuated glass container.

The honours for ultimate success in the production of an incandescent lamp must be shared fairly evenly between Thomas Alva

) Barton Street, Manchester, 1923
his shows an intermediate stage in the design which later led to a greater ratio between length
ıd diameter.
'hoto: Central Electricity Generating Board)

) Hackney 30,000 kW sets, 1930
igh- and low-pressure cylinders, 3000 r.p.m. Steam pressure, 350 lb per square inch.
'hoto: London Electricity Board)

(*a*) Baltersen A, First Stage, 1933
One set with an output of 105,000 kW was 120 feet long, the largest turbo-alternator in Euro
for many years.
(*Photo: Central Electrical Generating Board*)

(*b*) Meaford Power Station, 1948
Four sets, each 37,500 kVA.
(*Photo: General Electric Co. Ltd.*)

Edison, the able and versatile American inventor and, Joseph Wilson Swan, a Newcastle chemist. These two men, working independently for many years up to the point of establishing separate important manufacturing companies, ultimately resolved their commercial differences, after an important lawsuit, by combining in a joint concern, the Edison and Swan United Electric Light Company Ltd.

Joseph Swan (1828–1914) started his long and useful life in modest surroundings. His father, a maker and dealer in ships' anchors and chains in Sunderland, was of an adventurous and kindly disposition, and substituted for an unfulfilled ambition to lead a seafaring life an application to invention in connection with his business, a characteristic transmitted to Joseph. In addition to the incandescent lamp, posterity became indebted to his chemist son for important inventions in photography including the well-known carbon process, bromide printing paper, the use of cellular lead plates in storage batteries, and possibly most fundamental of all, the production of artificial cellulose thread by squirting a solution of nitro-cellulose in acetic acid through a fine orifice into a coagulating liquid.

After leaving school, Joseph was apprenticed to a firm of druggists and, being released from the indentures after three years by the death of both principals, set up a business with his friend Mawson in Newcastle at the early age of eighteen. During the next twenty years the firm of Mawson and Swan progressed and achieved a great reputation for its business methods and the integrity of the partners, due in no small extent to the way in which Swan had employed his time as a druggist's apprentice in Sunderland. In the Athenaeum of that city he had studied scientific books and journals and had attended lectures which turned his attention to the problems of the electric light. He had been particularly interested in the experiments of Starr and Staite already referred to. Writing later of his attendances at lectures by Staite illustrating the principle of electric lighting by heating to incandescence a length of iridio-platinum wire, he said: 'This arrested my attention and led me to ponder the question even at this early period, how to produce electric light on this principle, but so as to avoid the use of a fusible wire. *It was something like a seed sown in my mind, which germinated.*'* The seed certainly did germinate, slowly perhaps, but with great effort for, thirty years later, Joseph Swan solved the major problem in the 'subdivision of the electric light'.

* Present author's italics.

Swan's first experiments were made in 1848 after reading Starr's patent specification. He coiled narrow strips of paper in spiral form and packed them with powdered charcoal in a fireclay crucible and baked them at a high temperature in a pottery kiln. By saturating the paper with various liquids such as treacle and tar before baking, on the principle which had already been adopted in the manufacture of arc lamp carbons, he was able to produce some excellent fine filaments of carbon of a metallic strength and flexibility.

Having produced a filament which he could handle without fear of breakage Swan proceeded to mount various forms in glass bottles from which he exhausted the air, using a simple air pump with barrel and piston. By the year 1860, having interspersed these experiments with his other activities of running a druggist's business and investigating photographic materials and processes, he had succeeded in rendering incandescent a carbon strip $\frac{1}{4}$ in. broad, which he had made in the form of an arch $1\frac{1}{2}$ in. high. But the experiment, though promising after so many previous unsuccessful efforts, proved a failure. Although it justified the broad principle of the method of arriving at a strong flexible non-fusing filament, it failed because the vacuum was not good enough to prevent oxidation of the filament at the high temperature. Moreover Swan was ahead of the times: even the arc light was in its early days as a source of illumination and the dynamo was scarcely established as an economic alternative to the expensive battery as a source of current supply. So Swan, though still keeping the problems very much in mind, returned to his work on photographic processes and for the time being abandoned the pursuit.

As so often happens in the progress of science, an invention in another field removed the barriers which had halted Swan. Five years after he had given up hope Sprengel invented his air pump which was to become so famous. It made possible a vastly improved vacuum through a continual stream of discrete mercury pellets entering the upper end of a barometric column and on their way picking up the air from the chamber to be evacuated. Swan was slow to appreciate the significance of this invention, and it was another ten years before he realized the application it had to his own problem. In 1875 Mr. Crookes (afterwards Sir William Crookes) published the results of his researches into the electrical properties of gases at low pressures, while Geissler had produced his spectacular tubes with their beautiful colours, all the outcome of new vacuum technique. It was these announcements which led Swan to return to his attempts to produce a practical filament lamp and this time he succeeded, though not all at once.

Fortunately at this time Swan came by chance into contact with Charles H. Stearn, a bank clerk in Birkenhead, who as an amateur experimenter was investigating high vacua and was skilled in the manipulation of Sprengel pumps. Swan mounted a number of carbon filaments, similar to those he had made in 1860, in glass bulbs and got Stearn to evacuate and then seal off the bulb. As anticipated, improved results were obtained, but a new problem at once arose. On test at high temperatures, these filaments disintegrated and the inside surface of the bulb became obscured by a black film. Here was a new impasse as serious as that of fifteen years before, but after careful and prolonged thought Swan decided that the problem was still only a question of maintaining a perfect vacuum; the blackening—and therefore, presumably, the disintegration—was due to carbon being carried by the residual air to the glass bulb. He then discovered that no matter how good the vacuum was initially, it fell away as soon as the filament reached a high temperature, so here was the outstanding problem; the filament must be heated while the vacuum was still applied to get rid of the last vestige of air. The final hurdle was surmounted, and on 18 December 1878 Swan demonstrated at a meeting of the Newcastle Chemical Society a practical incandescent lamp in operation. This historic item consisted of a tubular bulb with conical ends carrying a straight thin carbon conductor one-twenty-fifth of an inch in diameter. During the demonstration, unfortunately, after giving brilliant light for several minutes the lamp burnt out through excessive current. Success had been achieved, however, and a month later the lecture was repeated with more precaution, and the world knew that Swan had at last solved the problem of the incandescent lamp. Great excitement followed and within the next few months he was called on to give lectures in many parts of the country.

Swan returned to his early U-shaped filament for which he was granted a patent in 1880, and the following year he formed the Swan Electric Lighting Company with Colonel Crompton as chief engineer, to manufacture lamps and carry out lighting installations. At an important lecture given by Swan in the Newcastle Literary and Philosophical Society on 20 October 1880 he gave the signal for the seventy gas jets which lighted the room to be turned out and then (with a suddenness which seemed like magic) transformed darkness into light by switching on twenty of his own lamps, producing a very brilliant effect.[8]

With Crompton's co-operation important installations were soon being carried out. Sir William Armstrong's house at Cragside, the first private installation after Swan's own home, was completed in 1880, the first hydro-electric plant in this country, while another famous one was the house of Sir William Thomson (afterwards Lord Kelvin), completed in 1881. In the same year the Savoy Theatre was equipped with 1,200 incandescent lamps as well as the first electrically equipped ship, the *City of Richmond* of the Inman Line, and a trial installation was carried out in the House of Commons. Within a few months Swan had read his famous paper before the Institution of Electrical Engineers,[9] lamps were being manufactured on an increasing scale and interest in new installations had spread across the Channel and indeed across the Atlantic.

So far we have traced Swan's progress independently of what was going on in the United States. In September 1877 Edison, who had been studying carbon in connection with telephone transmitters, turned his attention to the possibilities of producing incandescent carbon in an electric lamp. A company, The Edison Electric Light Company, was formed with a capital of $300,000 to permit research on a large scale and the famous laboratory at Menlo Park, New Jersey, became the centre of this activity. Edison's first experiments were made with strips of carbonized paper but, even under the glass bell of a vacuum pump, they were rapidly oxidised on the application of the current. After further experiments using chromium, boron and other metals, he became distracted by his phonograph developments, but a year later he returned to the attack, by which time he had concluded that the solution of the problems involved in the 'sub-division of the electric light' was to be found in the high resistance lamp on a parallel system.

In August 1879 Edison read a paper on his discovery of occluded gases and, abandoning his idea of employing metal filaments, returned a few months later to studying the possibilities of carbon. He engaged a German glass blower, Boehm, who had worked with the great Geissler in Bonn, and constructed a battery of Sprengel pumps as he also, in the same way as Swan had done a year before, saw that this new tool opened the way for the use of carbon filaments. Edison was soon satisfied by the results of his tests that carbon putty filaments would not be practical and turned his attention to the selection of suitable natural fibres. Horsehair, bamboo, fish line, an endless array of grass fibres collected from far and wide, fibres of trees, even hairs from the beards of some of his assistants were tested. Success arrived with the use of carbonized sewing thread filaments, for on 19 October 1879, a lamp continued to burn with every satisfaction for forty-five hours, the longest run so far. The day is recognized by Edison supporters as 'Foundation Day', for the incandescent lamp—ten months after Swan's public demonstration of his lamp in Newcastle.

Almost at once Edison turned over to carbonized paper and carbonized filaments and found them superior to the thread type, possessing as they did lives of over one hundred hours, and the laboratory quickly assumed the role of factory, turning out lamps in large numbers. Those for the s.s. *Columbia* constituted the first order.

By this time Edison's financial backers were pressing him for a public demonstration of the lamp which he had, in his various statements, promised the world, and all was bustle at Menlo Park. Some excitement was caused by the premature publication by the *New York Herald* on 21 December 1879 of a very well authenticated article prepared for a full-dress demonstration on New Year's Eve. But the event lost none of its importance as a result. Special trains came from east and west, old-fashioned wagons brought farmers and their families to see houses and streets of the village illuminated and the laboratory itself ablaze with this new marvel, the electric light.

Joseph Swan in England held the view that the broad features of an incandescent lamp could not be patented in view of his early disclosure in 1860 and consequently in spite of pressure from his colleagues in 1878 he took no steps to secure protection. In the copy of *Nature* for 1 January 1880,[10] the day after the great demonstration in New Jersey, a letter appeared over Swan's signature referring to Edison's published claims. In this he stated: 'Fifteen years ago I used charred paper and card in the construction of an electric lamp on the incandescent principle. I used it too in the shape of a horse-shoe precisely as you say Mr. Edison is now using it. I did not then succeed in obtaining the durability which I was in search of, but I have since made many experiments on the subject and within the last six months I have, I believe, completely conquered the difficulty which led to previous failure, and I am now able to produce a perfectly durable electric lamp by means of incandescent carbon.'

On 10 November 1879 Edison applied for and obtained British Patent 4576, which covered the lamp in the broadest terms, and a few months later Swan obtained a patent to cover the previously undisclosed but vital process of evacuating the lamp while incandescent to remove the occluded oxygen. Big issues threatened litigation, but in 1881 the two inventors and their respective English companies decided to merge and the Edison and Swan United Electric Light Company, Ltd., was formed. Of all the alternative substances considered for the filament Swan's extruded cellulose ultimately took the lead.

Although the two rivals were now reconciled, the new combination was at once threatened. The birth of the electrical era through the invention of the dynamo followed by the arc light and now by the versatile incandescent light created extensive interest in the financial world and new companies sprang into being to make and use lamps. The Company found it necessary to defend its position against the infringements of the lamp patents with the result that a *cause célèbre* developed. The weakest point in the company's case was the fact that Swan had exhibited his lamp in December 1878 in his Newcastle lecture, which might lead to the invalidation of the vital Edison patent of November 1879.

It was necessary therefore to jettison, to some extent, the claim for originality of some of Swan's invention in order to keep the more important Edison patent. The case occupied many days and finally went to the Court of Appeal, where it crystallized round the definition of a 'filament'. Was Swan's first and published carbon conductor a filament? If it was then it anticipated Edison's invention and anyone could make lamps in accordance with any of the ideas of either inventor. If, on the other hand, Swan had not made a 'filament' the Edison patent would stand. The Court after lengthy deliberation ultimately decided that no matter what Swan had used, he had not used a filament, and so the Company was saved.

In the United States there was even more litigation and rivalry between opposing interests in which Edison secured some important patent decisions, and in 1892 the position was so chaotic that, with the exception of a group headed by the Westinghouse Company, all the large manufacturers combined to form the General Electric Company—the American Company, of course, not the English one. One very interesting technical result of the patent situation at that time arose from a contract which the Westinghouse Company held for illuminating the World's Fair in Chicago, a very large contract. To avoid infringement of Edison's patent for the hermetically sealed bulb which had just been upheld in the Appeal Court the Westinghouse Company introduced a lamp which had a ground glass stopper. This device tided them over the two years until the Edison patent lapsed, when they, along with all other manufacturers, returned to the hermetically sealed glass bulb.

From this date the manufacture and use of carbon filament lamps went ahead by leaps and bounds. The average light output was 1·4 lumens per watt against today's 12, and with an expected life of 400 to 500 hours on 110 volts supply. By the end of the century it is estimated that the number of lamps in use in London alone was over 2½ million, while the provinces were using very many more than this.

About the time of Swan's success with the incandescent lamp, one of the pioneers in electrical development was R. E. Crompton. By the manufacture of arc lamps, dynamos and electrical fittings and by the carrying out of many of the early and famous installations in this country and abroad, he earned a world-wide reputation.

Rookes Evelyn Crompton was born at Thirsk in 1845 of cultured and widely travelled parents. His father, a country squire, was greatly interested in engineering, and his mother was an accomplished musician. R. E. was the fifth son and claimed direct descent from John Evelyn, the diarist (1620–1706). When he was six his mother took him to visit the great 1851 exhibition in London and he has recorded in his autobiography some of his recollections of those early days.[5] They travelled from Scotland to King's Cross in an extraordinarily long train drawn by six locomotives, and every seat was full. The first-class return fare was sixteen shillings.

With enthusiasm this child of six hurried his mother off to the machinery hall in Hyde Park, where he gazed in rapt admiration on the shining piston rods and all the other attractions of this remarkable display. But this first visit to London left even more lasting impressions on the young mind. A thing which impressed him was 'the extraordinary thunderous noise of the streets of London. The steel-wheeled and horse-drawn vehicles which then crowded the streets rumbled over the granite paving blocks with a noise so deafening that I, as a child walking by my mother's side, could not make myself heard until we got away into the side streets.'

The following year Crompton was sent to a small dames school where, among other impressions, were those of a sallow-faced youth named Dodgson who many years later from his study at Oxford delighted the world with his *Alice in Wonderland*. He also recalls how with horror he looked on the skeletons and skulls of the fugitives from the battle of Marston Moor which had been collected in a house opposite the school.

When the Crimean War broke out in 1854 his father, having obtained a commission in the militia, was transferred with his family to Gibraltar, a move which was followed later by a visit to Sebastopol. Here the curiosity of the boy resulted in his visiting the battlefield and, to his great amusement in after-life, to his receiving at the age of eleven the award of the Crimean Medal and the Sebastopol clasp.

From 1858 to 1864 Crompton spent an active school life at Harrow followed by a stay in Paris for languages and then at nineteen he was gazetted to the Rifle Brigade, which led to a period of military service in India. About this time he developed a keen interest in designing and constructing road engines, with considerable success. By 1876, however, he had decided to settle down to industrial life and secured a partnership in the firm of Dennis and Co., Chelmsford. After two years of selling steam valves, in which his interest and circle of friends were widened considerably, he succumbed to a growing passion for electrical development and started his own factory at Chelmsford.

As the use of incandescent lamps rapidly extended, the question of efficiency became a subject of research and doubts were cast on carbon as the final substance for the filament. At Siemens and Halske in Berlin experiments carried out on some of the rare metals led to the production of fine tantalum wire, which opened up the era of metal filament lamps and eventually ousted the carbon lamp from the field.[11] The use of tantalum enabled 110-volt lamps to be made down to a lower candle power and increased the efficiency, but experiments were already showing that tungsten offered even greater advantages. The major step in this direction was taken by Coolidge in America in 1908, who produced tungsten in a ductile form and made lamps with efficiencies of 1·2 watts per candle against the previous best of 1·6. By compressing tungsten powder in moulds rods were formed which after intense heating and hammering could be drawn to extremely fine dimensions with great tensile strength.[12] In comparison with carbon the tungsten filament for a lamp of a given candle power had to be made longer and thinner owing to the relative lower specific resistance of the metal and this factor introduced new problems in design and construction.

The next step in the development was the gas-filled lamp. It had been noticed that, with all the precautions which were taken in manufacture, the glass bulb of vacuum lamps slowly became obscured in use by an internal black deposit. This phenomenon was well known in the carbon lamps, but the same thing was happening with tungsten filaments. In 1913 Langmuir, of the American G.E.C., traced the effect to its cause and suggested filling the bulb with an inert gas as a remedy. This reduced the rate of evaporation of the tungsten and established what became known as the half-watt lamp, i.e. a lamp which required only half a watt per candle power emitted.

To bring the history of the incandescent lamp up to recent times a further major discovery was made on very simple lines in 1934. To reduce losses due to convection of the gas the coiled-coil lamp was introduced in which a coil is first made and then coiled on itself. This locks some of the gas layers and results in an efficiency varying from 9·7 to 12·7 lumens per watt.

Just before the Second World War the first steps were taken in what has proved to be a further revolutionary improvement in the electric light. Instead of the 11 or 12 lumens per watt reached by the improved gas-filled coiled-coil, metal filament lamp, we now get figures four times as good. This has been achieved after twenty-five years of intensive research based on a study of the visible glow which results from the discharge of an electric current through certain gases and vapours contained in glass tubes. The first-comer to the commercial field was the 'Moore tube', a hundred feet long, filled with carbon dioxide or nitrogen. The discharge resulted from the application of a high voltage between electrodes situated in the ends of the tube and in spite of rapid deterioration through chemical impurities these lamps were used both in this country and in the United States. Shortly afterwards the Cooper-Hewitt mercury vapour lamp appeared in which a tube less than four feet in length and about an inch in diameter gave a bluish light, which was, however, only suitable for certain special purposes.

When the development of gas discharge lamps gained impetus several distinct main types quickly emerged, hot and cold cathode lamps, mercury vapour lamps, sodium lamps, and fluorescent lamps. The problems involved centre largely round the nature of the gaseous discharge between electrodes. When the voltage reaches a certain value the stress in the atoms of gas results in ionization and the gas becomes luminous. Lowering the pressure of the gas and raising the temperature of the electrodes facilitate the process, which also depends to a great extent on the nature of the gas or vapour present. Another controlling factor in the practical application of the gaseous discharge tube as a practical light source is the well-known dark space near the cathode which constitutes a high resistance section of the circuit in series with the conducting luminous column stretching out from the anode nearly the full length of the tube.

A gas discharge lamp which has become very popular, largely because of its high efficiency, is the well-known low-pressure sodium lamp recognized by the orange colour of its light. At atmospheric temperature, sodium exists inside the glass tube of the lamp as a solid and to prevent the vapour condensing when the lamp is in use the glass tube is surrounded by a double-walled vacuum flask. It is, moreover, usual for the tube to be bent double to make it more compact.

The electrodes of a sodium lamp consist of short spiral wires coated with oxides of alkaline earths so as to emit electrons freely. These may be brought up to the condition for starting the discharge by the passage of a heating current through them or simply by the ionic bombardment due to the difference of potential between the anode and the cathode. To secure the initial ionization of the gas in the lamp tube while the sodium is still cold and solid, a quantity of the gas neon is introduced. This results in the characteristic red-coloured discharge which generates sufficient heat after a few minutes to vaporize the sodium and then the main orange-coloured sodium vapour discharge predominates. There are various ways of dropping the voltage in the lamp after the arc has been struck: ballast reactance can be included in the circuit or the lamp can be supplied from a specially designed auto-transformer. The latter gives a voltage of 350 to 400 volts for starting, which, through high-leakage reactance, is made to fall away, with increasing current, to the voltage required for running.

The second widely used low-pressure gas discharge lamp is that known as the fluorescent tube. This is primarily a low-voltage mercury vapour discharge lamp, the inside surface of the glass tube being coated with substances known as phosphors. The initial discharge through the mercury vapour generates ultra-violet light which, falling on the phosphors, causes them to fluoresce and become the principal source of light. Various inorganic compounds are employed as phosphors such as the sulphide and silicate of zinc to which various other metals are added to produce light of the required colour.

The electrodes in a fluorescent tube are in the usual coiled form with tungsten filaments coated with alkaline earth oxides having a high emissivity. To start up the lamp, these are heated by current supplied through a starting circuit which is afterwards cut out. A significant component in this auxiliary circuit is the Glow Starting Switch, a small, evacuated discharge tube containing helium. It also has two bimetallic strips which come into contact with one another when, through a discharge in the gas, they become heated. On switching on the tube, the mains current passes through the two filaments in series with the glow starter, and warms up the bimetallic strips, which then meet and short-circuit the glow current. The subsequent cooling of the glow switch causes the strips to separate again and by means of a ballast choke this sudden break produces an intermittent high voltage between the electrode of the main tube which strikes the arc. Other forms of starting switch have been designed as alternatives to the one described. In one of these, a thermal starter is substituted for the glow starter.

Another form of fluorescent tube has been developed known as the cold cathode tube, with a view to eliminating the starting mechanism. A high reactance transformer can be used to produce a starting voltage several times the running voltage required, which is ultimately assumed when the current starts. A fundamental modification has also been introduced in the artificial redistribution of the potential in the neighbourhood of the cathode. This in turn usually requires a longer tube and a higher overall voltage but there are associated advantages in long life and uniformity of performance, besides quicker starting characteristics.

It will be appreciated that in such a recent field of development as gas discharge lamps, many avenues for improvement are being explored. The use of higher pressures, different gases, variations in form and size of lamp have resulted in an avalanche of inventions which can only be followed by reference to recent books on the subject.[13]

REFERENCES

1. *Sci. American Supp.*, 21 Sept., 1878.
2. *Proc. Inst. Civ. Engrs.*, vol. 57 (1879), pp. 77 et seq.
3. Malcolm Maclaren, *The Rise of the Electrical Industry during the Nineteenth Century*. Princeton University, 1943.
4. Swinton, *The Principles and Practice of Electric Lighting*. Longmans Green and Co., 1884, p. 109.

5. R. E. Crompton, *Reminiscences*, Constable, London, 1928.
6. British Patent 14704. 17 July 1889.
7. *Electric Illumination*, vol. 1, p. 585.
8. M.E.S. and K.R.S., *Sir Joseph Wilson Swan*, p. 71. London, 1929.
9. *Journal Soc. Tel. Engineers*, vol. 9 (1880), p. 339.
10. *Nature*, vol. 21 (1 Jan. 1880), p. 202.
11. *Journal Inst. of Metals*, vol. 9 (1913), No. 1.
12. Harry Hewitt, *Modern Lighting Technique*. Arnold, London, 1952.
13. D. A. Clarke, *Modern Electric Lamps*. Blackie, London, 1952.

CHAPTER IX

Electricity Supply

The success which attended the application of electricity to lighthouse illumination following the pioneer installation at South Foreland in 1858 led rapidly to the improvement of both dynamos and arc lamps. The Gramme machine of 1870 was soon found to be a sound commercial proposition and by 1875 engineers in this country, on the Continent and in the United States, were tentatively putting down small installations for lighting open spaces such as railway yards, public squares, and large buildings. Several designers had discovered that inserting paper discs between the armature laminations would reduce eddy current losses and make water cooling unnecessary. The incandescent lamp was not to arrive for another five years, but with the arc and the Jablochkoff candle, electric lighting was already creating a great deal of interest.

In 1875 arc lamps were erected in the Gare du Nord, Paris, and soon afterwards 80 Jablochkoff candles were installed in the Grands Magasins du Louvre, supplied from a Gramme machine in the basement. The Avenue de l'Opera was lit by 46 lamps, the current supplied by three 20 h.p. engines and there were 22 lamps in the Place de l'Opera. In Germany a system of 12 improved arc lamps installed in the Kaiser Arcade in the centre of Berlin caused so much interest that the makers, Siemens and Halske, received orders for further installations in railway stations, Houses of Parliament and Royal Palaces.[1] In all these cases power was supplied by gas engines recently invented by Otto; accommodation for installing steam engines with their attendant boilers was usually difficult to find in urban centres.

In the United States Wallace and Brush had made dynamos and arc lamps and had installed systems for lighting railway yards, factories and public squares and Edison was hot on the trail experimenting at Menlo Park in New Jersey. In its first issue the new

English periodical, *The Electrician*, pointed out[2] that although Parisians were now enjoying the benefits of electric light there was not a single example to be seen in London. Soon afterwards, however, a French Company, the Société Générale d'Electricité de Paris, commenced operations in this country. The City Corporation placed an order for lighting Billingsgate Fish Market, the first example of a lighting system carried out by a public authority, and soon afterwards the same company installed electric light in the form of Jablochkoff candles on the Thames Embankment, on Holborn Viaduct and at the Mansion House.

The Embankment system extended from the present building of the Institution of Electrical Engineers in a westerly direction as far as Charing Cross Bridge. It had a 'power station' consisting of a wooden shed with a portable engine having two 10-in. diameter cylinders supplied from a boiler working at a pressure of 60 lb. per square inch and developing 60 horse-power at 160 r.p.m. The Gramme dynamo with a separate exciter supplied the lamps in groups of five in series and the current was carried by a multi-core underground cable. The Holborn Viaduct installation only lasted a few months as the costs were found to be excessive—four times those for gas—but the Embankment system proved a success and was soon extended. The Gaiety Theatre in the Strand had the distinction of being the first public building in London to have electric light: French engineers installed six Lontin arc lamps which caused great interest as 'half-a-dozen harvest moons shining at once in the Strand'.

Before the end of 1878 many installations had been carried out in industrial concerns both in London and in other parts of the country. *The Times* had the new light in its printing office while the railway companies had schemes in operation as far apart as the London Bridge Station of the London, Brighton and South Coast Railway and St. Enoch's Station in Glasgow. Sheffield footballers played by electric light before 30,000 spectators and a nearby picture gallery was illuminated by current generated hydro-electrically from a small stream. Both Gramme and Siemens machines were used and the makers of agricultural machinery were soon supplying complete portable steam sets suitable for lighting show grounds and public fairs. Siemens lit the Albert Hall with five lamps each of 6,000 candle power.[3]

In the year 1883 an enterprise was started in the West End of London which, although originally intended as a local private lighting installation, ultimately developed into an outstanding example of public electricity supply and one of great technical and engineering interest. As such the Grosvenor Gallery installation in New Bond Street will always rank as a pioneer in public supply systems. Two semi-portable steam engines were installed by Sir Coutts Lindsay with belt drives on to two separately excited Siemens single-phase 2,000-volt alternators. The lighting of the gallery was by arc lamps on a series circuit with an automatic regulator maintaining a steady current of ten amperes and gave such satisfaction that requests for supply soon began to come in from neighbouring residents and shopkeepers. Under a new Act the Grosvenor Gallery Company sought and obtained the permission of the local Authority to run overhead lines across the roofs of houses carrying the high tension current. Each consumer was provided with a small transformer, the primary windings of several being connected in series on the 2,000-volt supply circuit. The voltage was later raised to 2,400.

The demand from outside consumers grew to such an extent that the company had to establish a more permanent generating station in a large chamber dug out below the gallery and an adjoining boiler house. Two single-phase Siemens alternators each of 250 kW, the largest so far built, were installed and the series system of distribution continued. About this time the station and system was outgrowing its technical ability and a young man named Ferranti, who was already designing and building alternators, was called in to advise on the situation. He took charge in 1886 and initiated a series of changes and extensions which have had such an extensive influence on the progress of electrical engineering that they demand special consideration and a later chapter is devoted to them.

So far as the Grosvenor Gallery system was concerned Ferranti introduced electrostatic voltmeters and Thomson current balances for measuring the current and the undertaking grew rapidly under his direction. Within a very few months the system stretched from the Thames to Regents Park and from Knightsbridge to the Law Courts. Hundreds of iron posts fixed on the roofs of houses carried rubber insulated conductors supported on steel suspension wires by leather thongs.

During all this activity employing arc lamps on a series system the need for a source of illumination of smaller and more manageable character was the subject of wide public interest so that when Edison cabled from across the Atlantic in 1878 that he had succeeded in sub-dividing the electric light he created something of a scare. So long as the light of the powerful arc could not be subdivided the gas interests had little to fear, but smaller units of light would constitute a serious threat. Consequently Edison's announcement produced a startling result and gas shares experienced a sudden decline, though they recovered to some extent when it was found that his statement was premature and that his success in producing an incandescent lamp was limited to laboratory experiments.

In January 1880 Edison filed the basic patent for his 'system' and covered the field very extensively by some sixty subsidiary applications. Seven of these were for systems of distribution as such, 32 for lamps, six for dynamos and five for auxiliary equipment. One of the most interesting features, from the standpoint of modern practice, was the distinction which he made between 'feeder' and 'main'. He appreciated that if lamps were connected across the mains from the point of generation right through to the far end of the system, there would be excessive drop of potential and the far lamps would have insufficient voltage. Consequently he divided the system into sections, supplying each section by means of a feeder direct from the generating point.

Edison's main patent was not issued until 1887 but he lost no time in carrying his proposals into effect. He favoured New York as a suitable place to make a first attempt at public supply and in 1880 formed the Edison Electric Illuminating Company to operate as licensee for the local area under the original syndicate. This procedure soon became general and so was born the system of Edison Supply Companies which still cover the United States.

This was the year of Swan's announcement and in such promising surroundings the arrival of the incandescent lamp opened up the way for public supply. The speed with which Edison established his system by energetic and ingenious publicity methods was remarkable. The laboratories at Menlo Park became the mecca of such world celebrities as Sarah Bernhardt, and the first power station at 65 Fifth Avenue, New York, was chosen because of its spaciousness and dignity and as a suitable house to demonstrate the quality of the new light for interiors. For several years the rooms were crowded with sightseers and the wide interest created by Edison's methods no doubt accounted for his greater success compared with that of his competitors in other countries.

Edison's first real power station and system was in Pearl Street, New York, and the surrounding area where, in 1881 and 1882, he installed six of his 'Jumbo' dynamos with direct coupled Porter engines and laid cables in underground conduits. By the end of the

Modern 2-cylinder Steam Turbine under construction
The absence of the upper halves of the casings shows the blade assembly. The group on the right constitutes the high-pressure cylinder with the steam passing from right to left. The double group on the left is the low-pressure cylinder in which steam from the high-pressure cylinder enters at the centre and passes out right and left to the ports leading down to the condenser.
(*Photo: English Electric Co.*)

(*a*) Modern 60,000 kW Stator being Assembled
In this view the cylindrical portion carrying the windings in which the voltage is generat being inserted in the outside eccentric casing. Space is thus provided for cooling by hydrogen The conductors in the tunnel slots are brought out spirally to the connecting leads which c the current away. (*Photo: English Electric Co. Ltd.*)

(*b*) Modern 60,000 kW Rotor
The construction of stator and rotor in this view and the previous one shows the state of prog reached about 1945. The rotor is a 2-pole field magnet, the conductors being embedded in s and the magnetizing current is fed in through two slip rings. It runs at 3000 r.p.m. to gi frequency of 50 cycles per second. (*Photo: English Electric Co. Ltd.*)

year 193 buildings had been connected with a total of over 4,000 lamps installed; today a bronze plaque on the building at 257 Pearl Street commemorates this great achievement.

While these developments were taking place in New York, Edison concluded his arrangements with Swan and, through London agents, negotiated with the City Authorities for the establishment of a power station in Holborn Viaduct—the previous attempt having been abandoned—and a distribution system to supply incandescent lamps in the area. The primary idea was to light the streets with incandescent lamps instead of the gas lamps which had by now replaced the experimental arc lamps but it was understood that no objection would be raised to supply being given in addition to private consumers. In the power station, which occupied a site (No. 57) near the east end of the viaduct, was installed initially an Edison 'Jumbo' dynamo generating current at 110 volts and capable of running a thousand 16 c.p. lamps. The 125 h.p. horizontal direct-coupled Porter Allen steam engine was supplied from a Babcock and Wilcox water tube boiler.[4]

In addition to providing lighting for many buildings in the neighbourhood of Holborn Viaduct, the Old Bailey and Newgate Street, a supply was given to the General Post Office, and to the City Temple, which thus secured the reputation of being the first church to have electric light. The Holborn Viaduct station went into service a few months before the Pearl Street Station and so has the distinction of being the first public steam power station in the world. Another record was achieved about the same time for the first hydro-electric power station. Messrs. Siemens Bros. and Co. constructed a small pioneer station operated by the River Wey at Godalming which provided street lighting and a small amount of private supply. The system only lasted two or three years, however, and was then replaced by gas lighting.

In 1882, just when large sums were being poured into the newly formed companies for manufacturing dynamos, lamps and laying out supply systems, the first Electric Lighting Act was passed which proved to be a serious deterrent to development.

The opening up of streets and other public works placed those wishing to provide a public supply of electricity in a similar position to the gas and water undertakings who for many years had operated under Statutory Powers. Consequently, from the first, private bills were presented to Parliament for all new proposals. During the year 1878 there were no less than thirty-four such applications, a situation which in the absence of specific legislation on the subject, caused the government of the day some concern and a Select Committee under the chairmanship of Sir Lyon Playfair, F.R.S., was set up to consider the matter. The main functions of the committee were to investigate the powers of public authorities to supply electricity and under what conditions private companies should be permitted to enter the field if at all.

The Select Committee opened its proceedings on 31 March 1879 and heard evidence from many eminent scientists and engineers, including Professor Tyndall, Dr. John Hopkinson and Dr. Siemens, who indicated quite clearly their view that the development of electricity supply would be an important national activity. They commented on the rapid progress already being made and the report of the Committee expressed the hope that there would be no restrictive legislation which would in any way interfere with the development. They did, however, take the line that, although municipalities should have ample powers to open up streets for the purpose of distributing electric light, any such powers to private bodies should be subject to the consent of the local authority. Moreover, the 'public must secure compensating advantage for a monopoly of the use of the streets'. They also suggested that municipal authorities should have a preference in such matters and, even where they granted facilities to a private concern, there should be a restricted period after which the Authority should be able to take over the undertaking on easy terms.

In 1882 Mr. Joseph Chamberlain, President of the Board of Trade, possibly imbued with the merits of municipal enterprise after his success as Mayor of Birmingham, introduced an Electric Lighting bill into the House of Commons, which enthusiastically accepted the Committee's views and on 18 August 1882 the Act was passed. Anyone wishing to provide a public supply of electricity could obtain powers to break up the streets or erect overhead wires. Provisional orders given with the consent of the Local Authority were, however, subject to cancellation at the end of twenty-one years on terms settled by the Authority and moreover the area of the supply would be coterminous with that of the Authority. In spite of warnings the restrictive effects of the Act were not immediately realized by engineers who were dependent on securing capital support for their schemes. A rush of applications for Provisional Orders followed but consideration of the possible take-over terms damped the ardour of those who were invited to find the necessary funds. Two years after the passing of the Act a hundred and twenty applications had been made and seventy-three granted, but not one case existed of the supply of electricity having been commenced. The safeguards against monopoly had been too onerous and after six years, in 1888, it was admitted that there was not a single case of the powers available being exercised. In the same year the Government were forced to introduce a short amending Act in which among other ameliorating conditions was an extension of the period from 21 to 42 years.

The passing of the 1888 Act was sufficient to encourage the flow of capital again and within two years the arrested growth of the industry had begun to recover. The most promising central areas of London, Kensington, Charing Cross, Strand, Westminster and the City were quickly provided for and Local Authorities in St. Pancras, Hampstead, Shoreditch, Stepney and others followed during the next few years. Many provincial cities also followed suit; the local authorities usually got in first with provisional orders, but the companies made most progress in the actual supply of current.

The leading pioneer in the London development of electric supply was Colonel Crompton. In 1878 he improved the arc lamp by placing the mechanism overhead instead of underneath where it cast undesirable shadows, and he sold sets which included a Crompton-Burgin dynamo with one lamp per machine. He soon found that by shunting each lamp with a resistance he could produce stable strings of eight lamps for which he found a ready sale for use in railway goods yards and open spaces.[5] He became intensely interested in the subject through the work of Swan and has described how in 1880 at the urgent request of Swan he travelled to Newcastle to see the state of the incandescent lamp. He concludes: 'I saw at once that Swan's invention would open a new chapter in the history of electric lighting.' His own house in Porchester Gardens is claimed to have been the first private residence to be supplied effectively with electric light, as he started in 1879 using Grove cells and arc lamps. In 1882 he installed incandescent lamps in the Law Courts.

One of Crompton's more spectacular successes was the lighting of the Vienna Opera House in 1883. The building had been rebuilt after destruction by fire, and the dangers of gas having been emphasized, the new electric light was welcomed as a safe and more efficient alternative. A small power house was constructed on the nearest available site which was a mile away and this involved many new problems. Crompton had already been in close touch with Willans and had secured his co-operation in the development of high-speed vertical engines for driving dynamos. Consequently Crompton Willans direct-coupled sets were installed with an output of 700 to 1,000 kW. Direct current was generated at 440 volts and transmitted on a 5-wire system to the Opera House where a 200-cell battery was appropriately tapped to supply the 100-volt incandescent lamps on four separate circuits. The installation was a resounding success. Nothing had been attempted on this scale anywhere at the time and it was some years before it was equalled.

In 1886 Crompton became interested in the newly laid out Kensington Court Estate just south of Kensington High Street and persuaded the owners and residents to institute a system for the common supply of electricity for lighting. The Kensington Court Electric Light Company was registered with a capital of £10,000, and by the use of an extensive system of subways six feet high and three feet wide, originally constructed for hydraulic power mains, it was able to proceed in spite of the restrictive 1882 Act of Parliament. Crompton laid his cables in the ducts with which the power station was also in close proximity. After a preliminary run with a temporary engine a central valve high-speed Willans engine was installed, coupled direct to a 35 kW Crompton dynamo running at 500 r.p.m.

Within four years of its inception Crompton's Kensington Court station had been increased in capacity from 35 kW to 550 kW produced by four Willans sets, three of 100 kW each and working at a voltage of 200/240 on a 3-wire system. A 100-volt 600-ampere hour battery was used to help during the peak load and to replace the dynamos entirely after eleven o'clock at night.

By 1888 the Kensington Court Company had outgrown its resources and was taken over by the new Kensington and Knightsbridge Electric Lighting Company Limited, with a registered capital of £250,000, which built a new power station in Cheval Court and again Willans engines and Crompton dynamos were installed. A provisional order was obtained and an extensive system of mains laid down under the adjoining streets. Growth continued and in another ten years the capacity of the two stations on their restricted sites was so strained that the help of the similar undertaking which had developed in Notting Hill was called in. The two companies joined together in the building of a common power station at Wood Lane station, which had special new features referred to in the next chapter.

The release of capital by the 1888 Amending Act resulted in the inauguration of many new supply systems both in London and in the

Provinces. By the end of the century there were something like thirty power stations in London alone under the control of sixteen undertakings, most of them private companies. The progress during the two decades 1880 to 1900 was largely controlled however by the operation of the original Act. As we have seen, the provision of equipment for series arc circuits had made rapid strides through the enterprise of individual engineers and Swan's invention had opened up new avenues from 1880 which many were now following. This activity was, however, influenced by the restrictions imposed by the 1882 Act and resulted in some confusion which had repercussions on technical development. Many municipalities quickly sought and obtained Provisional Orders but took no steps to operate under them while Companies, avoiding the liabilities of Provisional Orders, and having no power to open up the streets for underground mains, carried their distribution circuits on overhead lines erected over the roofs of houses. In many cases the Company secured full facilities and there are noteworthy examples of such bodies developing in a most successful manner well into the present century.

A typical example of a system which made steady progress throughout the period under discussion was that which developed as the Charing Cross Electricity Supply Company Limited. In 1883 a well-known restaurant in the Strand, the Adelaide, under the direction of the famous A. and S. Gatti, put down a private plant to provide a supply for 330 incandescent lamps to light the restaurant. Two belt-driven Edison bipolar dynamos supplied the current and the result seems to have been a complete success. Three years later a supply was given by Gattis to the nearby Adelphi Theatre, and as the demands for current increased they soon afterwards laid down a small generating station in Maiden Lane. This had four Willans vertical high-speed engines, coupled to Edison-Hopkinson dynamos, two of 84 kW capacity and two of 50 kW, the voltage of the system being retained at 105 volts. It is interesting today to note that each dynamo had its own bus-bars to which any of the lighting circuits could be connected by hand-operated plugs; there was no common 'bus' for the station.

In 1889 Gatti's sold the undertaking to a newly formed company which became the Charing Cross and Strand Electricity Supply Corporation Limited with a capital of £100,000. The new company obtained a Provisional Order, reorganized the station, and by 1892 had five Willans direct-coupled sets working with a total capacity of 900 kW. They also adopted a two-wire system of distribution at

200/100 volts. In another four years the growth of the undertaking required more generating capacity and a second station was erected over the river, in Commercial Road, Lambeth. This plant started off with six Belliss high-speed direct-coupled 200 kW sets generating at 1,000 volts D.C. The current was taken across Waterloo and Hungerford Bridges in ducts and converted to 200 volts by means of 100 kW motor generators in two local substations. As progress continued towards the end of the century set after set was added until the total capacity had risen to 4,400 kW, which included two 1,000 kW machines.

Another interesting example illustrating the way in which successful supply systems developed over the twenty years from the introduction of the incandescent lamp was the Brighton scheme. This system originated in the enterprise of one man, Robert Hammond, whose name became well-known in electricity supply. With the co-operation of another pioneer, Arthur Wright, he built up one of the most interesting systems both technically and commercially, which survived without statutory powers until 1894, when the company was taken over by the Brighton Corporation.

The Brighton scheme originated in 1881 when Hammond staged an exhibition of the Brush arc lighting system there with a view to getting orders. Local shopkeepers became interested and at their request he extended the system for nearly two miles supplying sixteen dispersed arc lamps in series on an 800-volt circuit from a Brush dynamo driven by a 12 h.p. Robey engine. Encouraged by this display of interest the Hammond Electric Light Company was formed and consumers supplied at a cost of 12s. per lamp per week. As the demands increased a further dynamo was installed and an interesting method of charging introduced. For the supply of current from dusk to 11 p.m. the charge was 6s. per lamp-week plus an additional 1s. 6d. for every carbon consumed. Thus was born the two-part tariff which has subsequently assumed such prominence in methods of charging for electricity.

When Arthur Wright joined the concern his interest lay in providing the new incandescent lamps. In order to use the arc lighting circuits which carried a constant current of 10·5 amperes he connected ten lamps in parallel, thus giving one ampere per lamp and arranged these groups in series. To overcome the difficulties due to one lamp burning out and throwing more current on to the remainder, Wright used considerable ingenuity in devising automatic cut-outs which brought in spare lamps in such emergencies. By 1886 there were 1,000 incandescent lamps being run on the constant current arc lamp circuits, a record unique to Brighton. The overhead lines, consisting of No. 7 bare copper wire, extended over a distance of fifteen miles. The precautions taken to regulate the voltage, to keep the boy operating a rheostat awake, and to warn the driver that more or less steam was required, make interesting and amusing reading today.

The Hammond Company and its successor the Brighton and Hove Company continued to carry out extensions and tried to get the Corporation to transfer the Statutory Order which they had held since 1883 without any attempt to work under it. After many manœuvres on both sides the Corporation bought out the company and the first municipal supply was established. A two-wire 115-volt system was decided on and four high-speed Willans sets driving two 45 kW and two 120 kW dynamos installed. A standby battery maintained the service during the night when the engines were shut down. In 1893 more generating capacity was installed and the system converted to three-wire with 230 volts across the outers. Within a year or two after the end of the century the total installed capacity of the Brighton system was 6,000 kW supplied by nineteen Willans engines in a row.

Among the large number of systems which sprang into being at this time there was very little standardization in design either of system or plant: rather was there in most departments an interesting variety of technical detail. Except for a few stations in which gas engines were installed in the 1890's, the source of power in British power stations was steam produced from, first of all, portable boilers or, in the majority of cases, from Babcock and Wilcox boilers with here and there a Lancashire boiler installation. In the 1882 Holborn Viaduct station Babcock and Wilcox started its long tradition of association with the power station industry and very quickly assumed the lead. By the end of the period, about 1,900 Babcock and Wilcox boilers had assumed such pre-eminence that in the enlarged Bankside station there were over forty of this type installed.

In the very early installations engines of the semi-portable type were used, with the cylinders accommodated under the front end of the locomotive type boiler. Even when special engines were designed for power station work, the popularity of the horizontal low-speed type continued and the introduction of improved types such as the Corliss engine extended the period of their acceptability. Belt drives succumbed to rope drives, although this did not overcome the great disadvantage involved in the waste of floor space. In 1893 the rope drives in the Amberley Road station of the Metropolitan Company extended some twenty-five feet from the centre of the twelve-foot diameter engine pulley to the dynamo centres. This station exhibited a feature not uncommon at a certain period when the ropes returned to the engine pulley under the floor and the attendants walked about under the upper ropes. In the Sardinia Street station of the same company an effort to save floor space was made by installing the engines and boilers on the ground floor and passing the belt through holes in the floor to the dynamos on the floor above. With this system of belt and rope drive the dynamos were, of course, carried on slide rails so that the belts or ropes could be kept taut.

The advantage of direct drive from high-speed engines was early appreciated by Colonel Crompton, who influenced his friend Willans to design a vertical engine specially for power station work. In the first generating station of the Kensington Court Company a single-cylinder Willans engine drove a Crompton bipolar dynamo and this was probably the first time such a combination had been used, but the popularity of the Willans engine running at 350 r.p.m. increased rapidly and during the next twenty years large numbers were installed all over Britain. In 1892 the total power available in British stations of over 300 h.p. was about 33,000 h.p. and over two-thirds of this was being supplied by direct-coupled high-speed Willans sets.[6] The remaining third was represented by belt- and rope-driven sets in a limited number of stations. Improvements in design and extension to larger units continued the popularity of the high-speed vertical engine driving bipolar dynamos until the end of the nineties. By 1900 it was being threatened for larger power by the more efficient low-speed engine which could be built to much larger sizes and the effective speed of the dynamo increased electrically by the adoption of multi-pole machines.

The situation in this country was not representative of world practice as one example will show. In 1894 Guido Semenza, the President of the International Electrotechnical Commission and an eminent Italian engineer wrote *à propos* of British power station practice:[7]

'. . . They are reluctant to construct multi-polar dynamos, they have on the other hand abandoned the slow speed steam engine and transmission by belt and rope gearing and have adopted, in the Willans engine and the principle of direct-driving, two novelties which are slow to make their way in other countries.' The same critic made some disparaging remarks on the layout and cleanliness of our power stations. 'Let us visit one at random,' he said, 'no matter which, as they are almost all of the same stamp. An unimposing entrance, a gloomy little back alley, and at the end the workshop of the station. A covered passage, broad enough to allow a coal cart to pass, traverses the workshop. On the left-hand side, we will say, is the coal store, and on the right, up a few steps, we find the boiler room. I call it the boiler room for the sake of the word. There is a long row of Babcock and Wilcox boilers arranged along the passage with a few yards between them and the wall. Proceeding in the same direction we enter the engine room. Along the wall, separating it from the boiler room, there are installed parallel to the boilers, half-a-dozen or more Willans engines coupled to as many stout bi-polar dynamos mounted on the same bed plate. On the wall facing us is the switchboard. Into the engine room a grey light falls from a skylight, the walls are grimy and between one machine and the next, or the wall, there is hardly room enough to pass.'

This gloomy picture by such an authority as Semenza was, however, completely changed by the appearance of the Parsons turbines. This new prime mover, a British product, quickly solved the problem of the choice between increasing the number of high-speed engines, limited in individual size, and the adoption of larger and larger low-speed sets. It superseded both. By 1900 two 1,000 kW sets were running and the turbine rapidly gained a footing; but the story is proper to the next chapter as the development of the turbine is inseparably associated with the adoption of alternating current for electricity supply. Before passing on into the turbine/alternating current era ushered in by the twentieth century there are several further interesting features in the 1880–1900 period which must be disposed of.

As Semenza pointed out, the type of dynamo most popular in British power stations during the period was the bipolar machine of generous proportions arising largely from the design work of Hopkinson and Kapp. Crompton had made four-pole machines, some of which worked at the Eccleston Place station of the Westminster Company from 1891 for nearly thirty years. Generally speaking, however, the increase in number of poles on D.C. machines was very limited at that time.

In Germany on the other hand there was considerable misgiving on the employment of large numbers of small high-speed dynamos for electricity supply. As the load on the systems grew the position became even more acute. The high-speed direct-coupled Willans type of set had not established itself in that country as it had over here but 'the machine driven by flapping belt which purred like a spinning wheel while a bluish-green line of sparks under the commutator brushes showed that the dynamo was producing current' was anathema to the mechanical engineers. Among these was Emil Rathenau, who secured financial support for the development of electricity supply under licence from Edison and in association with Siemens and Halske. Rathenau was quite convinced that the steam engines to be efficient must be large, slow-speed units so designed with multi-cylinders as to secure the maximum advantage from steam expansion and with adequate condenser systems. They must be equipped with the latest and best precision valve gear like the American Corliss type and with an efficient governor. Such slow-speed engines should be coupled directly to slow-speed dynamos and so do away with all the overheated bearings and other troubles accompanying belt drive. This line of thought naturally led to dynamos having armatures and field magnets of greatly increased diameter which in turn meant more poles on the field magnet spread around the circumference of the air gap. This 'internal pole' machine was a great step forward and made possible the large machine which had become such an essential requirement of electricity supply. Its arrival on the scene was a vital factor in saving the Berlin electricity system at a time when technically and financially it was in low water. Within less than ten years Siemens and Halske manufactured more than five hundred of these machines for the home and export market. The Kapp six-pole machine driven by a triple expansion Willans engine, installed in the St. Pancras power station in 1891, was probably the nearest British approach to this innovation. This was the final stage in general dynamo design before the arrival of the changeover to turbine-driven alternators, although during the transition many multi-polar sets were installed in British and foreign stations driven both by vertical and horizontal medium-speed multi-cylinder steam engines.

The changeover from series arc lamp circuits to parallel operation for incandescent lamps, as well as the voltage limitation of the latter to about 100 volts and the considerations involved in the economy of copper, had resulted in an interesting transition period in which some ingenuity was displayed in the design of the distribution system. In most of the early systems the storage battery played an important and dual role. It was used to maintain the supply at nights while the engines were shut down but, more important still, it frequently enabled the current to be transmitted at a voltage several times higher than that required at the consumer's terminals, so saving copper in the mains. The magnitude of this use of the secondary cell opened up a big field for an old invention.

In 1859 Gaston Planté in Paris had carried out many experiments on voltaic cells and had discovered that, when a current was passed between two lead plates immersed in dilute sulphuric acid, certain chemical changes took place which gave the arrangement the character of a storage device for electrical energy. Oxygen was set free at the anode and, combining with the lead, resulted in a brown deposit of lead peroxide (PbO_2) on the surface of the plate. Hydrogen bubbles were released at the cathode and, rising to the surface of the electrolyte, escaped into the air. Planté discovered that if such a cell were disconnected from the supply of current and connected to a fresh circuit of wire, lamp, etc., it would act as a primary cell giving back on discharge the energy which it had received on charge. In this operation both plates became coated with a deposit of lead sulphate. When he repeated the process the white sulphate on the cathode was converted into lead again and the anode assumed its brown colour once more, the sulphate becoming peroxide. This process could be repeated time and time again, but the amount of charge stored was small. To effect an improvement Planté 'formed' the plates by repeated charge and discharge, but this was a costly process and in the absence of any commercial demand for a secondary cell no further progress was made.

About the time when Edison and Swan had produced the filament lamp and the need for a storage battery was making itself felt an important discovery was made by Camille Faure, another Frenchman, which revolutionized the Planté cell. He superseded the slow and expensive forming process by coating the positive plate with a paste of red lead (Pb_3O_4) and sulphuric acid and the negative plate with litharge (PbO). During charge the positive paste was oxidized to lead peroxide (PbO_2) and the negative paste reduced to spongy lead. Further improvements of a practical nature quickly followed Faure's idea, the most important of which was the construction of the positive plate (anode) in the form of a grid of lead with the red lead paste pressed mechanically into the interstices. Many detailed but important improvements followed, including the perforation of the negative plate to carry metallic lead in a form offering a large area in contact with the acid, and in the strengthening of the construction to withstand rough usage. With these technical improvements and the simultaneous demand for storage batteries in electricity supply, their manufacture on a large scale was quickly taken up in many countries. As an indication of the progress made in this field an American example may be quoted. In one particular substation of the Chicago Edison Company the installed capacity of the battery in the year 1900 was 45,000 amperes continuous discharge for a period of eight hours.[8]

REFERENCES

1. Georg-Siemens, *History of the House of Siemens*, 1957, vol. 1 p. 84.
2. *The Electrician*. 20 July 1878, vol. 1. Series of Article.
3. Adams Gowans Whyte, *The Electrical Industry*, London, 1904.
4. R. H. Parsons, *Early Days of the Power Station Industry*. Univ. Press, Cambridge, 1939.
5. R. E. Crompton, *Reminiscences*, Constable, London, 1928.
6. *Electrical Review*, vol. 30, p. 89. 22 Jan. 1892.
7. *The Electrician*, vol. 34, p. 758, 1894/5.
8. *Electrical World*, vol. 35, p. 735, 1900.

CHAPTER X

Alternating Current

The phenomenal growth of electrical engineering during the last two decades of the nineteenth century, which had followed the introduction of the incandescent lamp, left the supply industry in a curious state of uncertainty on an important and much debated issue. It had resolved the fundamental problem of 'the subdivision of the electric light' with complete success and had overcome obstacles placed in its path from the start by legislation. Some of the best brains in this country and abroad had devoted themselves to the many new scientific and technical problems involved in the design of generating plant while factories had been established for the production of lamps, dynamos and the rapidly growing range of items required for control of the power and in the equipment of mains and lighting installations. There was, however, one most fundamental question still unresolved, namely, should the current, when generated mechanically, be continuous or alternating? The experts, ranging themselves into two opposing camps, debated the question hotly for many years. As we know today, the exponents of the alternating system ultimately held the field and development followed this course. The direct current enthusiasts, however, made such progress over the years in carrying their ideas into effect that even now (1960), over half a century later, there are still vestiges of D.C. sections remaining, though they are rapidly approaching extinction.

From the early beginnings of public lighting supply alternating and direct current were adopted almost indiscriminately. Alternating current was used for the early arc lamps in lighthouses, in railway depots, for street-lighting, and elsewhere, while the use of the Jablochkoff candle demanded the use of alternating current to ensure equal burning away of the two carbons. About the time that Siemens Brothers were equipping the Savoy Theatre (1881) and Grosvenor

Gallery with 2,000-volt alternating current machines (1883), Edison was installing his 'Jumbo' dynamos in Holborn Viaduct and Hammond was laying down the beginnings of the Brighton system with a mixed load of arc and incandescent lamps supplied with continuous current. Gradually, however, a distinct preference developed for direct current which held the field for many years.

The most powerful argument in favour of direct current was that it permitted the use of batteries, by which reliability of service could be assured in case of breakdown of machinery, a frequent occurrence at that time. Moreover, the transferring of the night load to a battery permitted the shutting down of the generators and saving of labour. Repairs could thus be effected and the amount of stand-by plant reduced. The batteries certainly only returned some 80 per cent of the energy put into them and they were fairly costly to maintain, but most of the London companies adopted direct current and succeeded in providing a reliable standard of supply. Another argument which no doubt influenced opinion on this matter was in connection with the relative costs of running D.C. and A.C. machines. It was widely believed that the amount of coal consumed in an alternating current station was two or three times that in a direct current station working with a stand-by battery. Crompton gave an extensive analysis in 1891[1] and introduced the important term 'load factor' to show its influence on costs per unit. The costs were also of course a function of the type of prime mover employed: whereas the early alternating current generators were belt- or rope-driven from large engines, the direct current dynamos were usually coupled direct to high-speed engines. It is almost certain that the capital costs of an alternating current station per kilowatt of maximum demand would exceed that of a direct current station as the running plant would need to be large enough to take the whole load with no battery to fall back on. It must also be recalled that for a long time each alternator in a station was allocated to a particular part of the system and therefore a fall in total load effected all machines without making it possible to take one of them out of service. This defect was only overcome when paralleling of alternators was found to be practicable.

As the initial all-lighting era passed and the use of electric power developed, taking more and more of the load, the arguments of the direct current protagonists were strengthened. An efficient and reliable D.C. motor was quickly developed out of the theoretical and practical work put into the dynamo but there was no such suitable motor for operation by alternating current. This became a serious objection to the alternating form of current and was only removed at a later date by the invention of suitable A.C. motors.

As we saw in the previous chapter the direct current engineers, from the very beginning, appreciated that the extension of a system beyond the immediate surroundings of the generating station raised the question of economy in cost of the mains and the control of the drop in voltage at the end of long sections with increasing load. They saw that to overcome this difficulty they would have to raise the voltage beyond that at which the consumers' lamps would be connected, and Crompton, Hammond and others introduced three, four and even five-wire systems whereby the dynamos on a 100-volt system could generate at 200, 300 or 400 volts. In the alternative high voltage direct current system a number of consumers' batteries could be charged in series on a high voltage supply and discharged separately at the consumers' voltage. Ten consumers connected in this way enabled the current on the line to be only one-tenth of that required for the load but needed a dynamo for ten times the voltage. At a subsequent date motor generators or continuous current transformers, as they were called, were introduced. The double-wound dynamos with the separate windings brought to two separate commutators were supplied at the high voltage side direct from the central station and gave the consumers' voltage on the second commutators. High voltage D.C. transmission with voltages up to 1,500 was installed in this way at Chelsea (1889) and Lambeth (1896) and at several provincial towns from 1884 onwards.

British engineers must have been encouraged in thus using high voltage direct current by an experiment carried out in 1882 in connection with the Munich Electrical Exhibition. Marcel Deprez, a professor at the Paris Conservatoire des Arts et Metiers, and Oskar von Miller, a young engineer in the Bavarian Civil Service, made arrangements to transmit power by direct current from a coal mine at Miesbach, 35 miles away. They installed a small steam driven dynamo of about 1·5 kW output and 1,500 to 2,000 volts and employed a standard telegraph line which was made available for the purpose. A duplicate of the dynamo was used as a motor at the receiving end in the Exhibition and this drove a centrifugal pump which supplied a waterfall having a height of about 6 feet. The experiment caused a big sensation.[2] Transmission by high voltage direct current came much later and then overseas and will be discussed in a later chapter.

The year 1887 is a good date for assessing the stage reached in the development of alternating current. In August of that year the London Electric Supply Corporation Ltd., was formed to take over the Grosvenor Gallery system which was in charge of the young man Sebastian Ziani de Ferranti.

Ferranti, born in Liverpool of cultured parents[3] on 9 April 1864, developed an early fascination for scientific and engineering subjects. At St. Augustin's College, Ramsgate, his original outlook led him to take advantage of opportunities to visit the local electrician and learn how to use a lathe. In later years he expressed his indebtedness to Crompton for having allowed him, as an eager youth, to satisfy his curiosity in electrical matters by studying the portable generating sets at Alexandra Palace. At the age of seventeen he had made his first dynamo and sold it in the Euston Road for £5 10s. A few months later he joined Siemens at Charlton on the introduction of the electrical engineer at the British Museum with whom he had made friends. Very soon his outstanding ability resulted in his co-operating with Sir William Siemens in his research work which brought him into contact with many of the electrical pioneers of the day including Sir William Thomson (later Lord Kelvin). While not much more than a boy he was superintending the installation of electrical plant in different parts of the country. He formed a limited company with two acquaintances, an engineer and a solicitor, making himself chief engineer, had patented his first alternator and was paying royalty to the future Lord Kelvin on the Thomson Ferranti alternator.

The new company started a small factory in Finsbury to make dynamos and an electrolytic meter of his own invention but within a year it was dissolved and Ferranti, buying back his own patents, set up in business (now aged nineteen) on his own account, first in Hatton Garden and then in Charterhouse Square. Here the range of products, based on his own designs, extended to house-service meters. transformers for voltages up to 10,000 volts, oil-break switches and fuses and it was at this stage that he came into contact with the Grosvenor Gallery scheme.

Before proceeding with Ferranti's epic story a few lines are necessary to indicate the general progress which had been made in the use of alternating current. The Paddington scheme installed for the Great Western Railway by the Telegraph Construction and Maintenance Company provided for 4,100 incandescent lamps and 100 arc lamps and covered an area of some 70 acres. This installation had gone into operation on 21 April 1886 and was destined to give satisfactory service for twenty years.

Plate XXIII

(*a*) Power Station Switchboard about 1920
The board intended for low voltage has exposed live metal on slate panels and the switches are operated manually.
(*Photo: London Electricity Board*)

(*b*) Control Room in Modern Power Station
This installation at Llynfi, South Wales, is a good example of remote control, which was introduced about the end of the First World War for high-voltage operation. Only low-voltage circuits are brought into the control room. These bring in instrument readings via instrument transformers, and take out switching instructions to the high-voltage switches.
(*Photo: English Electric Co. Ltd.*)

(*a*) Switch Yard in Modern Power Station
The high-voltage circuits are made and broken in a switch yard outside the power station building through circuit breakers operated by low-voltage current.
(*Photo: Reyrolle & Co.*)

(*b*) Modern Boiler House
This boiler house at Cliff Quay demonstrated the remarkable change which has come about through the mechanization of stoking, and coal and ash handling. The operations are now controlled through an instrument panel as shown. (*Photo: Babcock & Wilcox*)

The most interesting feature of the Paddington scheme was the design of alternator employed. The designer, J. E. H. Gordon, who achieved a great reputation for his electrical engineering contributions, had joined the Telegraph Construction and Maintenance Company after a brilliant career at Cambridge. After trials in the Company's works at Enderbys Wharf, Greenwich, he installed at Paddington three machines, which were described as Brobdignagian machines to be driven directly by vertical steam engines. Weighing 45 tons each, they stood on bedplates 18 feet long. These Gordon alternators had rotating discs 9 ft. 8 in. in diameter, each carrying 28 horizontal electromagnets with cylindrical cores 6 in. in diameter and 3 ft. long. On each side of this central rotating ring were two fixed rings of boiler plate carrying 56 horizontal armature coils. These coils, connected to two separate circuits, virtually constituted a 2-phase arrangement although the two circuits were kept separate and served two distinct lamp circuits. As at first solid cores were used in these machines, excessive heating occurred which was only overcome by replacing them with a laminated construction. The machines gave a voltage of 150 volts and a frequency of about 68 cycles per second. As Gordon's installation ran satisfactorily until 1907, when it was superseded by a completely new power station at Park Royal, it may be considered as a landmark in the development of alternating current although he himself lost faith in A.C. and immediately joined the newly formed Whitehall Electric Supply Co., registered in 1887 with a capital of £200,000, to design and install a direct current station at Whitehall Court.

A few months later another activity developed which carried alternating current forward with a bound. The Metropolitan Electric Supply Co., with a capital of £250,000 at first, quickly increased to £500,000, purchased a small private alternating current station in Rathbone Place, Oxford Street, which had been in operation about a year. This station had already replaced a 32 kW alternator made by Ferranti generating at 100 volts (stepped up to 2,000 volts for overhead distribution by two Mordey alternators, the first of their kind) and in 1889 tried a further design—the Elwell-Parker alternator. The station was then laid out for generating single-phase current at 1,000 volts and 105 cycles, and was the beginning of a supply system covering a wide area stretching from Marylebone to Lincoln's Inn and Covent Garden. In a new station in Sardinia Street, near the present Kingsway, were installed American Westinghouse alternators brought over from Pittsburgh. Ten of these, each with an output of

125 kW at 1,000 volts, were belt-driven from Westinghouse Compound engines running at 250 to 280 r.p.m. Three separate belt-driven 100-volt dynamos supplied the exciting current. Here again there was no parallel running of the alternators, each machine having its own bus bars, the twenty feeders being connected at will by plug connectors.

A further alternating current station was established by the Metropolitan Company about this time in Manchester Square, and others followed, but these parallel activities do not in any way detract from the originality and foresight that Ferranti displayed in his work on the Grosvenor Gallery scheme. He was already justifying the assessment of his achievement made by Crompton many years afterwards, who said; 'It is only in recent years that I have realized how deeply electrical engineering and the world generally are indebted to him. I now believe that the problems involved in the successful application of the alternating current were far the most difficult.'[4] Under his energetic direction the system expanded rapidly over a wide area from the Thames up to Regent's Park and from Kensington to the Law Courts. The station was reorganized and two single-phase Ferranti alternators installed, each capable of supplying 10,000 lamps, probably about 400 kW. According to our modern views the rating of these machines was rather uncertain, but from overload experience quoted at the time[6] they must have been capable of producing an output not less than 700 kW. They were enormous machines for the time, standing 9 ft. 6 in. high and occupying an area of 9 ft. by 11 ft. Each alternator weighed 33½ tons. The field magnets were stationary and the armature, 8 ft. 6 in. in diameter, revolved. According to accepted methods of measuring frequency at the time these machines produced current at 10,000 reversals per minute, or as we should say, 83⅓ cycles per second. The method adopted for bringing the separate machines into service required both skill and courage. When an already loaded machine had reached the limit of its capacity a second set was run up to speed and, by means of dog clutches and a countershaft, was coupled up for use. When the speed was exactly right the operator brought the claws of the clutch into engagement, feeling his way in so as to effect a smooth transition. Ferranti later replaced the Gaulard and Gibbs scheme, in which the current from the station passed in series through the primaries of many transformers on the consumer's premises, by a full parallel scheme at 2,400 volts with step-down transformers.

Returning to the Grosvenor Gallery scheme, it soon became evident that with all the improvements possible the site imposed serious limitations on the development and in August 1887 a new company, the London Electric Supply Corporation, was formed with an authorized capital of £1,000,000 to take over and extend the system.

Ferranti for some time had held the view that the rapid growth of electricity supply called for the creation of much larger central power stations which should be sited where land was cheap, unlimited water available, and where water-born coal could be obtained at a low cost.[5] These conditions could only be met away from densely populated areas and transmission of the current at high voltage became essential. So far as London was concerned the gas industry had shown the way. They had installed their large gas plants away to the east of the City and on searching round Ferranti found a site at Deptford. He visualized an output of 12,000 h.p. and selected 10,000 volts as a suitable pressure at which to generate alternating current and transmit it by cable to the Grosvenor Gallery centre for distribution.

As such voltages with alternating current on this large scale were new and their practicability uncertain, Ferranti built an experimental transformer to step up the 2,400 volts at Grosvenor Gallery to 10,000 and gave a public demonstration. There were, of course, no suitable instruments available for measuring such a voltage but he connected a hundred 100-volt lamps in series and showed that it was a feasible proposition.

In the design of the building, boilers, engines and generators at Deptford, Ferranti entered new fields: quayside facilities, 40,000 square feet of concrete raft 4 ft. thick to carry the heavy machinery, and a layout for coal and ash handling such as had never before been attempted. The plan included 80 Babcock and Wilcox boilers to work at the unprecedented pressure of 200 lb. per square inch. To start with, two 1,250 h.p. Corliss engines were to drive two 5,000-volt Ferranti alternators by means of 40 cotton ropes from pulleys 21 ft. in diameter and provision was made for extension of capacity by means of four 10,000 h.p. engines each directly coupled to its own 10,000 volt generator.

So far as the transmission of the current to London was concerned the amended Lighting Act of 1888 posed certain practical problems. For instance, to open up streets over the distance of nearly five miles would involve many local authorities whose consent might be difficult to obtain or would at least result in delays and extra cost. The railways on the other hand were run on private property and soon provided the required rights of way for carrying the high voltage cables to London and across the bridges to Charing Cross, Cannon Street and Blackfriars. The underground railways moreover provided facilities for distribution through their tunnels to a wide area. The consumer was to be reached ultimately by running both overhead lines and underground cables in the streets and as the Provisional Orders sought were so extensive, involving some twenty-four local authorities, this one item created a furore in the form of an enquiry by the Board of Trade which imposed certain limitations on the scheme.

Sufficient was left, however, to ensure opportunities for a successful undertaking and by the middle of 1889 the power station was under steam with a total capacity of 414,000 lb. of steam an hour and two 1,250 h.p. engines completed and tested. The alternators with their 13-ft. diameter armatures proved to be satisfactory and all was ready for sending current to London and helping out with the growing load on the Grosvenor Gallery station when what was to prove to be the first of a series of tragedies interrupted the progress of the enterprise: the cables which had been made under contract and completed along the walls and through the tunnels of the railway companies were found to be incapable of carrying the 10,000 volts for which they were designed. Even at 5,000 volts they were not reliable and Ferranti and his directors were faced with the inevitable problem of replacing them with some more practical proposition.

In the engineering construction of the power station and plant Ferranti had displayed astonishing skill and enterprise. Failing other facilities, he had not only designed the immense alternators, but apart from making the necessary castings, had constructed them on the site. He had installed a lathe capable of swinging items 11 feet in diameter and of turning the massive engine shafts, as well as a wall planer with dimensions sufficient to deal with items 20 feet high and over 20 feet in length. Not daunted by these many problems, it is not surprising that the new situation arising out of the failure of the mains found him still more resourceful; he decided to make a new lot of mains himself! The setback imposed a delay of over twelve months but the ultimate success of his dogged determination in constructing this essential component of the system was outstanding and far in advance of anything that had previously been attempted. The results were of such permanent value that his high reputation today rests to a very great extent on this as one of the most brilliant of his achievements.

Although the construction of the 10,000-volt cable created a milestone in the history of cable making and as such is part of the story of Chapter XVI, Cables, it was so intimately related to the Deptford enterprise that a description is appropriate at this stage. Ferranti decided by an uncanny appreciation of factors which cable engineers only began to understand thirty or forty years later that success at this voltage most probably lay in a rigid concentric construction. His conductors for the transmission of the single-phase current consisted of two copper tubes each of 0·25 square inch sectional area one inside the other and separated by wax impregnated paper. The insulation between the two was applied to the inner tube by rotating it in a simple lathe, and applying the impregnated paper up to a thickness of half an inch. In the light of later experience it is clear that his success was due to the fact that by this method he was able to avoid the spiral spaces occasioned by lapping on narrow strips edge to edge with the attendant risk of voids, a necessary construction for cables which must be sufficiently flexible to wind on to a drum. The dielectric quality was moreover still further improved by slipping this covered inner tube inside a slightly larger outer one which was then drawn down mechanically so as to produce intimate contact between the two. In order to meet the requirements of the Post Office from the standpoint of interference with telephone and telegraph circuits a further lapping of waxed paper was applied to the outer tube to a depth of a quarter of an inch. This was then threaded into an iron tube, $2\frac{3}{8}$ in. external diameter, which was finally heated and the space between the core and the paper filled with molten wax from a hole at the centre of the length.

The cable was made in 20-ft. lengths and the ends turned down to a cone 6 in. in length, so that when laid one length fitted into the next. The inner conductor was jointed by a long copper plug forced into the two connected lengths as they were assembled and brought into intimate contact by hydraulic pressure. A copper sleeve previously slipped over one end of the outer tube was then made to bridge the joint and was forced down by a corrugating tool. The outer iron casing was similarly dealt with and a final application of hot wax injected to fill up any space left between the outer copper and the surrounding iron tube with its sleeve. Expansions due to temperature changes in service were catered for by giving the cable a slight wave here and there during the laying process. Over 28 miles of this cable were made and laid and gave completely satisfactory service at 10,000 volts for many years. More remarkable still, some of it survived in excellent condition more than forty years later.

With our more up-to-date knowledge of the precautions which must be taken to ensure safety in the use of high voltages, some of the experiences of those days may bring a shudder to the reader. It is recorded that the Board of Trade were rather concerned for the public safety in the use of underground cable operating at such a high voltage, but that Ferranti, on the other hand, pressed the view that the use of the earthed outer conductor made it perfectly safe. To satisfy the authorities, he arranged for the live cable to be spiked in their presence. Parsons records[6] that 'Mr. Harold W. Kolle, who had joined Ferranti's in 1888 after being for three years the Electrical Engineer at Eastbourne, volunteered to hold the chisel while the blows were struck by another assistant, Mr. C. Henty, who was the son of the famous war correspondent and writer of boys' stories, G. A. Henty.' Fortunately the test was successful in every way and convinced the Board of Trade who granted the permission desired.

Another example of the lack of precautions taken in those days is recorded by Crompton in his reminiscences. He received the Prince of Wales, afterwards Edward VII, and Princess Alexandra on their visit to the Electrical Exhibition at the Crystal Palace in 1882. 'I had the honour of conducting Her Royal Highness round', he writes, 'to point out and explain to her the electrical phenomena. I assured her that she could touch bare conductors carrying energy of 200 volts pressure without danger. She made the experiment and agreed with me that such a pressure was quite safe to use.' Today's comment can only be that the subsequent history of England must have rested to some extent at that moment on the dryness of the floor!

Before the new cables between Deptford and London were completed disaster again overcame the Corporation.[7] The Grosvenor Gallery premises, now a substation, were burned down. An attendant opening a high voltage switch on one of the feeders, was surprised by the strength of the arc and, instead of closing it again quickly, opened it still further and allowed the arc to persist, with the result that the roof caught fire and within half an hour the place was burned down and the plant put completely out of use. The combination of this unfortunate occurrence with the troubles at Deptford led to a complete cessation of supply for several months, with considerable financial loss to the Corporation. By dint of heroic efforts, however, the station was quickly re-established, this time with underground distribution cables replacing the former tangle of radiating overhead lines, and in February 1891 a 10,000 volt supply was restored by means of special generating and transforming arrangements installed at Deptford.

This unfortunate series of events led to a break between Ferranti and his directors. Besides the loss of consumers through the shutting down of the station it was realized that the hoped-for monopoly was no longer attainable in the light of growing competition from the neighbouring Metropolitan Electric Supply Company. Moreover Ferranti was under criticism from many quarters on the correctness of his views in favour of alternating current. Gordon, whose enthusiasm for A.C. had waned after several years' experience at Paddington, was now on the side of the battery men and Edison had moreover visited Deptford and strongly expressed his disapproval of the scheme. We read therefore with regret that the Annual Report of the London Electric Supply Corporation announced that the engagement of Mr. de Ferranti had ceased 'by effluxion of time'. In a stormy interview he had been told by the chairman, 'You are a very clever man, Mr. Ferranti, but I'm thinking ye're sadly lacking in prevision.'

As Ferranti turned to his manufacturing interests he carried with him his courage and a great reputation for the scheme which had evolved under his direction. The world of electricity supply turned to him for advice and within a few years the Hollinwood factory was established. Today it has, of course, become a great centre of electrical activity employing thousands of men and women under the direction of Ferranti's son, Sir Vincent de Ferranti, and with its products carrying the reputation of this great name to the furthest corners of the earth. In spite of continued ill luck at Deptford including destruction of all four cables on the way to London by a fire in one of the railway arches and an unusual spell of fog and frost which overloaded the plant to complete breakdown, the corner was turned at last and by 1905 the patient ordinary shareholders began to receive a dividend.

Among the items of outstanding historic interest in connection with the Deptford scheme must be mentioned the Ferranti effect and the nemesis of the large slow-speed reciprocating engine. While laying the cables from Deptford to London Ferranti carried out tests on sections and discovered an entirely unexpected phenomenon. The voltage on the cable remote from the station exceeded the generated voltage. A pressure of 500 volts was applied for the test and, in the absence of any voltmeter, 500 c.p. lamps were inserted in series as an indicator. As the length of the cable grew the lamps on test became brighter and presently an additional lamp had to be inserted to prevent the others being burned out. When the complete circuit was completed it was found that a voltage of 8,500 at Deptford resulted in one of 10,000 at Grosvenor Gallery. The phenomenon, which became known as the Ferranti effect, is now of course known to be due to the interaction of the induction of the transformer windings and the capacitance of the cable.

In Ferranti's original proposals for the Deptford station he included a pair of 10,000 h.p. vertical Corliss engines. Each engine had two high-pressure cylinders 44 in. diameter and two low-pressure cylinders 88 in. diameter, all with a stroke of 6 ft. 3 in. They were designed for steam at 200 lb. pressure and were to run at 60 r.p.m. The alternators weighed 500 tons each, the rotating armature, 46 ft. diameter, and shaft accounting for 225 tons. In the making of these sets some of the heaviest steel ingots ever made were employed but in spite of such an heroic conception the sets were destined never to be used. Their fate was being sealed, during the years of disaster, by the activities of Parsons, who was developing the steam turbine, which was to displace all large slow-speed engines for generating electricity as we shall see in Chapter XII.

One of the problems encountered in the adoption of alternating current for electricity supply was in connection with the running of machines in parallel. Dr. Hopkinson as early as 1883 had stated in a paper on 'Some Points on Electric Lighting' read before the Institution of Civil Engineers:[8] 'Now I know it is a common impression that alternate current machines cannot be worked together and that it is almost a necessity to have one enormous machine to supply all the consumers drawing from one system of conductors.' After some analysis of the conditions operating when alternators are run in parallel he concluded that stable running could be achieved. 'All that is required is that the incoming machine shall be thrown in with some care when it has attained something like proper velocity.' At a meeting of the Institution of Electrical Engineers a few months later[9] Hopkinson gave some experimental results obtained by running two lighthouse machines in parallel and then generously referred to the fact that it had only just come to his notice that as far back as 1869[10] Wilde had carried out a similar trial with complete success.

Notwithstanding these early and authoritative statements engineers were very hesitant about running alternators in parallel when larger machines were introduced for electricity supply. As the result of his experience at Paddington, Gordon, one of the few authorities on the practical running of alternators, referred to the fact that when two machines were connected in parallel they did not settle down until they had 'jumped for three or four minutes'. Such 'jumping' might be very costly in shortening the life of thousands of lamps and he concluded therefore that the only solution was one big machine for

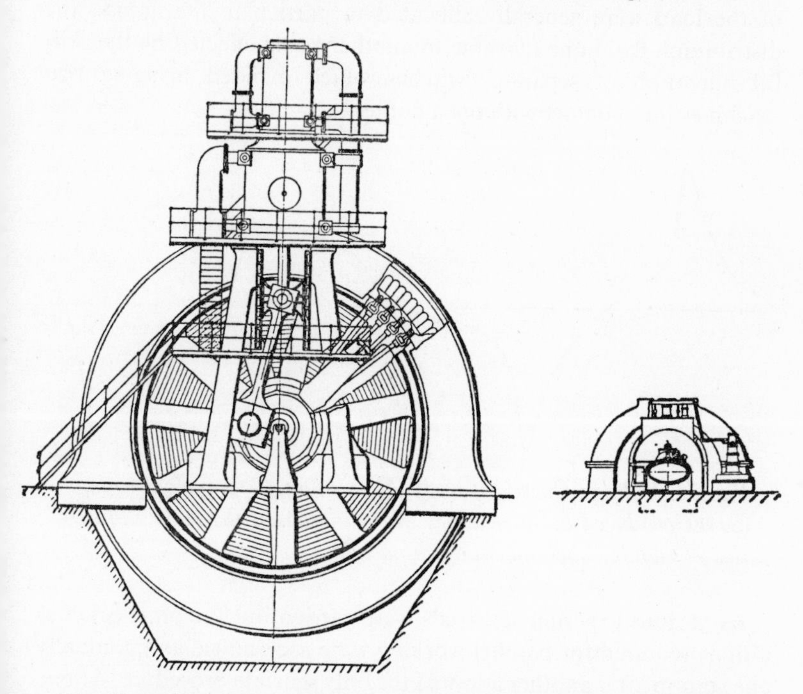

Fig. 21. Relative Sizes, Reciprocating and Turbine Units. This comparison of size illustrates the need about 1890 for a change from reciprocating engines in power stations to steam turbines which was inaugurated by Ferranti's abandonment of a scheme for large Corliss engines at Deptford.

the whole system. When we consider the difference in characteristics of drives which two alternators might have in the same station Gordon's conclusion is not surprising. A slow-speed engine coupled direct to the alternator could, of course, give wide fluctuation in speed over the revolution while a belt- or rope-drive would introduce some flexibility. Another factor would be the widely different electrical characteristics of different alternators. One could have a very peaky voltage curve while another might approach the rectangular. Two typical curves were given in the 1900 edition of Silvanus P. Thompson's *Polyphase Electric Currents*[11] for a Ganz and Co. alternator and one made by Siemens and Halske. Even if two such machines could have been induced to run in parallel the wasteful circulating currents must have been enormous. Consequently sections of the load were generally allocated to particular alternators, the distribution from one machine to another being effected by the skilful operation of separate switches which avoided bringing two machines into contact with one another.

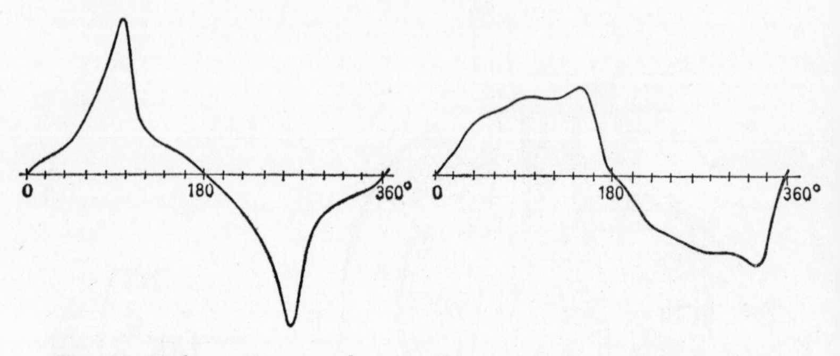

Fig. 22. Voltage Curves of early Alternators. Two typical voltage curves produced by alternators which increased the early difficulties of running such machines in parallel.

As various experimental trials were carried out the practical conditions required for parallel working were ascertained and gradually one station after another adopted the only sensible procedure. At one stage a Grosvenor Gallery alternator was run for a trial period in parallel with a Deptford machine. At the Richmond Road station of the Brompton and Kensington undertaking, three 100 kW 2,000-volt alternators, were being run in parallel about 1889, and by this date synchronizing methods using lamps had been introduced in the United States. Early methods of synchronizing two machines were rather crude. In one the cranks of the driving engine were observed in relation to one another and the corresponding arms on the alternators were painted white. Fortuitously the stroboscopic effect of the arc lamps illuminating the station enabled the correct paralleling condition to be observed and then the switch could be safely closed to bring in the additional machine. In another station the relative rotational position of the set on load and the incoming set was observed by opening the indicator cock on the two engine cylinders and noting the puff of steam which emerged. When the two puffs were happening together then the electrical sides could be safely connected.

The Battle of the Systems, as it was called, with Crompton, Hopkinson, Kennedy and others on behalf of D.C. fighting a rearguard action against the progressive Ferranti, Mordey and Professor Silvanus Thompson pressing the claims of alternating current, continued for many years, and did not die out suddenly but only through a process of attrition. In all this a paper read by Silvanus Thompson at the British Association in 1894 probably remains to this day the most complete and masterly vindication of the alternating current system.[12] In this statement he referred to the superstition that A.C. was less easy to manage, less economical, more dangerous, and less readily measured than D.C. It was well known, he wrote, that economies could be effected by raising the voltage of transmission, but pointed out the simplicity of a two-winding transformer compared with motor generators—no running machinery with sliding contacts, 'brushes, commutators and the like paraphernalia'. With the new alternators, then just introduced, there were not even slip rings in the high-voltage circuit but direct connection to stationary windings.

Silvanus Thompson's comments on the safety feature reveals that even so early there were many complaints from water and gas companies about the corrosion of their mains by leaks from the D.C. systems though his claims that A.C. shocks were less dangerous to human beings than those of D.C. were less reliable. In devices for regulating current the loss on D.C. systems by the introduction of resistances resulted in the waste of power and required the use of moving contacts, whereas for A.C. 'the use of choking coils is destined to find a great increase in services of this kind'.

In the systems described earlier the single-phase alternating type of current was usual though in one case two-phase current had been used almost inadvertently. Silvanus Thompson states the position quite clearly at the time of writing (1894): 'Two-phase and three-phase combinations of A.C. using three or four wires for their transmission instead of the ordinary 2-wire system of conductors are now quite well-known and their advantages fully recognized.' While admitting the complication of the multiplicity of wires he points out that in D.C. systems it had frequently been found necessary to employ three-, four- and even five-wire systems. In addition to the reduction in weight of copper required on a 3-phase 3-wire A.C. system over a 2-wire the former possessed one great advantage in the greater facility it offered for the use of the A.C. motor. On 2-wire single-phase circuits A.C. motors were not satisfactory, but the rotating field induction motor had provided a simple practical device with excellent starting characteristics.

At the time of Thompson's statement Scott had just invented the two-phase/three-phase conversion arrangement of transformers which has always carried his name and this he includes among the advantages of A.C. He also brings out some interesting differences between D.C. and A.C. magnets; with the former the pull on the armature falls off rapidly with distance, while with the A.C. magnet as the removal of the armature reduces the impedance the current is increased and the pull may actually be increased in consequence.

After pointing out that an A.C. synchronous motor possesses the advantage of fixed speed irrespective of load he goes on to explain how such a device can be employed as a condenser. Mordey had read a paper[13] on the operation of the synchronous motor and Thompson's remarkable summary of the state of knowledge at the time gave a detailed analysis of this subject; if *under* excited a synchronous motor acted as a choke whereas if *over* excited it acted as a condenser and could be employed at the end of a line remote from the supply point to compensate the inductive drop in the circuit. Elihu Thomson and Steinmetz had been led to the same discovery in America and had been able so to design the machine that it maintained the voltage on a circuit constant irrespective of the load. 'Thus', he continued, 'it would seem worthwhile to erect at one or possibly more points of the network, as remote as possible from the generating station, synchronous motors over excited, also to furnish the idle currents for magnetizing the transformer primaries.' He finishes with the remarkable prophesy: 'with such possibilities open in the future for alternate current working, and with such advantages in respect of motor power over continuous current working it can hardly be doubted that save in a few special cases, the vast majority of central stations will henceforth be operated by alternating current'.

The earliest example of three-phase generation and transmission in Britain was in the Wood Lane station constructed as a joint enterprise by the Kensington Court and Notting Hill Companies and put into operation in October 1900. The designers wished to adopt a voltage of 6,600 volts, but in spite of the existence of the Deptford

10,000-volt single-phase system the authorities would only sanction 5,000 volts and the system ran at this voltage for nearly forty years, when it was raised to 6,600 volts in accordance with other systems.

Abroad a spectacular three-phase transmission had been established some years previously in connection with the Electrotechnical Exhibition in Frankfurt. It was decided to transmit power to the Exhibition from Laufen on the Neckar, a distance of 110 miles.[14] A maximum load of 240 kW was transmitted at 15,000 volts by overhead line and used for both lighting and motive power. When the scheme was completed it caused a great sensation in the engineering world and no doubt influenced progress in this field in many countries.

This last decade of the nineteenth century saw much thought devoted to the design of alternators. Certain principles became established only to be abandoned after practical experience. At Deptford, for instance, Ferranti adopted a construction in which the field magnets consisted of two rings of double horizontal iron-cored coils supported in a massive circular shell. The armature was a thin ring of elongated flat copper strip coils supported on the periphery of what was virtually the engine flywheel. This form of armature had succeeded an earlier design in which Ferranti employed a continuous ribbon of insulated copper strip forming a circular zig-zag with alternate radial strips being in series with one another and spaced apart the pitch of the fixed field magnets.

In the Mordey alternator the armature, similar in general construction to that of the Ferranti machine, was fixed to the inside of the main iron casting; the field magnet consisted of a central coil concentric with the axis of the machine and had core extensions curved over from the end of this coil in a claw-like form. Thus nine pairs of pole pieces faced one another to provide the longitudinal field cutting the coils comprising the fixed armature.

A type of alternator which had a vogue for a short time was that invented by Kingdon and adapted by Fynn and others. In this machine the field and armature coils were both stationary and the rotating element carried masses of iron which came intermittently into such a position as to give the best possible magnetic circuit between the field and armature winding and subsequently moved away so as to increase the gap in the magnetic circuit connecting them. The repeated changes resulted in modification of the reluctance and consequently change of field in the armature circuit, which induced the required voltage. These were known as inductor alternators but as they possessed no real advantage over machines in which either the field magnet system or the armature coils moved they never became very popular.

Because of the advantages to be gained by keeping the coils in which the voltage was generated fixed rather than rotating this soon became the favoured practice. It avoided slip rings in the high-voltage circuit and simplified the dielectric problems. Very quickly the standard practice developed in which a multipolar electro-magnet rotated inside a surrounding iron core in the inner face of

Fig. 23. Mordey Alternator Rotating Field Magnet. In the Mordey alternator first produced in 1888, the current was generated in a fixed ring of flat coils projecting inwards from a surrounding cast-iron frame and the rotating field was provided by claw-like projections on an iron rotor which had one energizing coil concentric with the shaft of the machine.

which slots were cut to house the armature winding. This arrangement became very popular and survived in large diameter alternators until, at the beginning of the twentieth century, the introduction of the turbine-driven high-speed alternator revolutionized the whole practice.

The inception of high-voltage alternating current transmission resulted in the application of Faraday's transformer ring to the development of practical commercial transformers. About the middle of the nineteenth century, C. F. Varley had devised a construction in which the advantages of a subdivided iron core to secure the minimum eddy current loss were combined with a simple construction. He wound the primary and secondary windings over the centre one-third of the length of a bundle of iron wires, and then turned the ends, wire by wire, back over the windings to complete the magnetic circuit.

Thirty years later Gaulard and Gibbs introduced their system of distribution at 2,000 volts A.C. single-phase in which the primaries of a number of consumers' transformers were connected in series. They must be given credit for originality in this arrangement which converted the transmitted current of 13 amperes to 40 amperes. Their original transformer is on view at the South Kensington Museum. It has a core of soft iron wire with a primary coil of 3 mm. diameter

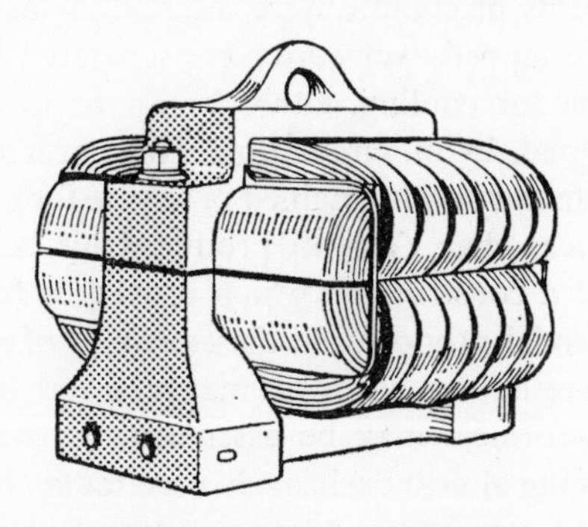

Fig. 24. Ferranti Transformer, c. 1890. The winding was applied over a number of iron strips laid side by side, the ends of which were subsequently turned over to complete the magnetic circuit and the assembly rigidly clamped together.

insulated wire surrounded by six equal bobbins of 1 mm. diameter wire. The secondaries are brought out to separate terminals on the side so that six separate sections can be used if required.

Ferranti followed the same general pattern but instead of using a magnetic core of fine iron wires he employed iron strips. In one of his early transformers designed for 15 kW he took six bundles of hoop-iron laid side by side and, over the centre portion, wound the heavy wire which was to become the low-voltage heavy current winding. The fine wire high-voltage winding was made up in sections carried by light frames and slipped over the first winding. After suitable application of insulating material the ends of the hoop-iron were bent over one by one so as to overlap one another and the whole assembly clamped together in a suitable cast-iron frame. This construction was adopted by Ferranti in 1887 for the Grosvenor Gallery system when he converted from series to parallel operation.

In 1891 Ferranti carried the design of transformers a stage further and constructed one with a capacity of 110 kW for transforming the Deptford current at 10,000 volts to the 2,400 required on the London system. This model consisted of three coils, a centre one for the incoming 10,000 volts and two adjacent coils, one on each side of it, for the 2,400 secondary. The coils were made up of copper strips separated by vulcanized fibre and wound over with varnished cloth and vulcanized fibre. Separately insulated component coils were built up one over another and connected in series. Considerable clearance was left between the central 10,000 volt coil and the two side coils and sheets of ebonite were inserted in the space. The flat iron bands of the lapped-over core were separated by half-inch air spaces to provide for cooling, either by air or in cases where oil cooling was adopted, by oil circulation. This particular transformer proved very satisfactory and remained in service for over thirty years.

In the same year that Ferranti produced his large transformer Mordey invented a construction which proved to be a sound technical and commercial proposition, the essentials of which have persisted up to the present time. For his magnetic core he used soft iron stampings. Transformers may be assembled as shell type or core type, the latter being almost exclusively used today. Mordey's was of the shell type, the iron sheets being assembled around the wound primary and secondary coils so that the windings, except for the end turns, are entirely surrounded by iron. He stamped a rectangular aperture in a larger rectangular sheet of such dimensions that the piece removed from the sheet could be used to bridge across the aperture at right angles, leaving two rectangular holes. The larger stampings were slipped, one by one, over the wound coils, each one alternating with one of the shorter pieces which was passed through the coil. In this way the coils were incased by the iron core from end to end and clamped together by iron bolts which, at one end, carried the terminal board.

About the turn of the century attention was paid to the characteristics of the material used for transformer cores with a view to securing high permeability accompanied by high electrical resistance, and silicon steel patented by Sir Robert Hadfield was found to meet the requirements. This alloy also retains its properties over long periods of service so that it soon became widely used and today 4 per cent silicon steel is standard practice, resulting in a total core loss in the best transformers of the order of only a few tenths of one per cent. Large modern transformers are discussed in the chapter on the

(*a*) First Nuclear Power Station, Calder Hall, 1956
This pioneer station had an output of 69 MW divided between two reactors.
(*Photo: Atomic Energy Authority*)

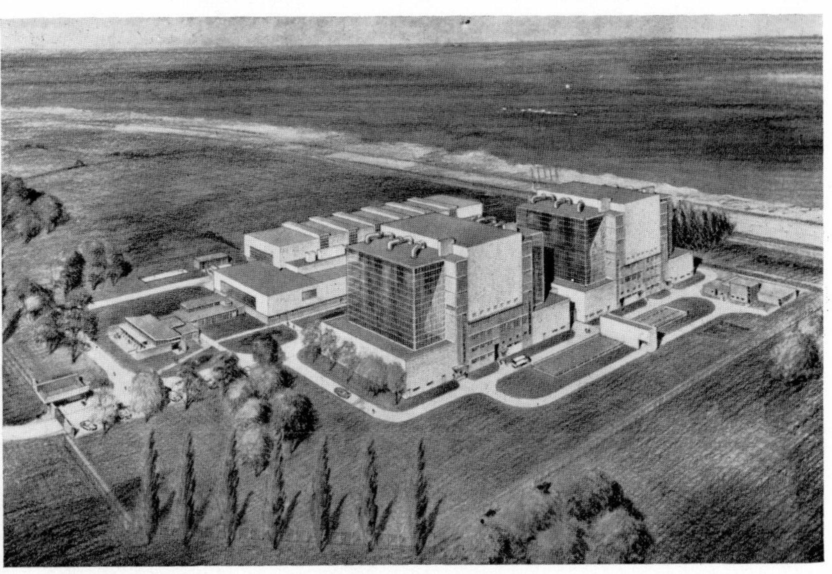

(*b*) Bradwell Nuclear Power Station, 1960 (artist's impression)
The heat generated by two reactors is passed to carbon dioxide gas which flows into steam raising units. Six Conventional type turbo-generators of 52 MW give a total station output of 300 MW. The generator's voltage of 11·8 kV is raised to 132 kV and is fed into the National Grid.
(*Photo: Atomic Energy Authority*)

PLATE XXVI

High Marnham Power Station. First 200 MW English Electric set. Brought into operation in 1959.
(*Photo: Central Electricity Generating Board*)

influence of Parsons, where it will be seen that the trend in design has led to the core type in which the iron core is first constructed with a removable link and the pre-wound coils slipped over before the link is closed.

REFERENCES

1. *Proc. Inst. Civil Engineers*, vol. 106, p. 2, 7 April 1891, and *Electrical Review*, vol. 28, p. 688, 1891.
2. Georg Siemens, *History of the House of Siemens*, vol. 1, p. 116, 1957.
3. W. L. Randell, *S. Z. de Ferranti.* Longmans Green, London, 1943.
4. Crompton, *Reminiscences*, p. 149.
5. W. L. Randell, *S. Z. de Ferranti.*
6. Parsons, *Early Days of the Power Station Industry*, p. 24.
7. W. L. Randell, *S. Z. de Ferranti.*
8. John Hopkinson, *Original Papers*, p. 67. Cambridge Univ. Press.
9. *J. Soc. Telegraph Engineers*, vol. 13, p. 496, 1884.
10. *Phil. Mag.*, Jan. 1869.
11. Silvanus P. Thompson, *Polyphase Electric Currents*, pp. 7 and 8, London, 1900.
12. *Electrical Review*, vol. 35, p. 199, 1894.
13. W. M. Mordey, 'On Testing and Working Alternators', *J. Inst. E.E.*, vol. 22, p. 116, 1893.
14. Georg Siemens, *History of the House of Siemens*, vol. 1, p. 122.

CHAPTER XI

Power and Traction

The credit of producing the first electric motor must go to Professor dal Negro, of Padua University, who, in the year 1830, obtained rotary motion from current supplied by a voltaic battery.[1] After showing that a pivoted magnet could be kept in continuous oscillation by current in an adjacent coil supplied through a simple contactor fixed to the magnet, he attached to one end of the magnet a light arm terminating in a pawl. This engaged with a cog wheel on a spindle carrying a flywheel. As the end of the oscillating magnet rose, the pawl engaged with the cog wheel and started off the flywheel. On its return stroke the pawl slid over the rotating cogs to come into action again on its next upward stroke. In this way the rotation of the flywheel was maintained.

Four years later Jacobi, a Russian professor, started a series of experiments which led to the first electrically propelled boat. In a letter to Faraday written in 1839 from St. Petersburg[2] he described how with a special battery constructed for the purpose and with ten people on board he had 'journeyed through many days' on the Neva. He was not satisfied with the trial, however, and estimated that to produce the force of one horse would require a battery of 20 square feet. In the face of such expenditure he reluctantly abandoned the experiments.

Jacobi's motor consisted of two wooden frames each carrying a series of horseshoe-shaped electromagnets with the pole pieces facing one another at a distance of a few inches. In the space between, a six-armed wooden spider rotating about a horizontal axis carried horizontal straight permanent magnets. The battery current supplying the electromagnets passed through a simple commutator which reversed the current in such a manner as to secure attraction and repulsion between the electro and permanent magnets always in the same direction. With a battery of 128 Grove cells a speed of 2·6 miles an hour was achieved, a remarkable performance for the time.

Other early examples of the application of the electric current to produce mechanical power date back to the times following Faraday's fundamental discoveries. In 1835, for instance, Francis Watkins,[3] evidently stimulated by consideration of Saxton's dynamo, mounted a bar magnet on a shaft so that it rotated with its poles sweeping past a group of stationary coils which were connected to a battery through a rotary switch.

In 1840 a Scot, Robert Davidson, equipped a car with an electric motor which consisted of eight coils on cylinders of wood with sliding cores of soft iron acting like engine pistons and connected to cranks on the car axles. The car, which had a loaded weight of five tons, attained a speed of four miles an hour and in 1843 made a number of successful trips on British railways.[4] The car created a great deal of interest when exhibited at the Egyptian Hall in Piccadilly.

Other early inventors produced crude electric motors about this time, but special mention should be made of one constructed by Elias in 1842. It consisted of two concentric wound iron rings, the outer one fixed and the inner one rotating. Each winding was divided into six sections, alternate sections being connected in reversed direction and two diametrically opposite points of the revolving ring were connected to alternate bars on a six-bar commutator. Froment, a Frenchman, made simple motors from 1844 onwards which consisted of 'paddle wheels' in which the paddles were bars of soft iron travelling past a series of horseshoe electric magnets. The magnets were energized through a commutator timed to cause attraction of the bars as they approached and to disconnect them while the bars were receding. This was virtually a precursor of the simple synchronous motor now used in electric clocks.[5]

From 1837 onwards, for a period of some ten years, Professor Page of Salem, Massachusetts, produced a number of interesting motor designs. Copying the well-known beam engines he substituted a solenoid for the steam cylinder and an iron core for the piston. Encouraged by a Congressional grant of $20,000, he built a double-acting beam type motor with flywheel and connecting rods which produced an output of one horse power and in 1854 a much larger one for traction purposes. A locomotive, which with motor battery and load weighed twelve tons, was tested on the Baltimore and Ohio Railroad and gave a speed of 19 miles an hour.

In a previous chapter Pacinnoti's iron ring construction was mentioned as anticipating by some years Gramme's ring armature dynamo. In 1860 Dr. Pacinnoti, an Italian, took a step in the development of the electric motor which resulted in a great advance. He was searching for an electromagnetic machine to give 'greater regularity and steadiness of action' than was achieved in the various attempts to make an electric motor. He had in mind, also, driving it not by battery power but by current from a magneto electric machine. The fundamental principle on which he based his idea was the creation of magnetic poles in the armature so controlled by the commutator that they did not move with the armature rotation but remained stationary as the iron itself moved on.

The armature of Pacinnoti's motor was an iron ring with 16 projecting teeth supported on a vertical spindle from a 4-armed brass spider. The winding was applied to the core in 16 sections between pairs of teeth and the 16 connecting wires were brought down to a 16-bar commutator with a wooden core. The field magnet consisted of two vertical wound coils standing on an iron bar to the upper ends of which were fitted pole pieces embracing the greater part of the armature with two gaps, diametrically opposite, each the width of two coils. The inventor said in a description of the motor,[6] published many years later: 'I like to consider such a ring as two transversal semi-circular electromagnets placed in juxtaposition, and having the poles of the same name in contiguity.' Thus, at all times, the exposed pole between one end of the magnet pole piece and the other had a polarity which was repelled by the pole it was leaving and attracted by the one it was approaching.

It is interesting to note in Pacinnoti's announcement of his motor some years afterwards a reference to the fact that a dynamo can be used equally well as a motor. This important fact seems to have dawned slowly. Lenz, as early as 1838, and Gramme, Siemens and others later, each made the discovery in turn. Probably the first public exhibition of one machine generating current which was then transmitted over a distance to reconvert the power back into mechanical energy on a similar machine was at the Vienna Exhibition in 1873. The operation of two Gramme machines in this way became the subject of a popular display at exhibitions in different countries about this time and outlets for practical application quickly developed. In 1879 a French farmer used electric power for ploughing and the advantages for industrial application were quickly realized. Both in Europe and in the United States many firms sprang into existence for the manufacture of motors of which a large number of types soon became available.

In this development there was, from the beginning, a close association between motor design and application to electric traction. The first electric railway in the world was constructed by Siemens and Halske in 1879. At a Trade Fair held in Berlin a demonstration line some thousand feet in length was laid down. Current at 130 volts was picked up from a centre rail clamped to a longitudinal sleeper and returned via the running rails. A train of five small cars with 'knife edge' seats was drawn along by a miniature electric locomotive propelled by a 3-h.p. motor. The insulation of the system was clearly insufficient for wet weather and the train carried passengers only on dry days, although during the short period of the exhibition it carried over 100,000 passengers.[7]

In the following year Edison, at Menlo Park, New Jersey, turned his attention to electric traction. He had already seen the possibilities of electric motive power and only a few years before had achieved great success with his famous electric pen driven by a tiny motor of his own design. The electric pen was a device for producing punctured copying stencils. A motor with dimensions about an inch by an inch and a half had a two-pole electromagnet and a flywheel armature carrying a short steel bar magnet. A small commutator reversed the current in the electromagnet coils as the magnet poles passed its cores. The motor ran at 4,000 r.p.m. and by means of a short crank caused a long needle in the pen-holder to vibrate so that the writing process pierced the paper with a large number of tiny holes. A two-cell battery supplied all the current necessary. The pen was widely advertised and when the Western Electric Co. took it up large numbers were sold. At one time over 60,000 were in use in the United States and elsewhere.

Because of its unique construction and resulting commercial success Edison's pen motor may be claimed as the first electric motor in history to be produced and sold commercially in large quantities. Apart from this example there were few electric motors in the United States but Edison quickly applied the knowledge he had gained in designing dynamos to produce one with superior qualities and applied it to running sewing machines. He arranged the armature between the pole pieces of the bipolar field magnet with its axis parallel to the coils. By providing a good cross section of iron and a small air gap and recognizing the importance of back e.m.f., he produced a motor which took only a small current on no load.[8]

In 1880 Edison laid down an experimental track at Menlo Park a third of a mile in length and supplied current from two of his Z-type dynamos rated at about 12 h.p. each. A third dynamo of the same type was fixed to a simple truck and used as a motor, the current being carried through the rails. In these experiments Edison used with success a series field coil for controlling the motor, so reducing the weight of resistance boxes carried. Speeds up to forty miles an hour were reached and wide publicity was obtained in the usual Edison style, but no commercial developments resulted.

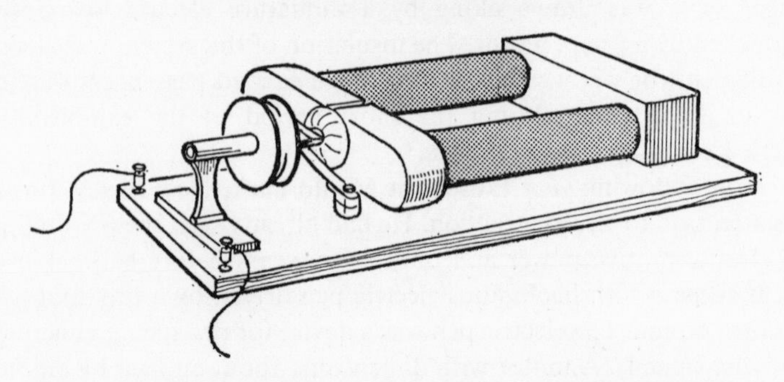

Fig. 25. Edison's Motor, 1879. This motor, designed by Edison following on his pen motor which had been such a commercial success, had two long magnet cores with inturned pole pieces between which the armature was housed. Characteristic of Edison's early machines, the field magnet cores were unduly long in proportion to the rest of the machine.

During the decades before and after 1890 increasing attention was given to the development of the D.C. motor. Although, fundamentally, motor design problems were common with those of the dynamo, the peculiarities of the situation in which the motor was to be used, and the type of service it was called on to perform, imposed special requirements. Speed variation, resistance to moisture in damp situations, peculiar shapes to fit in with other items of equipment, and such features, all called for special consideration. The relative advantages of shunt-wound, series-wound, and compound-wound machines were discovered and catered for while, all the time, efforts were made to improve the efficiency and reliability in operation. Larger sizes became available and very soon the electric motor established itself for industrial use as an essential piece of equipment in every factory.

The large central steam engine with its long cotton ropes to different floors in a mill, with other and smaller engines distributed over a large area, disappeared and were replaced by central generating stations supplying current to motors in separate shops. The long-established line shafting with leather belts to individual machines was one of the last features to go as separate motors for each machine became general during the past few decades. At first there were obvious applications for the electric motor, such as in the driving of fans, lifts and cranes, but it took longer to enter such fields as rolling mill drive and colliery winding gear when today electric drive is established practice.

About 1890 an important outlet for electric motors developed in an entirely different field. Since flanged wheels and grooved rails had been adopted for street tramways about the middle of the century many installations for horse-drawn vehicles had been laid down in this country and abroad, notably in the United States, and this form of transport for city dwellers held the field for many years.

In 1879, the year of Siemens' experiment with an electric train in Berlin, the British Parliament authorized the conversion of horse-drawn vehicles to mechanical propulsion. Steam locomotives were adopted in several towns from 1882, principally in the north of England, but were soon in competition with cable systems and electric trams. In 1891 the Thomson Houston Co. in the United States put on the market a slow-speed motor for street railway service which enabled single reduction gearing to be employed.[9] and the same year the first electric street tramway in this country was established at Leeds. This system pattern was set for a wholesale conversion throughout the country. The popularity of the electric tram continued until the First World War, by which time there were 264 undertakings in the United Kingdom. Street congestion, however, and the advent of the trolley bus and petrol motor bus with their greater mobility then resulted in loss of popularity for the electric tram car and by the end of the Second World War over 200 systems had been abandoned. During this half-century, however, the tramway was an important branch of electrical engineering.

Among the early motors designed specially for traction purposes was the Immisch (1886), which had double horizontal field magnets and an elongated Gramme armature. A peculiar commutator had two sections side by side in which the segments were staggered to short-circuit the coils during their passage through the neutral position in an attempt to reduce the distortion of the magnetic field and consequent sparking.

A much more revolutionary change was necessary, however, to

produce a design which could be accommodated within the dimensions imposed by track gauge, inaccessibility, weather resistance and the need to embody a large gear reduction between the motor spindle and the car axle. After friction drives, screw and worm wheel, and chain drives had been tried, the transmission system ultimately adopted was by spur reduction gearings. The Sprague motor, developed in the United States, established the construction in which the motor was supported at the end remote from the armature by a

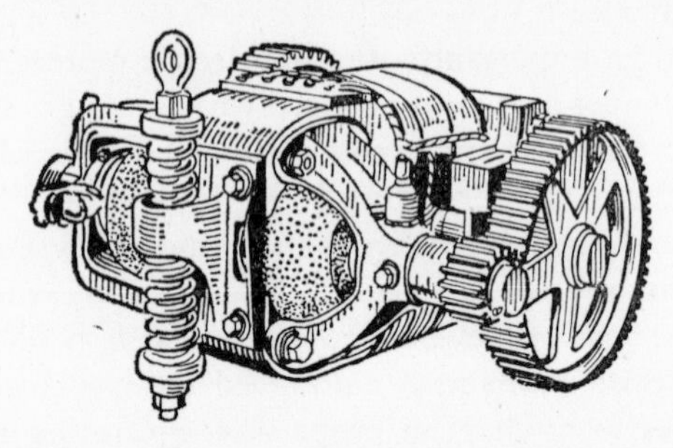

Fig. 26. Sprague Traction Motor. This motor, designed by Sprague in the United States, was carried on the axle of the vehicle at the end remote from the armature and was admirably adapted to the exacting limitations of space, dirt and moisture imposed by electric street-cars. It set a pattern which was widely adopted.

bearing on the car axle and connected flexibly through strong springs at the other end to the framework of the car bogie. In his original arrangement Sprague employed a horizontal bipolar magnet hinged to the axle at the yoke end. The motor pinion engaged with a large wheel on an intermediate shaft which passed through the motor between the field coils and at the other end of this spindle a second pinion engaged with a spur wheel on the car axle. The entire construction was an ingenious, efficient, and compact arrangement and rendered good service on street cars in America for many years.

In an improved motor for use in streetcars the Westinghouse Co. adopted the completely enclosed principle with a four-pole motor. The field magnet poles projected inwards from a cast-iron or mild steel yoke which formed the enclosing case of the motor, the upper half of which was hinged for inspection purposes. The armature was drum-type. A rectangular frame was hinged on the car axle and spring-supported to the chassis as with Sprague's design, but single reduction gearing was adopted, and the whole was protected from moisture and grit. By a special method of cross connection in the armature connections, it was found that although the field magnet had four poles, two brushes placed accessibly on the upper part of the commutator were sufficient.

The starting and control of speed with tramway motors presented the electrical engineer with many interesting problems. As no back e.m.f. is produced in a stationary motor, it was necessary in starting up to limit the current by placing resistance in series and to remove this, section by section, as the speed increased. Series-wound motors were employed and as there were usually two, one on each axle, a further device was introduced: the motors were first coupled in series so as to divide the voltage between them, and then ultimately in parallel to secure the maximum output. This sequence of connections required a complicated switching arrangement and after trials of a number of means which would be simple to operate and safe in an emergency, the cylindrical drum controller was adopted. In this device a vertical spindle fitted with an operating handle at the upper end carried a series of contact plates which came into contact with flat stationary springs connected in the circuit of the two motors. The usual order of procedure connected both motors in series with all the resistances at the first notch of the controlling handle. As this was turned around it removed the resistance section by section and ultimately connected the two motors in parallel, bringing back some of the resistance for safety during a transition stage. The usual handbrake was, at an early date, supplemented by an electric brake which applied a short-circuit to the motors and current to a magnet brake rubbing on the track.

Much ingenuity was displayed in the method of carrying current to the moving tramcar. The overhead trolley-wire was an obvious engineering solution, but although it was employed in 1883 at Portrush in Ireland and at Richmond in Virginia, opposition arose in other and older towns. Boston, Massachusetts, was able to take 9,000 horses off the streets but opposed the unsightly trolley wire. The result was that several alternative systems were tried out. The most important of these was the conduit system. In this the car carried a 'plough' collector which stretched down through a metal slot in the roadway and made contact with an underground conductor. The second, the stud system, depended on a collector rubbing on metal studs at the level of the road surface. These studs, fed from an underground cable, became alive automatically through the action of a magnet as the car passed over them. Neither system survived, however, and the overhead trolley wire operating at about 600 volts became standard practice. The return circuit was usually along the bonded running rail, though double-trolley wires were used at times to prevent interference.

The next stage in the application of electricity to traction was in hauling long-distance trains through tunnels at terminal stations as in New York and in the urban underground railways in London and other large cities. In the latter case the same relatively low voltage, 500–600 volts D.C. was adopted, as had become tramway practice, and the current was fed to the train along a third rail and a sliding metal slipper carried from the bogie axles. In the first London 'tube' railway the trains were hauled by electric locomotives, but on later systems they were assembled in units of a number of coaches, some of which were propelled and others trailers. The multiple control system developed by Frank J. Sprague in America in 1895 provided for motors distributed throughout a long train in this way to draw their current individually direct from the track and yet be under the control of a motorman operating a master controller at the front of the train. This has become standard practice. Many years later Sprague described his early experiences in his New Jersey shop in a most interesting review.[10]

The *Electrical Engineer* of 22 October 1890 reported that 'already a large number of electric railroads have been built as extensions of suburban lines', and progress continued. When, about the beginning of the twentieth century, the development of A.C. supply became so rapid, the question of employing alternating current motors on trains assumed considerable importance. Where the urban system was committed to D.C., as in London, the transmission was turned over to high voltage A.C., and the track circuits were fed from substations with rectifiers distributed alongside the track. For new traction schemes the possibilities of the A.C. motor were exploited and it is now necessary to retrace our steps in the story to introduce the A.C. motor which has been the subject of so much research and successful application.

Early in the development of the D.C. series electric motor it was discovered that although such a machine would continue to run in the same direction when both the armature current and field current were reversed, it would not run satisfactorily on alternating current. It was difficult to start and vicious sparking occurred between the commutator and brushes when it did run. This was due largely to the inductance of the field winding, which could be reduced by the use of lamination so cutting out eddy currents. Even then such motors could not be made to work on any but small sizes and many attempts were made to find a satisfactory A.C. motor suitable for general use.

Hopkinson in his speech in 1883 mentioned earlier observed that not only could two such machines be run in parallel on a common load but, if the power driving one of the machines was cut off, it would continue to run driven *synchronously* by the other. It had become a synchronous A.C. motor. The advantages of synchronous motors for certain duties were not at that time appreciated. Today, taking two widely different examples, we have valuable applications in the tiny synchronous motor used in every electric clock and at the other extreme the large multi-polar synchronous motors used for mine ventilating fans and for the correction of power factor of distribution systems. They have very definite disadvantages for some applications, however, in that special provision has to be made for running them up to speed and even while running satisfactorily a temporary overload may cause them to stop altogether. It is not surprising, therefore, that from the introduction of alternating current other types were sought.

The first practical type of A.C. motor was the induction motor which, in various forms, has become a valuable part of modern electrical equipment. For the origin of the induction motor we must go back to Arago's disc as both depend on the mutual force exerted between a magnetic field and the current which it induces in a conductor. Arago produced a rotating field by spinning a permanent magnet and the copper disc was dragged round after it, or vice versa, through the interaction of the induced eddy currents in the disc and the field of the magnet.

The first suggestion to produce a rotating field without the aid of mechanical motion seems to have been an exhibition before the Physical Society by Professor Walter Bailey in 1879. Bailey constructed a four-pole electromagnet in which the four poles stood up vertically from a common yoke. Designating the poles A, B, C and D, he sent two pulsating currents through the coils, one current through opposite coils A and C, and the other through B and D. The currents were supplied through a commutator which reversed them periodically and moreover reversed current B–D ninety degrees later than current A–C. Thus without any movement of the magnet itself he produced a magnetic field at the pole tips which rotated in a horizontal plane. On suspending a copper disc free to rotate over the four pole pieces he found that the rotating field carried the disc round with it.

In 1888 Galileo Ferrari of Turin fixed two coils at right angles to one another and sent through them two alternating currents 90° out of phase with one another. The lag was obtained by using a single-phase supply and passing one current through a highly inductive component. About the same time Nicola Tesla read a paper before the American Institute of Electrical Engineers[11] in which he described an important experiment. He produced a two-phase current from a dynamo with a drum armature by using two coils at right angles

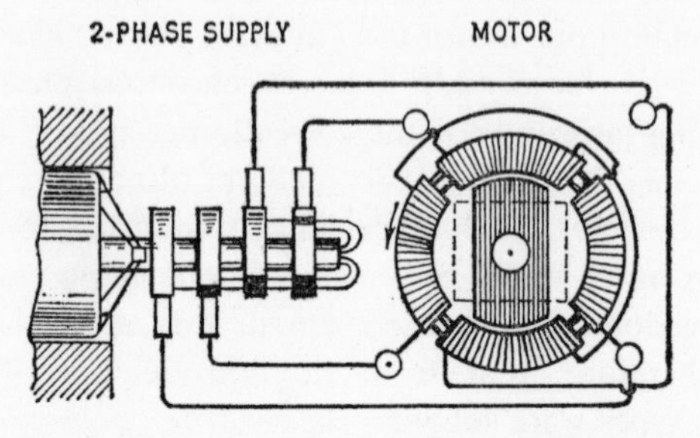

Fig. 27. Tesla's first Induction Motor, 1887. By feeding a two-phase current into two pairs of field coils at right angles, Tesla produced a rotating field. The rotor built up of iron discs carried externally closed circuit conductors.

connected to two pairs of slip rings instead of the normal commutator. On feeding the currents to two opposite pairs of windings on an iron toroid he produced a rotating field which would set suspended metallic discs spinning. This then led him to build the first practical two-phase induction motor by constructing an armature with two coils at right angles to one another rotating inside a four-winding ring. The current in two of the windings was out of phase with that in the other pair. Later he increased the number of coils in the rotating member and within a few months was exhibiting practical machines which were taken up by the newly formed Westinghouse Company, who showed their confidence in the invention by making a large number for the equipment of their new factory at East Pittsburgh.

The A.C. motor was soon in wide use in the United States and other countries, and at the Frankfort Exhibition in 1890 Dobrowolsky exhibited a three-phase machine to give 100 h.p. Great interest was aroused in the type and, as the use of alternating current increased, it became the subject of widespread technical and commercial exploitation.

In 1891 W. Langdon Davies produced an experimental single-phase induction motor. The ring stator with six small pole pieces was wound in six sections forming three groups of two coils each. One group had resistance in series, but the others were connected directly on the mains. Thus a lag occurred in the current of the one coil relative to that in the other, resulting in a rotating field. The rotor, a similar but smaller ring, was wound with heavy copper in six sections, each one being short-circuited.

This inventor soon improved on his original model and in 1897 was producing commercial machines. The stator was built up of slotted laminated sheets and carried two sets of windings, one displaced circumferentially relative to the other. The squirrel cage conductors were arranged askew with a view to improving the starting qualities. Early examples of these two machines are on display in the Science Museum, South Kensington, the latter machine partly sectioned to show the construction in detail.

Between the dates of Langdon Davies's two machines much development took place on D.C. machines. This is typified by Crompton's armature of 1905, also shown in the Science Museum, and indicates that already constructional practice was settling down to ideas which have persisted up to the present time.

Two main forms of rotor emerged. One of these, known as the 'squirrel cage', consisted of an assembly of iron stampings, forming a slotted cylinder, with a series of copper rods laid in the slots, their ends connected to copper rings so that all were short-circuited. In the other construction a wound rotor was employed, the winding being in three-phase formation with one end of each set being brought to a common point and joined together. The other three ends were brought to three slip rings.

In both types of induction motor the rotor revolves inside a stator provided with windings in slots on the inner face of a laminated core, and when these windings are supplied with two- or three-phase current they produce a rotating field which drags the rotor around with it. The simple squirrel-cage motor when started on load takes a momentary current many times its full load current, as the current induced in the rotor coil at any time is a function of the 'slip', that is the difference in speed between it and that of the rotating field. One of the main objects of the wound rotor is to enable adjustable starting resistances to be inserted via the slip rings and brushes.

The squirrel-cage induction motor in small sizes has become more and more popular because of its robust and simple construction, so that today it is used in very large numbers for such light duties as hair driers, floor polishers, vacuum cleaners, office machinery, machine tools and a host of other purposes. Under the general description of fractional horse-power motor it is employed in single-phase circuits, the rotary field being obtained by the use of capacity/inductance effects through specially designed windings.

An interesting development arising from the principle of the induction motor is the electrical coupling which has become an important feature in the driving of such items as mine ventilating fans. The driver and driven side of the coupling are connected by a controlled magnetic field in one member which induces a current in the other so long as the two speeds are different. This produces a torque between the two which can be controlled by varying the excitation. Another example is in the important precision device used for keeping two machine equipments in step, known as the 'selsyn'. Only an inch or two in diameter and three or four times as long, this motor, with a three-phase stator and a two-phase wound rotor with slip rings, can maintain automatic control between two industrial operations to within one-tenth of a degree.

When A.C. motors were introduced it was soon discovered that although a synchronous motor was difficult to start it had a great advantage over the induction motor when running, in that its power factor was very much better. A combination of the advantages of both types is obtained in the modern synchronous induction motor in which the machine is started as an induction motor and then, when its lag has been brought to a low value, the rotor can be speeded up to synchronism by the application of D.C. excitation to the rotor.

These and many other developments have taken place in the design of the A.C. motor so that now there is a particular type suited to almost every service requirement.[12] Generally speaking, D.C. series motors still hold the field for precision in such operations as the lifting of machine parts by crane in assembly shops. They gave ground to the A.C. motor slowly for the first quarter of the century but even the general supply of A.C. which came in between the two world wars did not diminish the demand for D.C. machines. For the many industrial applications requiring constant speed and torque as given by the shunt wound D.C. machine, the A.C. induction motor is cheaper and meets the requirements, but industrial development in the direction of automation has found an important field for the

D.C. series motor in which it will be difficult for A.C. machines to compete. In large complicated plant such as steel works, paper factories, etc., with the need for co-ordinating a number of different motors, the popularity of the D.C. motor has grown over the past thirty years. The introduction of Ward Leonard control, for instance, made possible applications of electric drive never before contemplated and legion in number. An up-to-date example is the use of a Ward Leonard control in connection with colliery winding gear. In 1957 a 2,500 h.p. D.C. motor was installed in a North Staffordshire colliery, the motor generator set being driven by a 1,800 h.p. induction motor.

It has sometimes been said that important development in electrical engineering stopped about 1900. Certainly a period of consolidation followed the astonishing spate of invention during the eighties and nineties, but sufficient has already been said in the present chapter to show that, although we still operate on the fundamentals laid down in those times, yet both in technique and magnitude the scope of electrical engineering has continued to widen without any slackening. A good example can be found in the production of electric motors. Never was there such a demand as today. After the Second World War this branch leaped forward to such an extent that the building of special factories followed one after another. One British firm has recently announced that after producing over three million motors in one factory, they have already reached the first million in a second one. The mass production of these essential and complex items is in itself a triumph for electrical engineering.

Considering the choice between A.C. and D.C. in the motor field it is not surprising that in the development of electric traction what might almost be styled a second 'battle of the systems' should have been fought. Indeed, it still continues, for countries even take sides in the battle between A.C. and D.C. traction and in this country the choice has several times been the subject of public inquiry.

Electric traction first appeared in Britain on 3 August 1883 when a quarter-mile length of railway was opened for traffic at Brighton by Magnus Volk. The current was supplied from a third rail at 140 volts D.C. A few weeks later the famous Portrush line in Northern Ireland began to operate at 550 volts D.C. carried by an overhead trolley wire. A most interesting account of this enterprise was written jointly by Alex Siemens and Edward Hopkinson.[13] The six miles of track had heavy gradients and sharp curves. The motors, referred to as dynamos, were made by Siemens Bros., and drove the wheels of the small 4-wheel locomotives through chain reduction gear. A speed of twelve miles an hour was achieved. The article is interesting as indicating the theoretical analysis which Hopkinson applied to the case.[14] In 1890 the City and South London tube was opened, employing electric locomotives, and the Central London tube between the Bank and Shepherd's Bush followed within a few years. Both were D.C. systems at 500 volts supplied from a third rail.

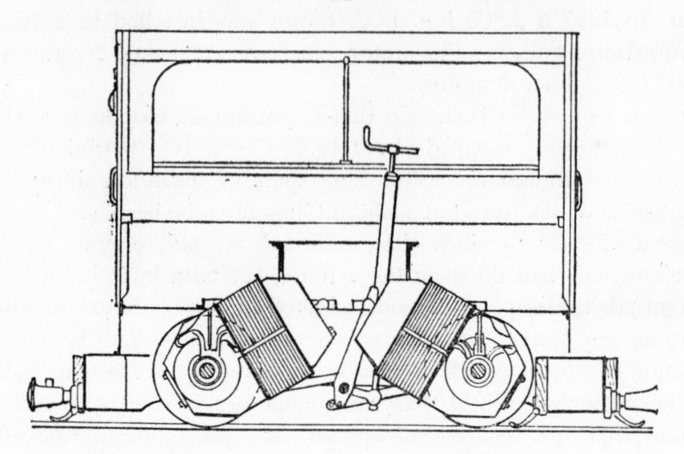

Fig. 28. Locomotive on City and South London Railway, 1890. The 50 h.p. 500 volt D.C. motors on these early locomotives were of the Edison–Hopkinson type with ring armatures fixed to the axle. For a train speed of 25 miles an hour, they ran at 310 revolutions a minute.

The City and South London system saw the first application of direct drive from the motors on to the wheel axles. Two series-wound Edison–Hopkinson type bipolar motors with Gramme ring armatures were mounted so that the latter were fixed to the axle and the pole piece end of the field magnets was carried by bearings on the axle. The motor speed was 310 revolutions a minute for a train speed of 25 miles an hour, and the combined maximum output was 100 h.p. The current at 500 volts was supplied through a third rail of steel channel supported on glass insulators from transverse wooden sleepers and picked up by cast iron slippers. The two motors were connected in series and controlled by a reversing switch and a running switch which cut in and out a set of resistance coils for starting purposes. Each locomotive hauled three passenger coaches.

(*a*) Ferranti Transformer, 1891
The core consisted of iron strips. Over the centre of ten bundles of such strips Ferranti wound the heavy wire which was to become the low voltage winding. The secondary winding of fine wire was made up in coils and slipped over the primary. The iron strip bundles were then bent back above and below and clamped at the joint.
(*Photo: Science Museum*)

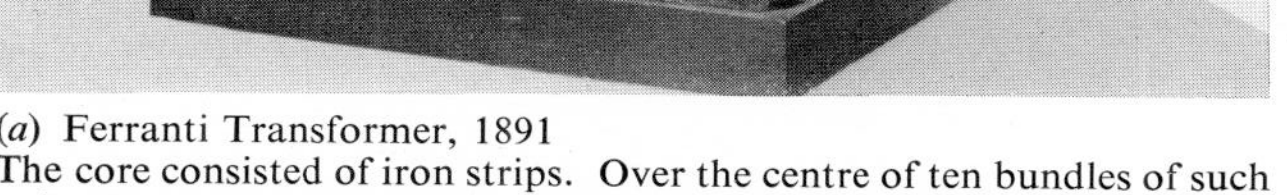

(*b*) System Load Curves
The photograph shows the daily load curves for London in 1954, each one on a separate card assembled over a period of 12 months.
(*Photo: London Electricity Board*)

(*a*) Large 3-phase Transforr
Core
Iron cores assembled ready
coils.
(*Photo: English Electric C*

(*b*) Large 3-phase Transformer
Cores complete with windings in position. Output 50,000 kVA 10·5/66 kV.
(*Photo: English Electric Co.*)

Plate XXIX

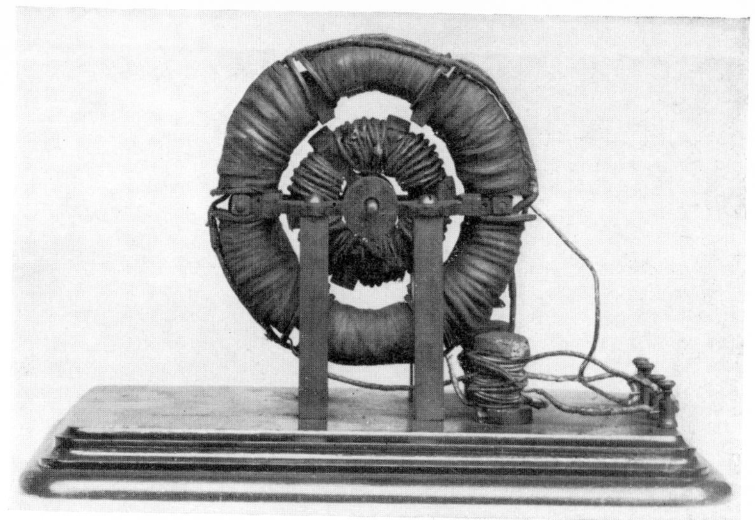

(*a*) Original Langdon Davies Induction Motor, 1891–2
The stator had three groups of windings each consisting of two coils in series, the groups being in parallel. By inserting resistance in series with one group a phase difference was introduced so resulting in a rotating field. This induced current in the short-circuited windings on the rotor and produced a turning torque.
(*Photo: Science Museum*)

(*b*) Commercial Langdon Davies Induction Motor, 1911 (partly sectioned)
Similar in principle to the experimental motor but with added refinements which made it a sound practical motor. (*Photo: Science Museum*)

PLATE XXX

The Central London Railway equipped by the British Thomson-Houston Co., started with locomotive haulage but each locomotive had two four-wheel bogie trucks. Direct current series motors, each of 120 h.p., were mounted direct on the axle shaft and were controlled by series parallel arrangements—all four in parallel for running—and resistance coils. The controller had sixteen steps. The weight of the train was 150 tons.

The rapid rate of development taking place at that time can be judged from the supply arrangements used in these two cases. The City and South London generated current at 500 volts by means of four dynamos each giving 450 amperes which passed to the third rail through lead-covered feeders in the tunnel. In the case of the Central London, three-phase alternators supplied current at 5,000 volts, 25 cycles, to transformers and motor generators, where it was converted to 500 volts D.C. for the track. In 1902 multiple unit train control was introduced on the Central London Railway and this type has replaced locomotive haulage almost exclusively for suburban traffic.

Other sections of electrified track appeared: Bow to Upminster in 1905, Paddington to Westbourne Park in 1906, Lancaster to Heysham in 1908. The London ones, influenced no doubt by the underground examples, kept to D.C., but for the Heysham line single-phase 6·6 kV at 25 cycles, carried on an overhead trolley wire, was adopted.

Wide differences of opinion arose as to voltage and system and from 1916 1,500 volts D.C. on an overhead wire became increasingly popular. In 1920 a Railway Advisory Committee issued a recommendation supporting this system which was endorsed by the Kennedy Committee in 1927 and legalized by a Ministry of Transport Order in 1932. Twenty-five years later (1956) the Transport Commission decided on a 25 kV 50 cycles single-phase system as national standard.

The introduction of the mercury arc rectifier in 1928 was a landmark in the history of electric traction in this country and when the National Grid was established the use of low-frequency supply with rotary converters gave way completely to a 50 cycle supply and the mercury rectifier. The perfection of the rectifier and the high cost of maintenance of single-phase A.C. commutator motors has now led to the use of rectifiers on trains and D.C. motors with their many advantages.

On the A.C. side different practices developed in different countries In 1930, for instance, the three-phase system disappeared finally from the Swiss Federal Railways and was replaced by single-phase 15,000 volt current at a frequency of 16⅔ cycles per second. About the same time there was one scheme in the United States employing 11,000 volts at 25 cycles on the trolley line with 300 h.p. A.C. motor, and another operating at 3,000 volts D.C. In 1892 Sprague had a 1,000 h.p. locomotive running in the United States and today there are electric locomotives operating on the Pennsylvania Railroad of 3,750 h.p. hauling 1,000-ton trains at 90 m.p.h. In the United States as in Europe, the use of rectifiers in traction substations became widespread in 1933. During the past few years electric locomotives of 4,000 to 5,000 h.p. have come into use, twice as powerful as most steam locomotives. Mention must also be made of the steady growth since about 1925 of the diesel-electric locomotive. By 1930 the Canadian National Railway was using this form of locomotive with two engines per train, giving a total of 2,660 h.p. The most recent development in this direction is possibly the use of the gas turbine electric locomotive, two of which with capacities of 2,500 and 3,500 h.p. are on trial in this country hauling heavy trains at 90 m.p.h.

The application of A.C. power for traction purposes has come along various routes over the past few decades: (*a*) conversion to D.C. in substations along the track and supply of D.C. motors through D.C. on a trolley wire, (*b*) the lowering of the transmission voltage to one suitable for the trolley wire with A.C. motors on the train, and (*c*) running the trolley wire at as high a voltage A.C. as possible with the employment of both transformers and rectifiers on the rolling stock to use normal type D.C. traction motors. Excellent discussions of the various systems appeared in papers by J. W. Grieve and Calverley and others.[15] The 1,500 volt D.C. overhead conductor system recommended by British Railways in 1927 was endorsed by a Government Committee in 1950, except for certain southern lines, and during the next few years the Liverpool Street to Shenfield and the Manchester–Sheffield–Wath lines were converted. In 1956, however, the 25 kV A.C. system triumphed and in April 1959 the first lines to be used on this system commercially in British Railways were taken into use on the Colchester–Clacton–Walton section.

A recent development (1960) constitutes a landmark in the electrification of railways. The first 3,680 h.p., 25 kV locomotives, with a top speed of 100 m.p.h. and a tractive effort of 48,000 lb., mark the real beginning of the disappearance of the steam locomotive on British Railways.

The logic by which railway electrification has been led through the various combinations of A.C. and D.C. to the present system in which national A.C. supply is supplied to the train at 25 kV, and transformed and rectified on the train for the use of D.C. motors, is indeed a story of a modern 'battle of the systems'. But the solution, compromise as it may seem, is established as is indicated by such large schemes as that in France with its 225 route miles joining Valenciennes to Thionville with extensions in contemplation. The whole development has been well recorded in the paper by Calverley and others.

The history of electrical engineering in the field of traction is one of great and continuous progress and one which no review can overtake. Even as the rectifier becomes an essential factor in solving one of the major problems so is the rectifier itself threatened by change. Germanium and silicon rectifiers already offer advantages over the mercury arc type though this movement has not so far reached the stage of history.

REFERENCES

1. *Phil. Mag.*, vol. 8, p. 45, 1832.
2. *Phil. Mag.*, vol. 15, p. 164, 1839.
3. *Phil. Mag.*, vol. 7, p. 107, 1835; vol. 12, p. 190, 1838.
4. *Electrician*, vol. 9, p. 400, 1882.
5. *Applications de l'Electricite*, pp. 341–420. Paris 1878.
6. Mullineux Walmsley, *Electricity in the Service of Man*, p. 588, 1911.
7. Georg Siemens, *History of the House of Siemens*, p. 85.
8. *Scientific American*, 18 Oct. 1879.
9. *Electrical Engineer of New York*, vol. 11, p. 141, 1891.
10. *J.A.I.E.E.*, vol. 53, p. 695, 1934.
11. *J.A.I.E.E.*, vol. 5, p. 308, 1888.
12. H. Cotton, *Alternating Current Machines*. London 1952.
13. *Electrician*, vol. 10, p. 545, 1883, and vol. 14, p. 303, 1885.
14. *Proc. I.E.E.*, vol. 103, Part A, p. 229, 1956.
15. *Proc. I.E.E.*, vol. 104, Part A, p. 15, 1957.

CHAPTER XII

The Turbine Era

From the early days of public electricity supply there has always been a close connection between the design of the electric generator and the development of the prime mover furnishing the mechanical power. Consequently, in this branch of electrical engineering, it has not been possible to separate mechanical and electrical considerations; rather has the electrical engineer been compelled to accept responsibility for the complete unit. This was so in the days when a choice had to be made between slow speed reciprocating engines with their rope drives and the direct coupled high-speed Willans and Belliss sets. The arrival of the steam turbine towards the end of the nineteenth century was a complete revolution for with it came a drastic change not only in design and size of generator but also in the layout of power stations and in the whole philosophy of the generation of electricity. Without the modern enormous and highly efficient turbines producing electrical energy on such a vast scale it is difficult to see how there could have been an electrical industry or indeed any industry at all on a scale commensurate with that which today flourishes throughout the civilized world. Within the supply system itself the increased availability of electric power and the larger amounts of energy to be controlled and distributed resulting from the adoption of the turbine have had considerable influence on the design of every item of equipment, transformers, switchgear and transmission line. Beyond question the invention of the steam turbine with its connected generator has proved to be the most potent factor in the history of electrical engineering throughout the world over the past sixty years.

The turbine story stems from two main roots, the interest displayed in A.C. transmission in the late 1880's, in which one young man, the enthusiast Ferranti, took such a lively and effective part and the arrival of another young man, Charles Parsons, a mechanical genius.

With his rapid developments at Deptford, Ferranti established the need for larger and larger prime movers. Parsons, working independently, provided the solution.

Charles Algernon Parsons was born in 1854 at 13 Connaught Place, Hyde Park, London, the sixth son of the third Earl of Rosse, Member of Parliament, who was an engineer and astronomer and President of the Royal Society.[1] Charles spent his boyhood largely in the family home, Birr Castle, Parsonstown, Ireland, under a strict educational regime with tutors, one of whom was Sir Robert Ball, the distinguished scientist, interspersed with happy workshop and yachting adventures shared with his father and brothers. His home was the rendezvous of the leading scientific men of the day. At seventeen he entered Trinity College, Dublin, where he quickly gave evidence of high intellectual endowments and two years later proceeded to St. John's College, Cambridge, to take the Mathematical Tripos. At that time there was no Engineering Tripos but Parsons displayed a keen interest in engineering problems and his study was littered with models on which he was working. At the age of twenty-three Parsons became a pupil apprentice at the Elswick works of Sir William Armstrong and continued to display his inventive genius and intellectual energy so much so that one of his devices, worked out in principle while at Cambridge, an epicycloidal engine, developing ten to twenty horse-power, was sold and about forty of them were made for driving dynamos.

Subsequently, after a brief period carrying out experimental work with Sir James Kitson at Leeds, Parsons secured a junior partnership with Clarke, Chapman and Co., Gateshead, where he turned his attention to the problem of the steam turbine and of a very high-speed dynamo and alternator to work with it. In about 1884 he produced a small model which worked well from the start and gave a steam consumption of about 150 lb. per kilowatt-hour. It ran at 18,000 r.p.m. and produced a direct current of 75 amperes at 100 volts. After a remarkably successful career in a lamp works at Gateshead this first turbine was ultimately placed in the Science Museum at South Kensington, where it records for all time the real beginnings of the power station industry as we know it today.

By the age of thirty Parsons was established in the career which led, before his death in 1931, to a remarkable world contribution of electric power, over thirty million horse-power.

The principle of obtaining mechanical power by means of steam issuing from a jet was, of course, not new. Where Parsons succeeded was in overcoming the limitations which had prevented other inventors carrying their proposals for a practical commercial turbine up to the largest sizes. With his mathematical knowledge he saw and solved the mechanical problems of turbine shafts rotating at very high speeds. He recognized also the value of adopting a multiplicity of wheels on one shaft, so distributing the drop in steam pressure along the length of the turbine. By alternating fixed and rotating rings so that the steam on its way through the turbine entered one set of blades direct from the previous ones, and by paying attention to the shape of the blades he obtained increased efficiency.

Parsons attacked the electrical problems of the turbo-alternator with equal energy. Iron core, conductors and dielectric all proved fruitful sources of inquiry and he was soon producing small commercial sets. In the year 1887 he made ten turbo-alternators with outputs up to 32 kW which supplied current through Swan's incandescent lamps at the Newcastle Exhibition. In 1888 the Newcastle and District Electric Lighting Company ordered the first of four 75 kW turbo-alternators. These ran at 4,800 r.p.m., producing single-phase current at 1,000 volts. In the following year his firm, Clarke, Chapman, Parsons and Co., installed two similar axial flow machines in a new power station at Forth Bank and the adoption of turbines for power station work had become firmly established.

At this time differences arose between Parsons and his colleagues and the partnership was dissolved, but he soon found other friends and with them in 1889 established the firm of C. A. Parsons and Co., with works at Heaton, Newcastle. Unfortunately Parsons was unable to take with him his original patents of the axial flow turbine but, undaunted, he set to work and produced an alternative on the radial flow principle.

Within three years of forming the new company Parsons, operating under the parliamentary powers granted to the Corporation of Cambridge, erected the first power station on the banks of the Cam. Here he installed three turbo-alternators of the new type, each of 100 kW capacity, running at 4,800 r.p.m., generating single-phase current at 2,000 volts and 80 cycles per second. Two interesting features of these sets from a modern viewpoint are that they were not bolted down, but stood on rubber blocks in a metal tray and that they exhausted direct to the atmosphere. The company adopted condensing at this time, however, with of course considerable improvement in efficiency.

In what subsequently became a traditional association between the

Engineering Laboratory of the University of Cambridge and the local power station, Professor Ewing in 1892 carried out comprehensive tests on the turbine installation. In his report he paid high tribute to this new arrival in the power station field, stressing its economy of steam under variable load, its exceptional lightness, steadiness and simplicity. His authoritative report did a great deal to meet current prejudice in some quarters and to establish the turbo-alternator.

Soon after the Cambridge installation Parsons secured the contract for a new generating station at Scarborough and installed two turbo sets each 120 kW to run at 4,800 r.p.m. for 2,000 volts single-phase, 80 cycles. Again the radial type of turbine was used and tests carried out by Sir A. B. W. Kennedy showed a consumption of 27·9 lb. of steam per kWH at full load, or 44·1 at quarter load. The results quickly led to additional and larger units.

The first municipal owned power station, to purchase steam turbines was Portsmouth, and here in 1894 Parsons installed a set, increasing the size once more, this time to 150 kW. This was the first occasion on which a turbine set was run in parallel with reciprocating sets. At 2,000 volts and 50 cycles, the results were completely satisfactory.

About this time Parsons turned his attention to the London power stations. In those employing alternating current, the Paddington power station of the Great Western Railway, with its two-phase alternators designed by Gordon, and the three stations at Rathbone Place, Sardinia Street and Manchester Square of the Metropolitan Electric Supply Co., reciprocating engines were in use and progress appeared to be steady and assured. A crisis arose, however, at Manchester Square. Here there were ten Willans sets of 120 kW capacity running at 350 r.p.m. which caused so much vibration that the local inhabitants took legal action to have the nuisance suppressed. Obviously something must be done and heroic measures in the form of underpinning to reach a firm foundation on solid ground were undertaken at great expense. This, along with mechanical modification of the sets, proved of no avail and the threat of having the station shut down was very serious.

While this trouble was causing the engineers much concern in London, the Parsons sets in Cambridge continued to run sweetly on their rubber pads. The solution was clear. Parsons built and installed a 350 kW set to run at 3,000 r.p.m., twice the size of any previous set, and the situation was saved. By this time Parsons had regained his patent position and had returned to the axial flow type of turbine in which he always had the greatest confidence. The two other stations of the Metropolitan Company were also changed from radial to axial flow soon afterwards.

About the end of the century the demand for Parsons turbines increased so rapidly that it exceeded the capacity of the Heaton Works, and licences were granted to other manufacturers both at home and abroad. In Germany an order was secured for the supply of two single-phase turbo-alternators to run at 1,500 r.p.m. for the city of Elberfeld. They were the largest to be built anywhere in the world and the results on test were outstanding, the steam consumption being only 18·22 lb. per kWH. The continental rights were at once taken up by the Swiss firm of Brown Boveri and those in America by the Westinghouse Company.

For some years the alternator remained of the conventional design with fixed field magnets and rotating armature, but in 1903 the first set with rotating field magnet, which subsequently became universal practice, was constructed and installed at Newcastle. At this stage the rotor had salient poles but very soon afterwards at the suggestion of Brown Boveri, Parsons sunk the windings in slots on the surface of the drum core. This distributed winding of the rotor has been adopted ever since.

As the years advanced more and more Parsons turbo-sets were installed. In 1905 the first designed for 11,000 volts were installed for the Kent Electric Power Co. at Frindsbury. These had an output of 1,500 kW at 1,500 r.p.m. and were followed two years later by 6,000 kW sets for Lots Road, London, and elsewhere. There was by now no question of the place of the turbo-alternator for generating electricity on the largest scale. The number of successful installations increased, as did the size of sets. Just before the First World War a Parsons 25,000 kW set was installed in the Fisk Street power station of the Chicago Commonwealth Edison Co. So successful was this, with its now-fashionable double-flow low-pressure cylinder and a steam consumption of 10·45 lb. per kWH, that soon after the war a spectacular repeat order was given: this time for a 50,000 kW set for the same Company's Crawford Avenue station.

In this set there were three turbines in series in the steam supply which entered the first at 550 lb. per sq. in. and expanded to 100 lb. It was then reheated to 700° F. and expanded to 2 lb. and passed through the low-pressure turbine out to the condenser. The speed was 1,800 r.p.m. and the alternator outputs 15,000 kW and 30,000 kW, with 5,000 kW on the low-pressure turbine.

Coming to more recent times many large and well-known stations were equipped—Carville, Barking, Brimsdown, Dunston and others —each with some improved feature. In 1928, for example, the first turbo-alternator to generate direct at 33,000 volts went into operation at Brimsdown and by 1930 50,000 kW sets at Dunston had achieved a record in efficiency. Using re-superheating the consumption had been reduced to 9,280 B.T.U. per kWH.

It is quite impossible here to follow the vast growth of power stations in many countries during the past two decades, but Battersea must be included in the record.[2] The station was planned from 1927 onwards as two separate stations, Battersea A and Battersea B. At that time the up-to-date British practice had adopted initial steam pressures of 350 lb. per sq. in. and total temperatures of 750° to 780° F. At the same time engineers in the United States were employing reheat cycles with pressures of 1,200 to 1,300 lb. per sq. in. and temperatures of 700° to 725° F. It was decided to adopt for the Battersea boilers a pressure of 600 lb. and a temperature of 830°F., and two sets were installed in 1928 each of the three-cylinder impulse type with double-flow exhaust and twin condensers. Each was coupled to a 64,000 kW main alternator running at 1,500 r.p.m., generating three-phase current at 11,000 volts.

Progress is again indicated by an extension to Battersea A station carried out in 1933, when a 105,000 kW set, 120 feet long, was added, the largest turbo alternator in any European power station at the time and for many years afterwards. At this period there were many sets on order for outputs of 50,000 kW to 100,000 kW, but the tendency to go much larger and to use multi-axis machines was changing. Difficulty in manufacture and transport was one factor, but the safeguards of a high-pressure governor were reduced when such a large quantity of steam beyond its control was already on its way through the turbine. General practice led to a maximum size of 75,000 kW at 15,000 r.p.m., though in France a tentative approach to twice this speed was being made. Steam pressures were, however, rising, and sets in Czechoslovakia and elsewhere reached 1,800 lb. per sq. in.

Alternator design was also developing rapidly. In the United States a 3-axis set of 208,000 kW was running in Chicago but 160,000 kW was still the largest single axis set with alternators running at 1,500 r.p.m. At Brimsdown, England, an alternator was generating direct at 33,000 volts, so obviating the need for a transformer between it and the grid, and South Africa followed with three machines at 36,000 volts on 20,000 kW sets; high voltage, however, pioneered by Britain, was not favoured in the States or on the Continent. Much attention was being given to the ventilation of large alternators and the use of hydrogen in place of air, a practice already established for synchronous condensers, was being seriously explored. By its use the windage losses were considerably reduced and output increased although, of course, the adoption of hydrogen demanded greater attention to precautions against explosion. One fundamental effect of the establishment of the grid on alternator design at this stage was the need for machines suitable for supplying the heavy light-load charging currents. Another was the adoption of 4½ per cent silicon steel on account of its low loss factor.

During the 1930's considerable progress was made in the United States with experiments employing mercury vapour instead of steam. A 10,000 kW set was put into operation at Hartford, and achieved an efficiency of 8,600 B.T.U. per unit and others were planned, but the capital costs were high and the movement had little effect on the continued rapid progress of the steam turbo alternator. During the year 1937 the output of British power stations was nearly 25,000 million units and was increasing at the rate of 12 per cent per annum. In comparison the output figure for the United States was over 114,000 million units.

Between the two world wars the great reorganization of the electric supply industry in this country which led to the construction of the 'grid' was formulated and carried out. In 1917 a government committee recommended that all supply undertakings should be brought under one central authority and in 1919 the Electricity Supply Act set up the Electrical Commissioners who started work in 1920. Districts were organized[3] under the title Joint Electricity Authority, but were often under suspicion from the undertakings and were not completely effective so that in 1925 reconsideration of the situation by the Weir Committee led to the formation of the Central Electricity Board. Its main functions were the construction of a nation-wide transmission network, the adoption of selected generating stations, and the standardization of frequency. At the time there were 17 different frequencies in use and 80 undertakings operating on other than 50 cycles.

This development was embodied in the important 1926 Electricity Supply Act and the now well-known steel transmission towers soon began to appear. A standard transmission voltage of 132,000 volts was adopted and the three-phase circuits were designed to carry 50,000 kVA on steel-cored aluminium conductors 0·77 in. in diameter having the equivalent copper section of 0·175 sq. in. The towers were designed for single and double three-phase circuits. In explaining in 1927 the eight-year programme for constructing the Grid, Sir Archibald Page said, 'The Grid will go a long way towards ensuring the universal availability of electric power throughout the country. It will bring the cost of production in the majority of supply undertakings down to the figure which has been attained by the few in whose areas electricity is already cheap and abundant.'

To continue the story up to the Second World War, great progress had been made in the quality of the steels for use at high temperatures, so that the second half of the Battersea Station, Battersea B, was put in hand on the basis of 965° F. steam temperature and a pressure of 1,420 lb. per sq. in. on the boilers. In this case the set was laid out in two lines; viz., an extra high pressure primary set of 16,000 kW at 3,000 r.p.m. and a high pressure secondary set of 84,000 kW running at 1,500 r.p.m. The station has since been equipped to a total capacity of 500,000 kW.

The general adoption of 3,000 r.p.m. for the speed of turbo-alternators for 50 cycles in Britain and 3,600 r.p.m. for 60 cycles in the United States has led to interesting developments in size. At the end of the Second World War the largest 3,000 r.p.m. sets were the two 40,000 kW sets at the Ferrybridge Station of the Yorkshire Power Co. A few years later two 60,000 kW sets were installed at Littlebrook in Kent, while sets of equal size were being installed in the United States and on the Continent. Today machines for 200,000 kW for steam conditions of 2,350 lb. per sq. in. and temperature of 1,050° F. are established, and plans are already (1960) well advanced for the installation of sets of 550,000 kW, or, in modern nomenclature, 550 MW. With units of this size an overall station efficiency of 36 per cent is expected.

By the outbreak of the Second World War there were 58 turbo-alternators operating in the United Kingdom at 3,000 r.p.m. and with individual outputs of more than 25,000 kW.[4] Hydrogen cooling was adopted widely and in the United States there were some thirty turbo-generators of sizes 30,000 to 150,000 kW working on this system. By 1942 the 1927 output of 5·2 million kWH had been more than doubled.[5] Without attempting to follow the complex post-war developments the subsequent progress may be indicated by the efficiencies now reached in modern turbo-alternators. The latest figure for efficiency given by the Central Electricity Authority in 1956 was one pound of coal per unit generated. Another sidelight on recent progress is the fact that during 1956 an American station put into service a large turbo-alternator, in which the stator conductors were liquid cooled, for which a current carrying capacity three times that of a hydrogen cooled stator is claimed. It is also interesting to note how the cost of generating stations has fallen with the increase in size. In a decade (1950–60) the capital cost of the complete station has fallen from over £60 per kW delivered to less than £50, and is still falling.

The continued expansion of the grid and the increased scale on which current is generated and distributed has naturally had considerable influence on the development of both transformers and switchgear. Larger and larger transformers have been constructed. One example is the 110,000 kVA at Battersea B, first installed for 80,000 kVA and modified for the higher output which converts from 11,000 volts to 66,000 and has forced oil cooling. Another is the 75,000 kVA transformer at Dalmarnoch, which is air cooled, while at Barking there are two, each of 93,750 kVA capacity. During the past few months a British transformer of 200,000 kVA has been supplied to the United States. Tappings on the windings of high voltage transformers for voltage adjustment on load have developed rapidly and are now in common use. Protection of transformers against damage through system surges becomes increasingly important on large networks and this has resulted in much greater reinforcing of end turns to withstand steep wave fronts. But possibly the most spectacular improvement in the modern transformer is in the efficiency achieved. Values of 99 per cent are now common and in the very large sizes 99·5 per cent is frequently reached.

A branch of electrical engineering which has grown enormously over the years and today assumes great importance is that involved in the design of switchgear for controlling the current in a circuit. The technical problems of switching were not serious on low-power circuits but as the magnitudes of current and voltage grew they called for increasing attention. The control of the outputs of large modern generators often requires the solution of engineering problems of the first magnitude. Increasing current values called for switches with large section and contact area to prevent overheating while, so long as the voltage remained low, there was not much difficulty in extinguishing the arc formed on opening the switch: a long enough air break was all that was necessary. Today the circuit breaker on a grid circuit, however, is a vital and complex piece of equipment, to the design of which much thought has been given.

The function of opening and closing a circuit for the purpose of normal control has usually been separated from that of opening the circuit in an emergency brought about by short-circuit or overload. In this case the situation has usually been met by inserting a fusible link or cut-out in the circuit. As loads increased, however, the two functions were combined in one device known as a circuit breaker, in which a main switch was released by an electromagnetic trip. The range of equipment thus stretches from the simple tumbler or press-button switch with a separate simple wire fuse, to the high-voltage oil-filled or air-blast circuit breaker in the switch yard of a large power station.

The first power stations were controlled by open knife switches on slate or marble panel switchboards and, as currents grew in magnitude, circuit breakers with magnetic blowouts to extinguish the arc were added. Well before 1900 the switchboard had taken its place as a vital component of a power station and improvements were introduced both into the arrangement of circuits and the design of individual switches to reduce the risk of failure and to ensure the continuity of supply. A major modification was in the adoption of the dead front in which no live metal was accessible, a condition which became essential on the adoption of alternating current and higher voltages. At one stage in this development remote mechanical control of the switches was resorted to, the switches themselves being housed in brickwork cubicles a short distance behind the control board.

During the first decade of this century the advantages of breaking a circuit between contacts kept under oil became appreciated and designs began to appear for metal-enclosed oil-immersed circuit breakers. A vast variety of switches of various kinds have been produced with a view to meeting increased loading requirements and notwithstanding similar activity in the air-break type the oil-filled type has held its own to a considerable extent. Immediately after the First World War great interest developed in research into the characteristics and requirements of high voltage switchgear, prominent in which was the then newly formed Electrical Research Association. There was, however, no facility anywhere to enable a switch to be tested for its breaking capacity. After the failure of attempts by Mr. H. W. Clothier to secure a national plant for the purpose, his firm, Messrs. A. Reyrolle and Co., installed the first short-circuit

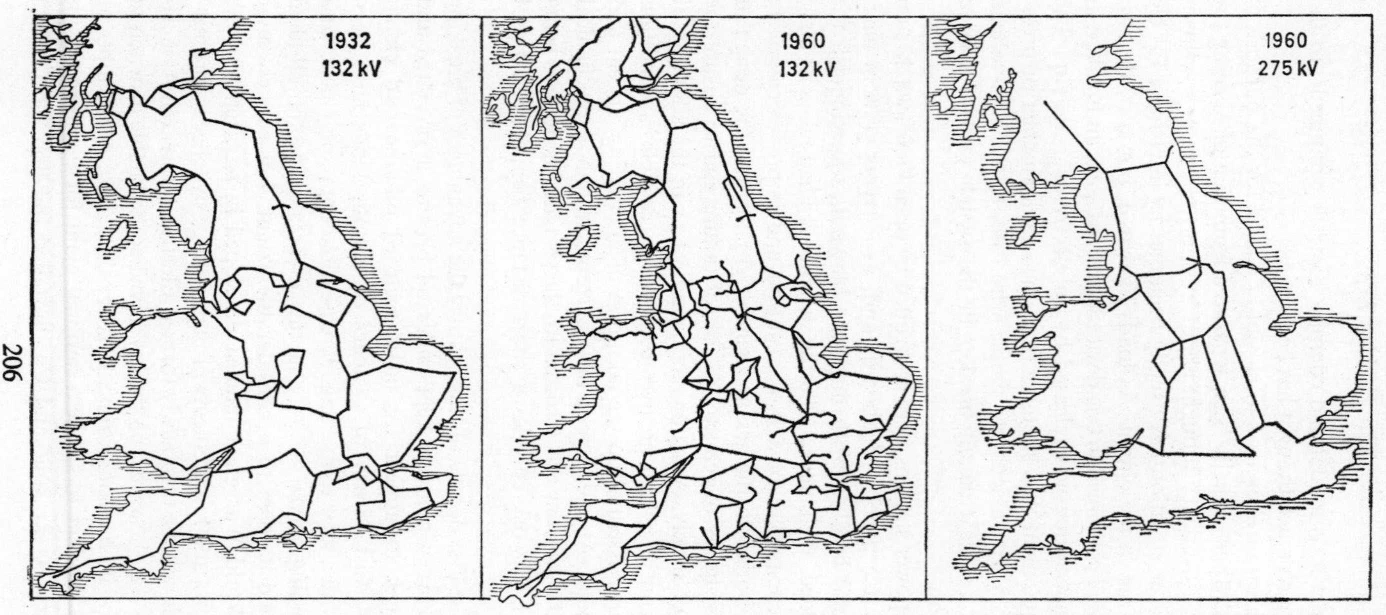

Fig. 29. British Power Grid. These three maps illustrate the intensive development in a highly electrified country. The entire country is covered by a three-phase 132 kV system on which a 275 kV was subsequently superimposed. (Acknowledgment to The Electricity Council.)

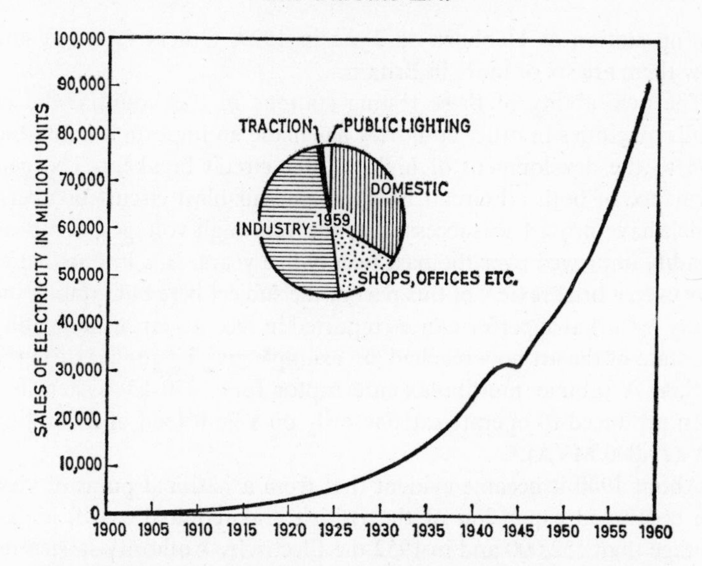

Fig. 30. Consumption of Electricity in Britain. The curve shows how insatiable the demand for electricity is and how the consumption has continued to grow apart from war years over the past half-century. (Acknowledgment to Garckes' Manual and the Ministry of Power.)

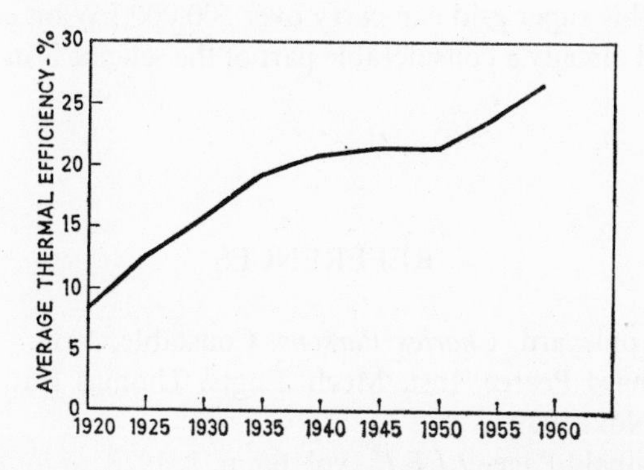

Fig. 31. Average Efficiency of British Power Stations. By continuous efforts to improve the design of power station equipment, the overall efficiency has continued to rise and individual sets have today efficiencies well above those indicated. (Acknowledgement to Electrical Times.)

testing station at Hebburn-on-Tyne in 1929. Others followed and now there are six or more in Britain.

The availability of these testing stations in this country and of similar facilities in other countries has made an important contribution to the development of high-voltage circuit breakers. The performance of both oil circuit breakers and air-blast circuit breakers, which have proved so successful at the very high voltages, has been steadily improved over the past twenty-five years. It is impossible to give even a brief review of this part of the subject here but, among the many remarkable performances reported in recent years as indicating the state of the art now reached, an example may be quoted from the U.S.A. A tubular multibreak interruptor for a 330 kV system has been produced to operate satisfactorily on a fault load of 25 million kW (25,000 MVA).[6]

About 1940 it became evident that from a national point of view the continued expansion of the British grid required a still higher voltage than 132,000 and in 1952 the Electricity Authority started on the construction of a super grid to operate at 275,000 volts. The two main objects were to interconnect the major groups of generating stations and to connect new generating plant located on the coalfields to bulk supply points in areas where there was a deficiency of fuel. In addition the superimposed grid would provide additional safeguards in times of breakdown or other emergency and so make possible an appreciable reduction in the provision of standby plant. One line of towers on this super grid can carry over 500,000 kW on each of two circuits and already a considerable part of the scheme is in operation.

REFERENCES

1. Rollo Appleyard, *Charles Parsons*. Constable, 1933.
2. Sir Leonard Pearce, Inst. Mech. Engrs. Thomas Hawksley Lecture, 3 Nov. 1939.
3. Sir Archibald Page, *J.I.E.E.*, vol. 66, p. 1, 1928.
4. Sir Leonard Pearce, *J.I.E.E.*, vol. 84, p. 342, 1939.
5. Sir Johnson Wright, *J.I.E.E.*, vol. 87, p. 1, 1940.
6. *Electrical Engineering*, vol. 76, p. 394, 1957.

Plate XXXI

(*a*) Electric Locomotive, City & South London Railway, 1890
The two motors, run at 500 volts, were arranged for direct drive on to the locomotive axles and developed 50 horse-power each. The current was supplied through a third rail.
(*Photo: Science Museum*)

(*b*) Crompton Motor for City & South London Locomotive, 1890
This drum-wound motor was substituted for the original ring-wound type soon after the railway was put into operation.
(*Photo: Science Museum*)

(*c*) 25,000 volt A.C. Locomotive, 1960
This locomotive, which develops 3600 horse-power, introduces the most up-to-date ideas for combining the advantages of A.C. transmission with D.C. drive. The power is supplied at 25,000 volts single-phase by means of overhead trolley wire and the locomotive carries transformers and rectifiers. Speeds of 100 m.p.h. are obtained.
(*Photo: Associated Electrical Industries*)

PLATE XXXII

The *Goliah* Laying Submarine Cable, 1850

The English Channel Submarine Telegraph Co., with a capital of £2000, arranged for the Gutta Percha Co. to make 25 miles of No. 14 gauge

CHAPTER XIII

Submarine Telegraphs

In an earlier chapter we have reviewed the beginnings of telegraphy and seen it established for commercial purposes both in Europe and America. At various points in this development proposals were made for underwater lines.

In 1811 Sömmering and Schilling had carried out trials across the River Isar near Munich. Colonel Pasley, of the Royal Engineers, had crossed the Medway in 1838 and Dr. O'Shaughnessy (later Sir William O'Shaughnessy Brooke, F.R.S.), the Director of the East India Company Telegraph Co., in 1839 had laid lengths across the Hooghly.[1] It was not until 1840, however, that serious official consideration was given to a practical scheme. In this year a Committee of the House of Commons held an inquiry on the feasibility of laying a cable from Dover to Calais, and Wheatstone gave evidence. He had evidently been studying the problem for some time previously as he produced a design for the cable with drawings of proposed cable-making machines, details of arrangements for loading the cable on to the ship, a chart of a proposed cross-channel route and proposals for joining lengths together.

Wheatstone suggested a cable consisting of seven copper wires each lapped with hemp twine saturated with tar. The seven lapped wires were to be laid up together and lapped overall with similarly treated twine. Some months later he carried out successful trials in Swansea Bay. The most comprehensive and reliable account of the events of this period appears in the record of an inquiry held in 1861[2] after the troubles with the Atlantic cable. It is clear from the evidence given by Mr. Charles West associated with Messrs. S. W. Silver and Co. (whose name was the origin of Silvertown) that there was some considerable competition to be the first to provide such a cable. Mr. West was offering a rubber insulated conductor and, having obtained permission from the British Government to proceed, he laid out a length in Portsmouth Harbour and transmitted messages before a large concourse of spectators. At this stage Paxton, the original designer of the Crystal Palace, and Charles Dickens were both interested spectators, but West failed to obtain the necessary financial backing.

Experimenters in America, Professor Morse and Ezra Cornell, were carrying out trials in New York Harbour and about the same time, Dr. Montgomerie, a Scottish surveyor with the East India Company, brought to England samples of a new gum, gutta-percha. Faraday and Wheatstone saw that this substance had merits as a cable dielectric and trial lengths were laid by Mr. C. V. Walker, the electrician of the S.E. Railway Co., in shallow water at Dover. While all this was going on West and the Silver Co. were forestalled by the Brett Brothers, who obtained Treasury consent here and went over to Paris and secured an exclusive concession to lay a cable connecting England and France. A company was formed, the English Channel Submarine Telegraph Company, with a capital of £2,000 and arrangements made for the manufacture of 25 nautical miles of cable by the Gutta-Percha Co. The cable, which had a single conductor of No. 14 gauge copper with a coating of gutta-percha half an inch in thickness, was made in 100-yard lengths. These were jointed together by twisting and soldering the conductor, each joint being subsequently covered with hot gutta-percha pressed on by means of a wooden mould. The lengths were subjected to a very crude test with 24 cells: if the rough galvanometer used showed no leakage they were jointed to the previous length and wound on a large wood and iron drum which revolved on a horizontal axis. To ensure that the cable lay on the bottom, leaden weights of 10–30 lb. each were bolted to it at intervals of 100 yards. The *Goliah*, a small Thames steam tug, took the drum to Dover and after shore ends had been laid, the paying out commenced at 10 a.m. on 23 August 1850 in fine weather. Late the same evening signals were exchanged between huts on the beaches at Dover and Cape Grisnez by means of a Cooke and Wheatstone needle instrument. Unfortunately, after complimentary messages had been sent, including one to Napoleon Bonaparte, communication ceased and was never restored. The trial had, however, proved an epoch-making event—communication was possible across twenty miles of sea—and the way was opened up for further attempts. To complete this story, Charles Bright[3] describes how a Boulogne fisherman had—accidentally or otherwise—raised the cable in his trawl and on finding gold at the centre of this unusual seaweed, had cut out a considerable length.

The publicity resulting from the 1850 failure brought suggestions for improving the construction of the cable. The Bretts obtained an extension of their concession and with the financial and technical aid of Mr. T. R. Crampton, made a second attempt. This time the cable contained four No. 16 copper wires, each one coated with gutta-percha, laid up with tarred hemp filling the interstices, followed by a lapping of tarred spun yarn. An armouring of ten No. 1 gauge galvanized iron wires was then applied with a long lay to provide the mechanical strength found so lacking in the first cable. The finished weight was seven tons to the mile.

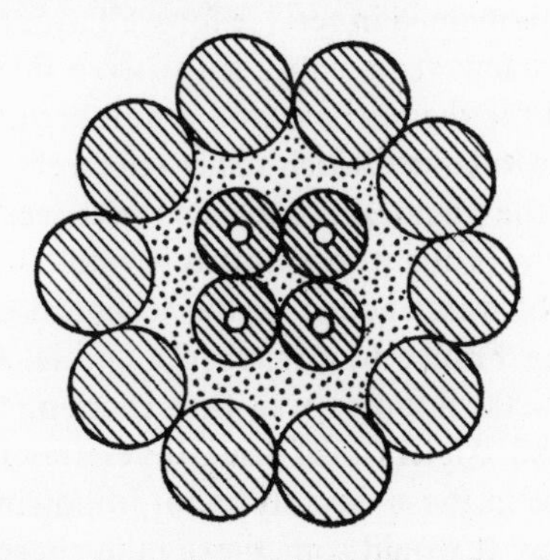

Fig. 32. 1851 Cross-Channel Cable. The original 1850 single-wire cable having failed, this one was constructed with four separately insulated wires laid up with tarred hemp and protected with an armouring of ten No. 1 gauge galvanized iron wires. It operated satisfactorily for many years.

When this cable was laid from the South Foreland lighthouse bad luck overtook the enterprise in the way of stormy weather and currents which carried the vessel off its course. The result was that when still a mile from the French coast no more cable was left on board. Through the dogged determination which has always characterized the development of submarine telegraphs, three temporary gutta-percha wires were connected to fill the gap temporarily, later to be replaced by a new piece of armoured cable and on 19 October 1851 service was opened to the public. Despite corrosion and bad distortion of the dielectric this veteran lasted for many years.

The success of the 1851 Channel cable established submarine telegraphy as a sound engineering proposition, and many further schemes were quickly proposed. Within two years, the traffic between

England and France having increased so rapidly, a second cable was laid across the Channel, this time from Dover to Ostend, a distance of 70 miles. Attempts to establish communication between Scotland and Ireland by a cable from Portpatrick to Donaghadie were not so successful. Three different attempts failed either through faulty design of cable or bad weather during the laying operations. The problems were very different in a channel 180 fathoms deep from what they had been in 30 fathoms, but undeterred by repeated failures the Magnetic Telegraph Company ultimately succeeded, in 1853, with a six-core cable very similar in construction to Crampton's 1851 cable, which remained in service for well over twenty years.

During the next few years cable ships were fitted out with gear specially designed for the purpose and a number of cables were laid from England to Holland. British cable engineers went further and further afield and the Thameside cable factories secured many orders, but greater depth brought continued trouble. Time after time increased depths of water added to the technical difficulties. In the scheme for linking France to Algiers via Spezia, Corsica and Sardinia, a depth of 300 fathoms was first encountered and then between Sardinia and Algiers over four times this depth. It was no unusual experience in those days to find it impossible to restrain the cable during laying. It would sometimes take charge and run out at terrific speed, leaving insufficient length to reach the destination. The cable makers no doubt recouped themselves, but the financial backers of these early schemes must, at times, have lost heavily, though technical success, when it came, did result in profitable business.

Interest was soon aroused in the possibility of linking Europe with America. The Western Union Co., on the success of its vast network of land-lines over the United States, planned to construct an overhead line between America and Europe via British Columbia, Alaska, the Aleutian Islands, and Siberia, which involved a cable across the Behring Sea. The expedition, comprising a fleet of thirty vessels, started off and in spite of many difficulties had erected a considerable part of the line when talk of an Atlantic cable brought the proposal to an end, involving a loss of some 300,000 dollars. The famous telegraph trail through the forests of British Columbia is a reminder of the scheme.

Many minds were soon directed to the possibility of linking the extensive telegraph systems which had grown up on both sides of the Atlantic, but the length of cable required was far in excess of anything so far attempted and the depths of water much greater. The engineering problems were clearly of an entirely different order from anything so far achieved, but another and even more serious one caused much concern. How long would it take a current of electricity to traverse such a distance and would the speed of working a telegraph on such a circuit prove to be economic? Professor Wheatstone had carried out an ingenious experiment with a long wire circuit connected to a Leyden jar through two spark gaps, one at each end of the circuit. By viewing the two sparks through a tiny rotating mirror fixed to the works of a watch he had calculated that electricity travelled 288,000 miles a second.[4] Practical tests on a made-up circuit on the underground system of the Magnetic Company 2,000 miles in total length were much more illuminating. What was important was not the 'velocity of electricity' but the rate at which signals could be transmitted. What mattered was the influence of the capacitance or, as it was then called, the induction of the circuit, a phenomenon by then well known and provided for by double current working. The tests showed that a speed of up to 270 signals a minute could be expected on this length of cable, which would meet commercial requirements.

In 1854 an English engineer, F. N. Gisborne, working for an American Company engaged in constructing a line to Nova Scotia, had discussions with a Mr. Cyrus W. Field, an influential business man, with the result that a concession was obtained from Newfoundland for the landing of an Atlantic cable to link up with landlines just laid. Mr. John Brett joined the group and a length of 85 miles of single conductor cable made by Glass, Elliott and Co., on the Thames, was sent out to connect St. John's with the Canadian and American networks. In 1856 the same partners, along with Mr. Charles Bright, who became a famous figure, formed the Atlantic Telegraph Company.

The first task was to determine the nature of the bed of the ocean. Soundings showed a gently undulating plateau between Ireland and Newfoundland varying in depth from 1,700 to 2,400 fathoms. Many specimen cables were made and considered. A public subscription list was opened in Lancashire, Yorkshire and Scotland and amid great enthusiasm a sum of £350,000 was raised. The promoters were to receive nothing until the shareholders had been paid 10 per cent on their investment. Professor William Thomson, an enthusiast for the Atlantic cable, was one of the directors and proved to be a source of great scientific strength throughout the enterprise.

The father of Professor William Thomson (later Lord Kelvin) was a professor of mathematics at Glasgow University and his mother was the daughter of a Glasgow merchant. William Thomson, the second of six children, was born on 26 January 1824 while the parents were living in Belfast, and was eight years of age when they moved to Glasgow, which then became his home virtually for life. William was a child of quick perception and wide knowledge, due to a large extent to his father, who taught him and his brother and sisters English, geography, mathematics and classics. The extent of his early progress may be judged from the fact that, at the age of ten, William matriculated at the University and commenced a course in arts subjects. His ability astonished everybody and his older brother James, who eventually became a professor at Glasgow, was in the same class, usually with second place to William in the examination results.

In 1841, not yet seventeen years of age, William Thomson was entered at Peterhouse, Cambridge, where he soon became popular with the other students, distinguishing himself in mathematics and geology, and acquitted himself as a capable oarsman. He also showed a great love of music, played the cornet and French horn, and was one of the founders of the University Music Society. Nobody was surprised when the results of the Tripos showed that he had taken a high place—second wrangler, and first Smith's prizeman.

During and after his Cambridge course, Thomson made many scientific pilgrimages at home and abroad and at the age of twenty-two, already widely known, was elected Professor of Natural Philosophy at Glasgow University. The enthusiasm with which he and his father collected testimonials is told in an amusing way by Silvanus Thompson in his two-volume life.[5]

In 1846 he gave his first lecture to the Natural Philosophy Class from a manuscript which, it is said, he used for fifty years and never reached the end, notwithstanding the fact that the interest of his students was such that a one-hour lecture was often extended to two or three times its normal length.

From the early days of submarine telegraphy Kelvin became interested in the scientific problems involved as well as in navigation and designed many instruments for use in both. He wrote a vast number of scientific papers. Honours crowded in on him including Presidency of the Royal Society from 1890–5 and of the Institution of Electrical Engineers three times. He had honorary degrees conferred on him by twenty-two universities, honorary membership by some ninety learned societies, a Baronetcy in 1892, and the Order of Merit ten years later. Lord Kelvin died on 7 December 1907 and was buried in Westminster Abbey near to the grave of Sir Isaac Newton.

The construction of the Atlantic cable in which Kelvin became interested in 1854 was as follows. The conductor consisted of seven No. 22 gauge copper wires weighing 107 lb. per nautical mile, insulated with three coats of gutta-percha to a diameter of $\frac{3}{8}$ in. and weighing 261 lb. per nautical mile. This was covered with hemp saturated with a mixture of Stockholm tar, pitch, linseed oil and wax

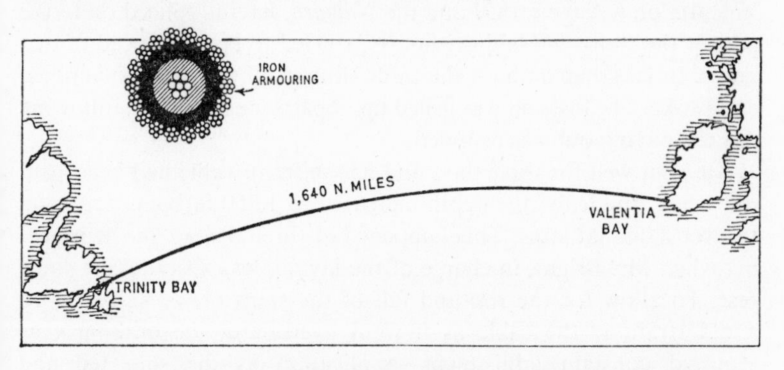

Fig. 33. First Transatlantic Cable (1857–8). The first attempt to bridge the Atlantic was made with a cable having a conductor of seven No. 22 copper wires insulated with three coatings of gutta percha. A bedding of tar-impregnated hemp followed and an armouring of eighteen strands each of seven No. 22 gauge iron wires.

to form a bedding for the armouring, which consisted of eighteen strands each of seven bright iron wires of No. 22 gauge. The complete cable was drawn through a compound of tar, pitch and linseed oil. It weighed one ton per nautical mile and had a breaking strength of 3 tons 5 cwt., or five miles of its weight in water. To protect the cable against rocks on the coasts at Valentia and Trinity Bay special shore ends were made with an increased armouring in the form of twelve iron wires, No. 0 gauge.

Unfortunately insufficient time was allowed for the manufacture of the cable which had to be delivered in June 1857. From February there was a race against time in the drawing of over 20,000 miles of copper wire, over 380,000 miles of iron wire and the preparation and application, in three coats, of 300 tons of gutta-percha. The laying machinery had to be designed and made and the two battleships, detailed for the task of laying, required considerable modification to make them suitable for the purpose. The crowding of this vast project, so far in advance of any such previous undertaking, into a few months, no doubt contributed to the ultimate failure which overtook the attempt.

The British ship *Agamemnon* (3,200 tons) and the United States *Niagara* (5,000 tons) were to carry half the cable each, and the lengths were to be joined together in mid-Atlantic. They assembled at Valentia on 6 August 1857 and the *Niagara*, having spliced on to the part of the shore end laid previously, started to lay at a speed of two knots. In less than an hour the cable slipped off the sheaves, jammed and broke. The lost end was fished up, repairs made and the following day the paying-out was resumed.

All went well for three days and 334 miles of cable had been paid out when, suddenly, the depth increased to 1,750 fathoms and soon to over 2,000 fathoms. This happened at three o'clock in the morning when Mr. Bright, in charge of the laying, was away taking some rest. To allow for the rise and fall of the stern of the ship and to avoid straining the cable as it went overboard, the braking gear required constant adjustment—application as the ship fell and release as it rose. Unfortunately the man left in charge, surprised by the speed at which the cable was racing away, applied the brake at the wrong moment and the cable broke again.

This disaster terminated the expedition for the time being and the ships returned to Plymouth. The remainder of the cable was taken ashore for storage and the paying out gear was completely redesigned in the light of the experience. Trials were in due course carried out in the Bay of Biscay at 1,800 fathoms, both in paying-out and picking up cable. The following year it was decided to make a second attempt. Additional cable had been taken abroad by the *Niagara* and both ships proceeded to mid-Atlantic with the intention of splicing the cable there and paying out in opposite directions. On 16 June 1858, after a fearful storm in which the *Niagara* nearly foundered, the splice was made and the ships separated. Unfortunately the lay of the two lengths of cable were found to be in opposite directions, so that to prevent the wires being pulled out straight a special rigid jointing device ten feet long, consisting of wood and iron, had to be inserted. Three times the cable broke and jointing was repeated, but on 29 July 1858, after a return to Queenstown for stores, the paying out proceeded, the *Niagara* towards America and the *Agamemnon* towards England. The Newfoundland end was landed on 5 August with current still coming through the cable from the European end. The *Agamemnon* had a more eventful time. She ran into a storm and at a critical moment, a fault was discovered in the cable at a point shortly due to go overboard. The paying-out was slowed down as far as was safe and the jointers, racing against time, completed their work just as the ship was hanging on to the cable. In another few minutes the violent pitching would have snapped the cable and brought the laying once more to an end.

On 17 August 1858 the first message was transmitted across the Atlantic and there was great excitement both in this country and in America. Congratulations were exchanged between the Queen and the President. But within a few weeks a fault developed, the signals became more and more unintelligible until after some 700 messages had been transmitted the cable failed completely. Even so the short period of operation had demonstrated the great value of such a cable—by one message alone the Government saved £50,000 through being able to transmit some instructions regarding troop movements.

Various spasmodic but unsuccessful attempts were made to resuscitate the cable, but in the meantime failure had overtaken several schemes, in particular one of major importance from Egypt, through the Red Sea to India, and losses amounting to more than a million sterling had been incurred. The Government were involved in guarantees so that in 1859 it set up a Joint Committee, consisting of four representatives nominated by the Board of Trade and four by the Atlantic Telegraph Company to carry out a complete investigation. During its twenty-two sittings extending from 1 December 1859 to 4 September 1860 the Committee heard evidence from over forty witnesses, including engineers, professors, seamen and manufacturers. Outstanding names among these were Latimer Clark, Sir Charles Tilson, Bright, William Thomas Henley, Professor D. E. Hughes, Fleeming Jenkin, Charles William Siemens, C. F. Varley and Professor William Thomson (later Lord Kelvin).

The Report,[2] which extends to over 500 pages, is one of the most remarkable statements ever made on the position reached in any branch of electrical engineering and today makes fascinating reading for the historian in this field. A tabular statement gives a list of submarine cables already laid during the previous nine years. Thirty-one shallow water schemes totalled in length over 3,000 miles and fourteen schemes, including the Atlantic cable, 2,200 miles, and the Red

Sea, 3,500 miles, totalled 8,290 miles. Out of the 11,364 miles laid only a little over 3,000 miles were actually working.

In their analysis of the evidence regarding the Atlantic cable the Committee stated: 'We attribute the failure of this enterprise to the original design of the cable having been faulty owing to the absence of experimental data, to the manufacture having been conducted without proper supervision, and to the cable not having been handled, after manufacture, with sufficient care. We have had before us samples of the bad joints which existed in the cable before it was laid; and we cannot but observe that practical men ought to have known that the cable was defective, and to have been aware of the locality of these defects, before it was laid.'

The Report contains an interesting account of theoretical and practical points affecting the quality of the conductor of a submarine cable and a useful statement of the art as it existed at that time. The method of jointing was to solder the overlapped wires and wrap with small binding wire which was then soldered with silver solder. The joint was always much more brittle than the wire itself and liable to fracture. Moreover the wire itself was not homogeneous. Little was known at that time about the conductivity of copper and the Committee commissioned Dr. Matthiessen to investigate the matter with the result that the Report contains the first comprehensive statement ever made on the effect of the admixture of different impurities on electrical conductivity. It also gives some enlightening figures on the values obtained in various coppers available commercially. Lake Superior had a conductivity 92 per cent of pure copper while a sample of bright copper wire gave 72 per cent. In a range of samples taken from the 'Gibraltar core' furnished by C. W. Siemens the conductivity on Matthiessen's standard ranged from 67·4 per cent to 90·7 per cent, all at 15·5° C. In his book[3] Bright states that the purity of copper in those days was so low that an electrical conductivity of 40 per cent was as much as was ordinarily obtained for telegraphic purposes and that the primary cause of the failure of the first Atlantic cable was the fact of the core being insufficiently large coupled with this low specific conductivity.

There is much also in the report on the design, methods of manufacture and precautions to be taken in laying submarine cables, but possibly the subject of greatest fundamental value touched upon was the electrical theory of signal transmission. In experimenting with his early telegraph as early as 1823 Reynolds had observed the delay occasioned in passing a signal through an insulated wire and later

Faraday had likened such a wire to an attenuated Leyden jar, the conductor taking the place of the inner coating of the jar and the earth or surrounding water the outer coating. The laws of charge and discharge were not understood, however, until Professor Thomson studied them[4, 5] and came first of all to the conclusion that the velocity of charge or discharge was inversely proportional to the square of the length of the cable. In 1856 as the result of criticism by Whitehouse he sent his views to the Athenaeum.[6] In this letter he gave suggested dimensions for a transatlantic cable which he calculated would give a speed of three words a minute over a distance of 2,400 miles. These dimensions were ultimately adopted and as a result of the interest which he displayed in the subject the directors of the Atlantic Telegraph Company invited his co-operation. He supervised the electrical arrangements throughout the laying of the cable and his evidence, given before the committee of inquiry on 17 December 1859 covering eighteen pages with many diagrams, makes most illuminating reading. In this he reaffirmed his inverse square law and further enunciated the formula for calculating the capacitance of the cable from the specific inductive capacity of gutta-percha and the ratio of conductor and dielectric diameters. He also described in some detail his marine galvanometer which consisted of a very light steel magnet cemented to the back of a tiny mirror, the weight of the whole being from a grain to a grain and a half. This was first of all attached to a platinum wire suspension but later to a stout bundle of silk fibre. The needle and the mirror were balanced by their centre of gravity and therefore unaffected by gravity or by the inertia called into play by the pitching and rolling of the ship. Thomson, in a further modification, substituted a magnetic zero control for the torsion control and this became general practice.

From the date of the failure of the 1858 cable Cyrus Field and his associates never gave up hope of ultimately bridging the Atlantic with a reliable commercial cable and gradually overcame the natural reluctance of the investing public to provide funds for a further attempt.[7] After discussion of the technical problems by a committee including Professor Thomson and Professor Wheatstone and full consideration of all the findings of the Select Committee on the previous failure a new cable was designed and adopted. The conductor was a strand of seven copper wires, each of No. 18 gauge, and weighing 300 lb. per nautical mile. The gutta-percha was to be applied in four separate coats, to each coat being applied Chatterton's compound, the dielectric weighing 400 lb. per nautical mile. The centre wire of the strand was also coated with compound to prevent air occlusions and creepage of moisture. This time the copper wire was to be tested and to give at least 85 per cent conductivity compared with pure copper. With this dual increase in conductivity of the conductor a speed of seven words a minute was anticipated. The ten armouring wires were to be of No. 13 gauge iron wire laid over a cushion of tarred jute and separately lapped with compounded hemp to prevent corrosion. The weight was 1 ton 15¾ cwt. per nautical

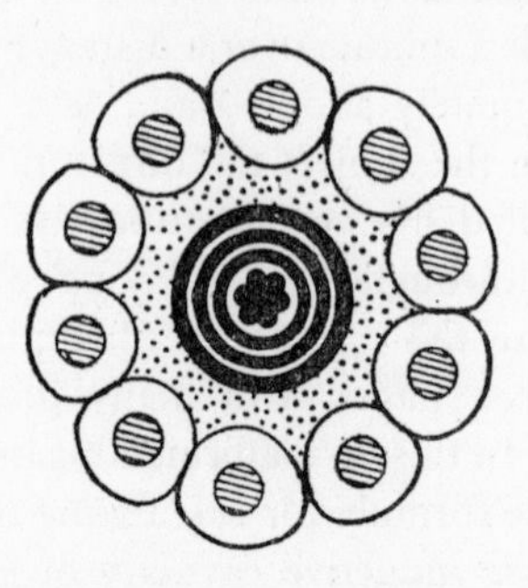

Fig. 34. 1865 Transatlantic Cable. In view of the 1858 experience and that on other ocean cables supported by the important 1859 enquiry a new cable was designed with a stranded conductor of No. 18 gauge wires with four coats of gutta-percha and ten solid armouring wires of No. 13 gauge. Many other refinements were added but only after a second cable had been laid in 1866 and the 1865 one repaired was communication satisfactorily established.

mile in air and in water 14 cwt. The overall diameter was 1·1 in. and the breaking strain 7 tons 15 cwt. or 11 miles of its own weight. Specially reinforced shore ends were, of course, again provided.

By this time the importance of keeping lengths of the cable under water during manufacture and of testing for a high insulation resistance was appreciated. It soon became clear that applying the gutta-percha as a multi-layer coating was an important safeguard against faults. The *Great Eastern*, a vessel of unusual size for the time—22,500 tons—which had been constructed some years before and had been a commercial failure owing to lack of suitable freight, was commissioned for the task of laying. After suitable modification and design of a special paying-out gear based on previous experience, the cable was taken aboard. Professor William Thomson and Mr. Varley accompanied the expedition on behalf of the Atlantic Telegraph Company.

Several unfortunate incidents occurred as the cable was being paid out including faults due to pieces of iron wires having been forced through the dielectric and, when 1,186 miles had been laid, the cable parted. Many attempts were made at grappling but one after another failed when within a measured distance of success, so that on 11 August 1865, all grappling rope having been exhausted, the expedition was abandoned and the ships returned. But the promoters, though discouraged, were not defeated. More lessons had been learned. By amalgamation of companies a total capital of £600,000 was made available. The scheme now was to lay a complete cable all the way and to fish up and complete the broken 1,186-mile length. The only change in design was the adoption of galvanized wire for the armouring.

On 30 June 1866 the *Great Eastern* set off once more. Again Sir William Thomson was abroad: Mr. Latimer Clark and Mr. C. F. Varley were at Valentia. The new complete length was successfully laid in fourteen days and, after many days of alternate hope and despondency, the formidable task of bringing up the end of the 1865 cable from a depth of two miles was accomplished.

News of the success was sent back to Valentia over the recovered cable and thence to America on the completed cable. Transatlantic telegraphy was really established. The broken length was soon completed and within a few days two cables were working between Europe and America.

A dramatic demonstration was made by Mr. Latimer Clark on the two completed cables. He had the two ends joined together at Newfoundland and connected one of the Valentia ends to a silver thimble containing sulphuric acid and a tiny zinc plate and the other to a mirror galvanometer. Within a little more than a second the signal crossed the Atlantic twice and gave a strong deflection of the spot of light of twelve inches or more.

The successful outcome of the transatlantic cable gave a great impetus to the industry and many other deep-sea cables were laid all over the globe. Very little further modification of the cable design was necessary. In 1879, trouble having been experienced on cables laid in warm waters, through the ravages of the teredo and other boring insects, the Penang–Malacca cable was furnished with a brass tape protecting the core and this became standard practice for such situations. Attention was now paid more to obtaining higher speeds of working and with the extension of telegraph service came a variety of telegraph instruments designed to increase the amount of traffic which could be passed over a particular cable.

In the forefront of these must be placed Sir William Thomson's siphon recorder invented in 1867, the advantages of which led to its replacing the mirror receivers. The early siphon recorders, while not possessing much greater sensitivity, had the advantage of producing a permanent legible record. In time they also became much more sensitive.

Fundamentally the siphon recorder consisted of a light rectangular coil of wire suspended centrally between the poles of a powerful electromagnet so that a small current passed through the coil caused it to rotate in the same manner as the mirror galvanometer. An extremely fine glass tube bent into the form of a siphon was suspended in such a way that its upper end rested in a vessel of coloured liquid and the lower end, slightly bent inward, could move transversely across a strip of paper which was driven electrically at the requisite speed. Silk threads connected the siphon with the coil in such a way that as the coil oscillated backwards and forwards it moved the lower end of the siphon backwards and forwards. By means of a light lever the motion was magnified so that the end of the siphon moved further than the coil to which it was attached. Thus, as the coil responded to the current received from a cable, and the paper strip travelled forward, the ink from the siphon drew a wavy line on the strip. In Thomson's early recorders he electrified the ink so that a mark was made on the paper by the ink being attracted to the paper without actually touching it. Thus he was able to eliminate any friction between the siphon and paper. He generated the charge by the use of a 'mouse mill' influence machine driven by a small motor. In later forms the electrified ink was abandoned and satisfactory marking obtained, without friction, by the use of vibration on the siphon.[8] Towards the end of the century Muirhead introduced an improved type of siphon recorder and large numbers were subsequently used throughout the cable systems of the world.

In 1862 Varley patented the use of condensers at each end of a long submarine cable to sharpen the signals and so increase the speed of working and in 1866 a speed of twenty-five words a minute was obtained on the Atlantic cable using manual transmission and mirror instruments. The condenser provided what had become known as a curbing current to speed up the discharge of the cable after each signal. Bright's *Submarine Telegraph* gives an excellent analysis of the action of such condensers (p. 542).

To avoid manual translation between the terminal of a submarine cable and the continuing land-line, relays such as the Siemens polarized relay of the type employed already were brought into use. The next major step in speeding up the output was the adoption of duplex working. Two fundamentally different systems were available for sending a signal in one direction along a line without interfering with the received signal travelling in the opposite direction. In one method known as the differential system the sending current was split between two opposing coils in the receiving instrument, one half going to the line and the other half to earth through an artificial line. The opposition of the effect of the two thus did not interfere with the received signal. In the second method—the bridge method—the receiving device was connected across the vertical pair of terminals on a Wheatstone bridge, the upper one of which was connected to the cable and the lower one to earth through a balancing artificial cable. The sending current was fed into this bridge system through two equal resistances forming the left-hand arms of the bridge. Thus when the bridge was correctly adjusted for the cable constants the sending current, again as in the differential system, passed half to the cable and half to earth leaving the receiving circuits across the vertical terminals of the bridge unaffected. Some difficulty was at first experienced owing to capacity kick as the theory of long lines was not yet understood, but Stearns discovered in 1872 the advantages of inserting small condensers in the artificial cable line, a practice which was, of course, continued. Muirhead and others gave considerable attention to the construction of artificial lines and important refinements were later introduced.

The next significant improvement in submarine telegraphy was the adoption of automatic working, as had already become wide practice on long land circuits in the Wheatstone system. Perforated tape was used in the ordinary way but an additional mechanism and contacts were added for curbing the signals by sending a charge of opposite sign at the end of each signal. The use of the cable code called for modification of the ordinary Wheatstone transmitter and this again resulted in much ingenuity in the design. More sensitive repeating relays became available which led to other major developments alongside those which were taking place in land telegraphs. The need to translate code messages into straight Roman script was recognized by Hughes as early as 1854 and he invented a system which came into commercial use and was widely employed for over fifty years until its supremacy was attacked by other more efficient systems.

An excellent description of the Hughes system is given in the book by Crotch.[9] Fundamentally the Hughes transmitter had a keyboard similar to that on a piano, the separate keys being connected to vertical rods arranged in a circle. Pressing a key raised the tip of the associated rod above the normal level. Concentric with this circle of rods was a rotating vertical spindle which carried an arm called the 'chariot'. This swept round over the tips of the rods and was engaged by any one protruding. The engagement was thus transferred mechanically to an arm with contacts in the transmitting circuit so sending out a momentary signal current. The spindle rotating the chariot was geared to a printing wheel having raised characters on its outer periphery so that as the chariot rotated one revolution so did the printing wheel which presented in turn the individual printing characters to a strip of paper. It was raised into contact by a further simple mechanism and released by the signal current. The sending and receiving instruments had, of course, to be synchronized but this was achieved by a simple technique of interchanging recognized signals.

The Hughes system was applied to duplex circuits and used extensively between England and the Continent with relay repeater circuits every few hundred miles, though Murray in 1911[10] had stated that its use was confined to long land-lines and that there was little scope for printing-telegraphs on long ocean cables. He estimated that at that time there were over 3,000 Hughes machines in use carrying the bulk of the telegraph traffic on the Continent of Europe but that it had only a very limited application outside Europe.

A major development in the direction of increasing the capacity of a single telegraph circuit was in the adoption of the multiplex principle which became the basis of a number of well-known systems, some of which have continued in one form or another to the present day. In the multiplex arrangement a number of transmitting sets and their corresponding distant receiving sets are given access in turn to one and the same line for a short period of time. Thus the basic principle of multiplex differs essentially from that of duplex, whether differential or bridge. As the individual periods of access to the line are short in comparison with the duration of a dot or dash the operators need adopt no special procedure: all that happens is that the signals are broken for a negligible interval of time. The first to

Plate XXXIII

(*a*) Hughes Type Printing Telegraph, 1854
In this early automatic telegraph which was widely used for over fifty years, a row of keys, similar to piano keys, operated a series of vertical pins arranged in a circle. A "chariot" rotating above these pins stopped when it encountered a pin which had been raised by the depression of the associated key. By simple mechanism this action was conveyed to a printing wheel at the sending and receiving ends of the circuit.
(*Photo: Science Museum*)

Early Bell Telephone
's early telephone passed through many stages. He pted a thin iron plate for the diaphragm and used a nanent polarizing magnet in conjunction with the coil ying the speech current.
to: Science Museum)

(*c*) Early Telephone Switchboard
The first telephone exchange was established by Jones, at New Haven, Connecticut, in 1878. Drop indicators were employed but there was nothing approaching an operator's cord circuit as developed later.
(*Photo: Science Museum*)

PLATE XXXIV

(*a*) Manual Telephone Switchboard, 1940–50
While retaining manual operation for many decades, great improvements were introduced in switchboard techniques to meet the increase in number of circuits. In this illustration can be seen the bank of incoming jacks and, above, the subscribers' multiple. The set of plugs and cords assigned to the individual operator is shown low down on the right.

(*b*) Modern Telephone Exchange, Crossbar Canyon
This illustration gives an idea of the extensive provision in a large exchange which has to be made for making circuit adjustment to suit traffic requirements.
(*Photo: American Telephone & Telegraph Co.*)

propose multiplex working was Farmer of Boston in 1852[11] and the principle has been adopted by several subsequent inventors, conspicuous among them being Baudot in 1874. It depends solely on a rotating switch known as a distributor in which the arm connected to the line makes contact in turn with any number of segmented contacts, connected to the separate transmitting and receiving sets. A manual key operator could transmit at the rate of 30 to 40 words a minute, but the current in an ordinary telegraph line could rise and fall much more rapidly than was necessary for this speed so that a four-segment switch and four operators enabled the same line to carry 120 to 140 words a minute and the Morse multiplex was used in this way for many years.

Before the multiplex system was brought to complete success it was necessary to substitute a more satisfactory code in place of the dots and dashes of the Morse code and this was achieved by the five-unit code. In this code there is no discrimination on duration of the elements of current as between dots and dashes but only in direction, either 'spacing' or 'marking'. All characters occupy the same length of time in transmission and consist of five units so that by arranging these in different ways 32 combinations are obtained. Thus letter A is transmitted by a marking interval followed by four spacing intervals, B by two spacing intervals, then two marking intervals and finally a spacing interval, and so on.

In the Baudot transmitter the operator had five keys which were connected to a distributor sweeping over contacts connected with them. The contact arm then moved on to a second five controlled by a second operator, then to a third and a fourth group. To prevent the operators pressing the keys for the next letter too soon there was a telephone signal sent by a contact on the distributor at the end of each group of five. The message was received in Roman characters on a paper strip as in the Hughes system. From about 1880 the Baudot system and its modification extended throughout France and by 1910 practically every important telegraph line was equipped with this apparatus, a practice which soon spread to other European countries.

The adoption of printing-telegraphs was not so rapid in Britain, although some interest was displayed about 1901 in the Murray automatic, while in the United States the Buckingham system was being developed by the Western Union Company. In the Murray system the message was punched on tape in the five-letter code. The tape was passed through a transmitter operating on the Jacquard loom principle in which selective mechanism is controlled by the movement of metal pins or fingers engaging with holes in a paper tape.

Key levers were arranged to select power controllers which consisted of distance pieces and these became interposed between the heads and the hammer so that only those punches which had the distance pieces inserted were driven through the paper. Rates of 100 to 120 words a minute were obtained and the final message was received in 'clear' on a paper tape.

The Murray–Creed instrument which replaced the original Murray had a differential feed which enabled it to work on the ordinary Morse perforated tape. The many ingenious modifications of these multiplex-type printing-telegraphs which were introduced early in the past half-century are much too complex to be described here, but several excellent papers are available on the subject.[12, 13]

Multiplex systems have been adopted for submarine cables for the same reasons as for land-lines, but the difficulties are much greater. The greater distortion of the signal due to the distributed capacitance is a disadvantage with the Baudot type of system depending as it does so much on the time of arrival of the signal. Such apparatus has had to be restricted to submarine cables of not more than 200 or 300 miles in length. Satisfactory transmission was obtained in an experiment in 1903 from Porthcurnow to Gibraltar, a distance of 1,190 nautical miles, but much more development of both transmitting and receiving apparatus was necessary before the speeds on ocean cables could be raised. The greatest step forward was taken after the invention of the thermionic valve when voice-frequency telegraphs employed trains of alternating impulses.[14] The science of telephony had not only made the telephone an ever-increasing competitor to the telegraph but had made available to it certain new techniques such as valve circuits and continuously loaded cable conductors. By 1929 the Western Electric Company in the United States were operating an A.C. multiplex system which had twelve channels with an overall speed of 480–600 words a minute over a single pair of land wires. By the 1930's such techniques had made it possible to transmit on an Atlantic cable at 400 words per minute. But about the same time gutta-percha, the long-established dielectric for submarine cables, found for the first time a new competitor. The chemists of Imperial Chemical Industries had discovered that when the gas Ethylene (C_2H_4) was compressed to 1,000 atmospheres it became a waxy solid and remained so. As a cable dielectric polythene had vastly superior properties to anything previously tried for communication cable and it will be referred to more fully under 'Science in Telecommunication'.

REFERENCES

1. *Journal Asiatic Society*, September 1939.
2. *Report of the Joint Committee appointed by the Lords of the Committee of Privy Council for Trade and the Atlantic Telegraph Co. to enquire into the Construction of Submarine Telegraph Cables*, 1861.
3. Charles Bright, *Submarine Telegraphs*. London 1898.
4. *Royal Soc. Phil. Trans.*, vol. 124, p. 583, 1834.
5. Silvanus P. Thompson, *The Life of Lord Kelvin*, 2 vols. Macmillan, 1910.
6. *The Athanaeum*, 1 Nov. 1856.
7. E. and C. Bright, *The Life Story of the late Sir Charles Tilson Bright, Civil Engineer, with which is incorporated the story of the Atlantic Cable and the first telegraph to India and the Colonies*. London, 1898. Abridged edn., 1908.
8. *J. Soc. Tel. Engineers*, vol. xiv, p. 243, 1885.
9. Arthur Crotch, *The Hughes and Baudot Telegraphs*. London, 1908.
10. 'Practical Aspects of Printing Telegraphs', *J.I.E.E.*, vol. 47, p. 450, 1911.
11. Stone, *Textbook of Telegraphy*. London, 1928.
12. Donald Murray, *Setting Type by Telegraph*. *J.I.E.E.*, vol. 34, p. 555, 1905.
13. H. H. Harrison, 'The Principles of Modern Printing Telegraphs', *J.I.E.E.*, vol. 54, p. 309, 1916.
14. W. Cruickshank, 'Voice Frequency Telegraphs', *J.I.E.E.*, vol. 67, p. 53, 1929.

CHAPTER XIV

The Early Telephone

At the beginning of the nineteenth century several simple devices such as speaking tubes and megaphones had been used for conveying speech to a distance. Transmission of musical sounds through solid rods had also been tried with success, notably by Wheatstone, who in 1831 described his experiments with the optimistic prediction: 'Could any conducting substance be rendered perfectly equal in density and elasticity so as to allow the undulations to proceed with a uniform velocity without any reflections and interferences, it would be as easy to transmit sounds through such conductors from Aberdeen to London as it is now to establish a communication from one chamber to another.'[1] There is some uncertainty as to the exact date on which the word telephone was first employed, but it was well before the time of the electric telephone, for in the famous arbitration case on the electric telegraph involving Cooke and Wheatstone, in 1841, the word was frequently used in connection with the transmission of speech.

The inventor of the telephone, as we know it, using an electric current for the transmitting medium, was Alexander Graham Bell, and his fundamental discovery was made on 2 June 1875. Bell did not stumble over the secret of the telephone in the traditional manner of many discoveries but arrived at it after years of patient study and deliberate step by step progress to the ultimate goal. Not only did he, himself, devote nearly twenty years of his adult life in pursuance of his objective, but the foundations on which he built had been laid during two previous generations. A subject which had been the absorbing life interest of both his father and his grandfather became a passion with him and proved to be the avenue which led direct to his great achievement.

Bell's grandfather had been a professor of elocution, who wrote books on the subject—which in the 1830's became a very popular branch of self-education. His father, following the same line, had specialized in the mechanism of speech. As a professor at University College, London, he in turn wrote books on the subject aiming at the analysis of sounds for such purposes as teaching the deaf and dumb to speak, and, by a system which he called 'visible speech', expressing sounds graphically.

Alexander Graham Bell was born in Edinburgh on 3 March 1847, and after attending the Royal High School he studied classics in the University there and subsequently anatomy at University College, London, where he matriculated in 1867. At an early age he developed a keen interest in the family subject and read everything written by his father and grandfather, as well as the works of Wheatstone and Helmholtz. As a boy he developed a flair for practical experiments and at eighteen had studied the formation of vowel sounds in his own mouth and the effects of cavity shape on resonance. Finding at twenty-three that he could read Helmholtz in French, but that he could not understand the work through a lack of electrical knowledge, he set himself to study electromagnets, the telegraph, tuning forks, etc., with a view to producing vowel sounds artificially. In 1870 the family moved to Brantford, Ontario, Canada, where Bell continued his activities and he was soon absorbed in experiments in electric telephony employing groups of electrically controlled tuning forks.[2] The vibration of a fork was made to operate a wire-in-mercury contact by which an interrupted current was transmitted to a similar tuning fork at a distance, operated by means of an electromagnet. Thus he produced what was virtually an electric telegraph based on a musical code of signals.

The next step in the development of the telephone idea came when Bell recalled that a permanent magnet when moved towards the poles of an electromagnet produced a current in the coils of the latter and that on withdrawing the magnet a current was generated in the reverse direction. If, therefore, a permanent magnet was fixed to one of his vibrating reeds and held in front of an electromagnet a current would be induced of the same frequency as that of the reed. He wrote, 'In this arrangement voltaic batteries, current interrupters and induction coils became unnecessary and I was fascinated by the simplicity of the arrangement of circuit.'

In 1874 Bell conceived the idea of the electric harp, which consisted of a large number of reeds operating alongside one another in front of the common core of an electromagnet. In describing the proposal he said: 'Utter a sound in the neighbourhood of the harp H and certain of the rods would be thrown into vibrations with different amplitudes. At the other end of the circuit the corresponding rods of the harp H would vibrate with their proper relations of force and the *timbre* of the sound would be reproduced.' In his account of the matter later Bell explained that the expense of constructing the apparatus deterred him from making the attempt. In spite of this explanation the electric harp has frequently been referred to in the literature as a practical achievement and as the real invention of the speaking telephone, but to be historically accurate we must await Bell's next step in the development.

Studying the form and construction of the human ear and noticing the disparity in size between the membrane and the small bones moved by it, Bell then conceived the idea of making a telephone with a large membrane carrying a small piece of steel, but this time it was lack of confidence, not of finance, which held him back. He had no faith in the power of the small induced currents to transmit the necessary amount of energy.

Proceeding by careful logical steps Bell was moving slowly along when accident, that frequent ally of the discoverer, stepped in and took a conspicuous part. One day Bell and his assistant were conducting their experiments with vibrating reeds fitted with coils of wire in an attempt to send more than one telegraphic message on the same circuit by means of different frequencies. Now and then one of the armatures would stick to its reed and had to be released by hand and set into vibration by plucking. On one of these occasions Bell, at the receiving end, noticed something peculiar; his own armature was vibrating in unison with the remote plucked reed. In his famous deposition later he said,[3] 'I called out to Mr. Watson to pluck his reed again. At every pluck I could hear a musical tone of similar pitch to that produced by the instrument in Mr. Watson's hands and could even recognize the peculiar quality—or *timbre* of the pluck. These experiments at once removed the doubt that had been in my mind since the summer of 1874, that magneto electric currents generated by the vibration of an armature in front of an electromagnet would be too feeble to produce audible effects that could be practically utilized for the purpose of multiple telegraphy, and of speech transmission.'

On 2 June 1875, Bell wrote a letter from Salem, Mass. 'Dear Mr. Hubbard, I have accidentally made a discovery of the greatest importance in regard to the Transmitting Instrument. Indeed, so important does it seem to me that I have written to the Organ

Factory to delay the completion of the Reed arrangement until I have had the opportunity of consulting you.

'I have succeeded today in transmitting signals *without any battery whatever*! . . .'

After a day of plucking and observing Bell made up a system of two instruments in each of which a small electromagnet had a hinged armature, the moving end of which was attached to the centre of a membrane closing one end of a cone. But the parts were too heavy and insensitive and in the following month a more delicate construction was tried, this time with a stretched membrane. In the account of the experiment Bell says the assistant rushed upstairs in great excitement to tell him that he could hear his voice quite plainly and *could almost make out what he said.* This achievement later stood the test of high tribunal on the validity of the patent and was universally accepted as the origin of the electrical transmission of articulate speech.

In spite of personal financial difficulties Bell persisted in the development of more effective receivers and the form and size of the electromagnet went through various stages. A thin iron plate was adopted for the diaphragm and the coils were carried as extended pole pieces on a permanent magnet of which both rod-type and U-shaped were used. Within a few months the equipment had been assembled in a long box with a mouthpiece at one end. A fundamental difference then developed between transmitter and receiver. In the former the box shape persisted but the pole pieces of the magnet were turned at right angles to the permanent magnet so that the box with the long magnet could lie flat against a wall with the mouthpiece standing out convenient for speaking into. The receiver, on the other hand, soon took the form of a tube for convenience in holding to the ear.

Early in 1877 the telephone had created a world-wide interest. Speech had been transmitted over the existing telegraph lines and the time had come for commercializing the invention. Bell offered pairs of instruments on lease 'for social purposes' at $20 a year, and 'for business purposes' at twice this rent. Very soon the instruments were being produced in a small factory in Boston, Massachusetts, and supplied over a wide area, including New York and London, where Bell read a paper before the Society of Telegraph Engineers.[4]

Before following the steady course of Bell's progress it is appropriate at this stage to refer to another name in the history of the telephone. Philip Reis, a German school teacher, invented a telephone in 1863, which was exhibited at the British Association. It depended on the magnetic tick occurring in an iron rod standing on a sounding board when the current in a surrounding coil is interrupted. This device took the form of a receiver and the transmitter was a diaphragm to which was connected a mercury contact. Reis seems to have followed very much the same line of reasoning on the analysis of sounds as Bell did, but from the standpoint of producing practical speech transmission there seems to be a very weak case for considering him as having anticipated Bell as the inventor of the telephone.

The idea of a central switchroom by which subscribers could be interconnected was suggested in October 1877 by an enterprising journalist in Boston and quickly took practical shape based on the telegraph system which had already operated to a limited extent in the United States, England and France. At first a ticket system enabled an operator to receive requests from individual subscribers for a specified line between certain times and to issue instructions by means of the ticket to another operator, who inserted and withdrew plugs as required.

The first telephone exchange on a commercial basis was installed at New Haven, Connecticut, in January 1878.[5] Drop indicators were used with a call bell and two-way lever switches enabled the operator to connect his own telephone to any line and obtain instruction. Another switch enabled him to send out a signal in the form of a loud buzzing sound on the subscriber's receiver. A few months later an exchange in which cords were used was set up in Chicago. The subscribers' indicators were arranged along a wall and a series of horizontal metallic rods grouped in pairs as connecting racks. These had clearing-out drop indicators associated with them.

In the early eighties, several businesslike attempts were made to produce practical switchboards. The Western Union already had the key switch for telegraph purposes and the Bell Company generally followed this design. One of their licensees, Williams, of Boston, made a board with an upper upright section and a lower sloping one with two rows of annunciators between. The horizontal bars were connected to vertical rows of spring jacks behind the panel by the insertion of long plugs. The operator's set was inserted at a horizontal series of spring jacks along the front edge of the sloping panel.

The first telephone exchange in England was established by the Telephone Co., Ltd., at 36 Coleman Street, E.C., in August 1879. On the Williams principle two vertical panels were arranged with three rows of drop indicators, and three rows of flat spring jacks, one for each subscriber's indicator. On the upper part of the board were 24 pairs of coloured interconnecting horizontal metal bars provided with holes to take circular pegs connected to cords for coupling up to selected subscribers. In front of the board were two small tables reminiscent of the old-fashioned treadle sewing machine, one for the answering operator and the other for the calling operator. One operator having taken the call, intercepted the line and connected it via the jack cord to the coupling strips; the answering operator passed the details to his colleague who then called the wanted subscriber and also walked to the board to complete the connection. Other exchanges followed and soon the number of subscribers was measured in hundreds. It is interesting to recall, however, that the circuits at this period were single-wire earth return. The transmitter was electromagnetic, as the microphone had not arrived, and clearing of a line depended on the operator listening in to ascertain when the conversation had finished.

The microphone came very quickly. Both Bell and another American, Gray, conceived the idea of modulating the current by a variable resistance in a battery circuit and both filed patents within a short time of one another for almost the same method of carrying out the idea. They devised a variable resistance transmitter formed by a rod attached to the sound diaphragm dipping into a liquid. Wires of platinum and liquids of dilute sulphuric acid or a salt solution were found suitable and capable of replacing the inductive transmitter already widely established, but in 1877 Edison appeared with his carbon transmitter, a much more practical proposition.

In America the Western Union Co. saw that the telephone now had big commercial possibilities and commissioned the great inventor, Thomas Alva Edison, to carry out experiments with a view to producing improved equipment. He soon hit upon two joint ideas, the use of a step-up transformer at the sending end of the line and substitution of the variable resistance between carbon contacts for the transmitter. There was some conflict with part of Bell's patent but sufficient was left to enable Edison to carry out an experiment in England by the transmission of speech from Norwich to London, one of the earliest long-distance circuits set up in this country. Edison, however, used a receiver of the Bell form and an injunction was threatened. Nothing daunted, Edison set to and in the course of a few days produced a receiver free of the Bell patent. Although the device never reached wide commercial application, it was such an ingenious arrangement as to be worth a little detailed description.

Edison knew that if a metallic point was drawn over a chalk surface moistened with potassium iodide, the passage of small currents between the point and the chalk affected the amount of drag. He therefore made a chalk cylinder which could be rotated by hand and arranged a mica diaphragm with a platinum point connected rigidly to its centre. When the received undulating telephone currents passed through the moving contact they caused a vibration of the diaphragm which reproduced the speech. In this way the ground was laid for an Edison system clear of the Bell patents and the Edison Telephone Company of London, Ltd., was formed with a capital of £200,000. Exchanges were established in Victoria Street, Eastcheap and Chancery Lane, with switchboards possessing new features, but essentially of the pattern having vertical and horizontal metal strips so that connections could be made between subscribers with plugs passed through holes at the points of intersection. Only 24 subscribers! could be accommodated on one board and when the numbers increased additional boards were erected.

We have seen in a previous chapter how David Hughes, a Londoner who spent the greater part of his life in America, made an important contribution to the development of machine telegraphy. It is appropriate at this stage to show how again his name appears in the progress of the telephone. On 8 May 1878, Professor Huxley communicated to the Royal Society[6] a description of Hughes' investigation into the effect of contact pressure between elements of an electric circuit on the value of the current flowing.

He gave the name *microphone* to the instrument he was using because it magnified weak sounds. As was pointed out by Wheatstone and others it did not in fact respond to sounds in the surrounding air but only when in contact with a vibrating body. Thus at the Royal Society meeting the members 'heard' the tramp of a fly: in reality what they heard was the effect of a fly walking about on a loose contact, a rather different thing from reception of airborne sound waves. Nevertheless the use of a loose contact between granules of carbon stemming from Hughes' discovery led to a host of types of transmitter, one of which was attributed to Hunnings, a Yorkshire clergyman, and another to Blake, an American. Hunnings was the first to use particles of carbon loosely assembled between two conducting electrodes, but he did not carry his idea forward to a commercial transmitter. Blake, on the other hand, in association with the Bell Co., had his design accepted and it was used extensively by early telephone subscribers for many years.

The first transmitter of the granular form to be brought into common use in this country was the Deckert in which carbon granules were disposed between a carbon diaphragm and a carbon block. The surface of the block had pyramidal projections, the tips of the pyramids having tufts of cotton wool to prevent 'packing' of the grains. The construction replaced Blake's transmitter and was assembled in its place in the same wooden cabinets. The Deckert transmitter was found to be so efficient that the speech currents resulting soon caused intolerable interference between the lines of separate subscribers and double-wire working had to be resorted to. So important was this that in the case of the National Telephone Co., the former Blake transmitters were retained for some time to avoid the expense of doubling the circuits.

The outstanding step made in transmitter development was in the invention in 1892 of the 'solid back' instrument by White for the American Bell Telephone Co. A cell, partially lined with carbon, was carried on the front side of a rigid support and the front side of the cell was closed with a small mica diaphragm lined with carbon and attached at the centre to a large acoustic diaphragm facing the mouthpiece. By careful selection of the carbon for the granules and screening for uniformity in size, as well as by the introduction of precision in the manufacture, the White solid-back transmitter quickly established itself. It could handle larger currents than previous designs of transmitter and enabled satisfactory conversation to be carried out on longer lines.

In 1880 the instruments standardized for subscribers' use were Telephone No. 1 and Telephone No. 2. The former, the wall type, was made of wood having a projecting ebonite mouthpiece on the transmitter with the tubular receiver and switch hook at the side and a double magneto bell in front. The desk type, constructed mostly of metal, carried the transmitter with its ebonite mouthpiece on a tubular pillar with the receiver and switch hook on the side, but the bell was accommodated in a separate wooden box. Solid back transmitters were adopted for both types of instrument.

Subscribers' apparatus has been subject to continuous improvement right up to the present day, and the efficient, robust hand set receiver and transmitter of aesthetic appearance converted for automatic working by the addition of a calling dial is a marvel of engineering development. Today's receiver has a sensitivity twenty times that

Fig. 35. Early Wall-type Telephone Instrument.

of the instrument used fifty years ago with a magnet weighing only one-hundredth that of the Bell receiver. With the latest balanced armature not only has the sensitivity of the receiver been increased but greater purity of tone achieved so that the voices of different speakers can be clearly distinguished through a more uniform response over the frequency range. This improvement, in addition to

Fig. 36. Early Table-type Telephone Instrument.

giving better service to the telephone user, has had an important effect on the overall economy of the system: it has made possible the use of much smaller wires in the subscriber's circuit: conductors only 0·012 in. diameter may become standard.

On the amalgamation in 1889 of the several telephone companies in this country in the £4,000,000 National Telephone Company attention was turned to the development of the best form of switchboard suitable for handling the large numbers of subscribers which it had become clear would require switching facilities. As the numbers increased the Edison system, with its arrangement of subscribers' terminations on vertical bars and connecting circuits on horizontal, became altogether too unwieldy and the plug and socket arrangement adopted by Bell—elementary jacks for line connections and plugs with cords for interconnecting—provided the only possible solution. Some form of supervisory signal to indicate the termination of a conversation was found to be necessary and at first a galvanometer was used. This was soon followed by a clearing indicator on the principle just adopted in the Chicago exchange on a patent granted to Horace H. Eldred in 1884.

A major event in the development of the telephone exchange was the introduction of the subscribers' multiple. At first there was only one point of access to a subscriber's line and, as one operator could not deal satisfactorily with more than fifty subscribers, it meant that in an exchange with over fifty incoming lines two or more operators would be concerned in each call. This involved special arrangements for connecting from one operator to another as well as a considerable increase in the cost of operating. The problem was solved for some time by so designing the board that any one operator could reach the boards on either side of her but even this arrangement was soon inadequate. Each operator with not more than fifty incoming lines, must be able to plug into any number of outgoing lines, in fact, to place her own subscriber in contact with any other subscriber on the exchange without the intervention of a second operator. The solution was provided by the multiple switchboard in which every line was looped into a jack at every operator's position, a solution without which larger telephone exchanges would have been quite impossible. To avoid the confusion which would result from two operators plugging into the same subscriber many devices were patented and tried, involving engaged indicators of various kinds. In many exchanges the operators called out to one another causing pandemonium at busy hours. At others a large display board which all operators could see, had connected subscribers shown by numbers and a special operator covered up the numbers of subscribers actually speaking by continually hanging up small boards. Engaged indicators on the switchboard were tried, but the ultimate solution was the modification of the jack, so that an operator could touch the tip of her calling plug on a ring surrounding the multiple jack of the wanted subscriber and so receive an 'engaged' click in her receiver.

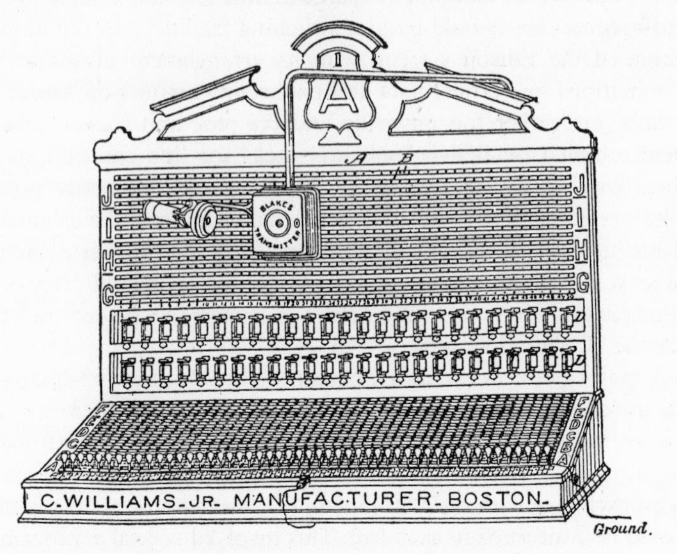

Fig. 37. Early Cordless Switchboard.

During the last decade of the nineteenth century there was an extensive development of the magneto system with drop indicators on the switchboard; but lamp signalling, found to save space and enable more equipment to be brought within the compass of an operator, gradually came into use. In 1896 the major trunk lines of the National Telephone Co., were purchased by the Post Office and at the turn of the century common battery (CB) working was introduced. Many inventors and many committees contributed on both sides of the Atlantic to this important innovation. In the famous Hayes circuit, based on a patent granted in 1892, the subscriber received current from a large common battery located in the exchange and signalled by lifting his receiver from a switch hook. He was called by the exchange operator sending out a magneto current which passed through a condenser to the subscriber's bell. Improvements by various inventors in the U.S.A. and the U.K. followed and the system settled down into one in which the exchange operators' cord circuit had a 24-volt common battery connected through repeating coils to the calling and answering plugs, speaking and ringing keys, supervisory relays, operating lamps, and a meter circuit. After a long investigation had been carried out in the United States by a 'switchboard committee', and experimental exchanges equipped, the first large commercial common battery installation was completed at Worcester, Massachusetts, in June 1896. Several others followed quickly in the United States and in 1900 the first installation in England was completed in Bristol for about this time the British Post Office had extended its interest from trunk operation to include subscribers. By 1906 there were 26 central battery exchanges in the London area, the early ones being equipped largely by American apparatus. In 1912 the Postmaster General took over all the National Telephone exchanges, of which there were 1,565. Of these 231 had over 300 subscribers, 68 were central battery exchanges, and the remainder magneto.

When the first telephone exchanges were projected the telegraph system with its ramification of circuits was already established, and it was natural that practice should follow precedent. There was, however, a basic difference in the requirements of the two services; telegraphs were usually long-distance circuits with lines from town to town, whereas in telephony, initially at any rate, the main problem was to connect many subscribers in a congested urban area with one another through a local exchange. Consequently bare wires strung over rooftops became the usual practice. Baldwin has given an excellent description[7] of the early history of overhead telephone line plant in which he traces the development from the practice of the telegraph engineer.

Telephone exchanges were conveniently housed in high buildings surmounted by a derrick from which lines radiated in all directions. The most commonly used wire to start with was bare galvanized iron, No. 12 gauge, with other sizes depending on the whim of the particular engineer. To prevent corrosion in industrial towns and large cities the wires were often protected with an impregnated cotton covering. Although the high electrical resistance of iron wire was a great disadvantage it required many years of testing and consideration before its place was taken by copper. Because of their increased tensile strength phosphor bronze and silicon bronze were favoured by some but hard-drawn copper held the field until the brittleness of bronze was overcome. Soft copper binding wires were introduced for holding the wire to the insulator of porcelain of various shapes, and gutta-percha covered wire was used for leading into premises. By 1890 silicon bronze had been so much improved that it had been brought into general use for subscribers' circuits throughout Britain in a standard weight of 40 lb. per mile.

The congestion resulting from the large number of wires leaving the structure over an urban telephone exchange soon called for a bunching of the circuits into groups and this was done by the use of multi-core aerial cables. Early forms consisted of a number of gutta-percha insulated wires bound together with tape and associated with a stranded galvanised wire to give supporting strength. Later rubber was substituted for the gutta-percha which cracked when exposed to air and sunlight. This also enabled more conductors to be accommodated in a cable of the same weight and standard types were adopted with 26 and 52 pairs. As an example of the magnitude of the task of getting the circuits away from an early telephone exchange the case of Sheffield may be cited. There the angle iron derrick erected in 1891 accommodated over 1,000 wires.

A problem which faced the early telephone engineer was the prevention of overhearing between circuits when so many had to be carried in close proximity to one another. The impracticability of the single-wire earth circuit which had been suitable for telegraphs was soon appreciated and in 1892 expensive schemes of doubling, or making the circuit 'metallic' as it was called, were carried out. Even then induction effects were found to be serious and twists were introduced. The credit for this idea goes to Professor Hughes, who referred at a dinner of the National Telephone Co. in 1895 to his paper[4] in which he had first announced the idea. For long-distance overhead lines, then growing rapidly along the main roads of Britain, the system of twisting a pair between poles and of inserting crosses was adopted extensively.

One of the most important improvements in the provision of circuits between subscribers and exchange and between one exchange and another was the invention of the dry-core paper cable. The first lead-covered cables were manufactured by the Western Electric Co., of Philadelphia in 1884, but the dielectric consisted of two wrappings of cotton giving a total thickness of 0·125 in. In the discussions of a special conference of specialists held in New York in 1887,[5] the importance of keeping the electrostatic capacity at a low value was emphasized and at a similar conference two years later, John A.

PLATE XXXV

o Dutch Telephone Cable,

struction of submarine cable
ythene dielectric and spiral
tion to minimise capacitance.
Submarine Cables Ltd.)

(*b*) Strip System for Underground Distribution, 1889
System used by Latimer Clark & Muirhead on the Pall Mall and St. James's system. Copper bars carried on porcelain bearers in cast iron troughs.
(*Photo: British Insulated Callendar Cables*)

(*c*) Ferranti 11 kV Concentric Cable, 1890
Ferranti lapped paper in the form of wide sheets on copper tubes and impregnated it with wax. This was then slipped into an outer copper tube which was drawn down with intimate contact. After a further application of paper a protective iron tube was applied and the space filled with hot wax. Over 28 miles of this cable gave satisfactory service for many years.
(*Photo: Associated Electrical Industries Ltd.*)

(*a*) High Speed Stranding about 1930
Great increase in the efficiency of the stranding operation for conductors was obtained b abandonment of the slowly rotating bobbins and the substitution of concentric reels from w the wire was "peeled" off and passed through to the stranding die.
(*Photo: Associated Electrical Industries Ltd.*)

(*b*) Longitudinal Covering of Rubber Conductors about 1920
The application of two strips of rubber, one above and one below the conductor, and the u steel rolls to form a complete sealed coating was in use long before the First World War. S quent progress in this field has been the increase in the number of conductors simultane covered on one machine. (*Photo: Associated Electrical Industries Ltd.*)

Barrett, electrician, American Telephone and Telegraph Co., announced that he had 'been engaged in an effort to reduce specific inductive capacity of the wrapping used upon the conductors so as to secure a considerably lower limit for static capacity while still using the same dimensions for the cable. We have had an almost unlooked for success in this direction in the employment of manilla paper in the place of cotton as the wrapping of the conductors.'

Dry-core cable was thus established and has held the field for telephone work both on short subscribers' circuits and for long-distance transmission right up to recent times, when its supremacy has been challenged by the coaxial cable. Early in the 1890's two 41-pair lead-covered dry-core cables with 20 lb. conductors were installed in the Mersey railway tunnel between Liverpool and Birkenhead and by 1905 the National Telephone Co. had 500,000 wire miles of underground cable, mostly with 20 lb. conductors. Similar cable but with 10 lb. conductors was also being used on rawhide suspenders carried by suspension wires to relieve exchange line congestion. In 1898 the first underground telephone cable of any appreciable length was laid with 38 pairs of 150 lb. conductors. For long distances the multi-twin, i.e. one pair twisted round another, and the star quad, i.e., four wires laid up together with diagonal wires forming pairs, were adopted. The cables were at first drawn into cast-iron pipes and later earthenware single and multiple ducts.

The heavy overhead long-distance lines continued to grow in number and size until the First World War but, as even then they failed to meet the demand, more and more cables were laid. For subscribers' lines 1,000-pair cables with 6½ lb. conductors became common. After the war much attention was given to the production of a star quad cable with improved symmetry. This was achieved by the use of a spiral thread under the paper and a thread down the centre of the quad assembly.

In 1925 the first quad cable of this kind was laid between the City and Ealing, and had 254 pairs of 40 lb. conductors. By 1930 the type had become standard and known as star quad cable. It accommodated 40 per cent more circuits in a given diameter than the multi-twin cable and was a possible construction up to 1,400 6½ lb. pairs.

Throughout the period already covered in the development of the early telephone, some thought had been given to the possibilities of automatic switching of lines. As early as 1879 the brothers Connolly of Philadelphia proposed an automatic telephone switch in which pulses of current were transmitted by the rotation of a letter pointer on the principle of the Froment telegraph. The possibilities of this system were strictly limited and others were suggested, but in 1889 Strowger, an undertaker of Kansas City, U.S.A., dissatisfied with his telephone service, made the first tentative step towards what has become a practical system applied on a world-wide basis.

The basis of the Strowger system[8] was the operation of a ratchet and pawl by means of current impulses sent out by the subscriber on a counting system. To send the figure 1 he pressed a key once, to send 5, five times, and so on. Several keys were provided, one for units, one for tens, one for hundreds, and so on, and each was connected by a separate wire to the central exchange. The contacts were contained within a hollow cylinder and a radial arm, moved step by step by incoming impulses, could rotate over the contacts. It could also rise and fall. Various modifications were followed in 1895 by a design very similar to that which has survived, i.e. ten rows each of ten contacts arranged in the form of an arc. The upright shaft had both vertical and rotary movement. In 1896 the subscribers' dial-sending apparatus was invented and with modifications has become the standard equipment.

The next step was the provision for larger numbers of subscribers, without extravagant increase in size of the multiple, on the rotary switch and this was achieved in 1897 when a 1,000-line system was produced for Augusta, near New York, in which one selector was provided per subscriber, 100 groups of 10 selectors, with 10 groups of 100 selectors, a total of 1,000 lines. By 1898 there were 22 automatic exchanges in the United States. Progress continued in the simplification and reliability of the equipment. In this country the first automatic exchange was opened at Epsom in 1912 with, at first, 500 subscribers, increased later to 1,500. Other exchanges quickly followed and in 1918 Siemens entered the automatic field with important improvements in the circuit arrangements. The 10-pt. first and second selectors were combined with the 100-pt. two-motion group and final selectors. Great progress also followed in the design of relays and other equipment and at this stage the subject becomes much too complex to cover even cursorily in a single chapter.

The dramatic growth in the use of the telephone over the past eighty years can be seen from the graph, Fig. 71. When Sir William Preece visited the Philadelphian Electrical Exhibition in 1884 the number of telephone stations in the United Kingdom was already 11,000. He found that the comparative figure for the United States was 148,000. He explained the discrepancy to his hosts on the grounds that in London an errand boy cost 2s. 6d. a week whereas a similar service in New York cost 12s. to 15s. It paid to use the telephone. Today (1960) there are still about nine times as many telephones in the United States as there are in the United Kingdom.

In his inaugural address as President of the Institution of Electrical Engineers[9] Sir Gordon Radley summarized telephone progress in this country. The inland system had more than doubled in fifteen years, cables with one conductor were carrying 600 simultaneous conversations with a proposal to increase the number to 1,000, and altogether some seven million telephones were in daily use. The story of some of these remarkable achievements over the past few decades is that of the application of science in its most advanced state to the work of the telephone engineer and calls for a separate chapter.

REFERENCES

1. *J. Royal Institution*, vol. 2, 1831.
2. *J. Soc. Tel. Engrs.*, vol. 6, p. 387, 1877.
3. *The Bell Telephone*. The deposition of Alexander Graham Bell in the suit brought by the United States to annul the Bell Patents. The American Bell Telephone Co. 1908.
4. *J. Soc. Tel. Engrs.*, vol. 8, p. 169, 1879.
5. Kingsbury, *The Telephone and Telephone Exchanges*, 1915.
6. *Proc. Royal Soc.*, vol. 27, p. 362, 1878.
7. F. G. C. Baldwin, *The History of the Telephone in the United Kingdom*. London, 1925.
8. *J.P.O.E.E.*, vol. 49, p. 173, 1956–7.
9. *Proc. I.E.E.*, vol. 104, Pt. B, p. 1. Jan. 1957.

CHAPTER XV

Science in Telecommunication

In no branch of electrical engineering has there been such a dramatic development in the direct application of science as has taken place during the past few decades in the fields of telegraphy and telephony. In a rather remarkable manner this has resulted in a complete integration of the engineering development of what grew up as two quite separate activities and today the telegraph and the telephone are technically inseparably associated with one another. It was during the inter-war years that a number of striking innovations finally opened the way and these constitute a vital link in our history, although the seeds were sown much earlier.

In January 1886 Professor Hughes, President of the Society of Telegraph Engineers, adopted as the subject of his inaugural address 'The Self-Induction of an Electric Current in Relation to the Nature and Form of its Conductor',[1] and this resulted in considerable interest being taken for the first time in the theoretical considerations governing the transmission of a fluctuating current through a long line. Within a few months Preece followed with a paper on 'Long Distance Telephony'[2] in which he indicated broadly the relatively greater importance of the characteristics of the conductor in comparison with those of the transmitting and receiving apparatus, and in January 1887 Silvanus Thompson wrote on his own investigations.[3]

Two very important matters arose from these contributions. During the discussion on Silvanus Thompson's paper Preece formulated the theory that the limiting distance for coherent speech through a current was settled by the capacity and resistance of the circuit, his K.R. law, and Oliver Heaviside, that remarkable genius, laid the foundation for a full mathematical analysis. In pointing out the severe limitations of Preece's employment of capacity and resistance only, Heaviside showed in his famous *Electrical Papers* in 1892[4] that a circuit had four fundamental electrical constants, the resistance, inductance, capacity and leakance. He pointed out that in an Atlantic cable the attenuation is very great and increases with the frequency, 'thus leading to a most prodigious distortion in the shape of irregular waves as they travel along'. 'Of course', he said, 'we may *send* as many waves as we please per second but they will not be utilisable at the distant end. This distortion is a rather important matter. Mere attenuation if not carried too far, would not do any harm. . . . Within the limits of approximately constant attenuation the distortion is small. This is what is wanted in telephony to be good. Lowering the resistance is perhaps the most important thing of all. Increasing the inductance is another way of improving things.'

In 1899 Michael Idvorsky Pupin read his paper before the American Institution of Electrical Engineers on the propagation of long electric waves[5] and took out a patent, for distributing inductance along the length of a conductor, which made telephone history. In a fascinating autobiography[6] Pupin describes how he was born in a little Serbian village of peasant parents, who could neither read nor write, and how, in 1874, he arrived in New York at the age of fifteen, an immigrant with only five cents in his pocket. In search of knowledge he had left his birthplace first for Prague and then, having saved the cost of a steerage passage, reached the promised land of so many of his day. By dint of five years' hard work reading while earning a living, he had become a student at Columbia University, passed out successfully, and turned his thoughts to the mathematical analysis of electricity. Returning to Europe he spent some time studying at Cambridge and Berlin Universities, returning to Columbia, where he was appointed to a chair. It was here that during twenty-five years he made his inventions, the most significant of which, the Pupin Coil, brought him in $1,000,000. The patent was bought by the American Telephone and Telegraph Co.

Thus was laid the foundation stone of 'loading' an electrical transmission line and the first practical application was in August 1902 when loading coils were inserted in a ten-mile length of telephone cable between New York and Newark, New Jersey. By this modification the grade of transmission was improved to that obtainable on a five-mile length of unloaded cable.[7] In 1905 a long-distance cable between New York and Philadelphia, a distance of 90 miles, was loaded by the insertion of toroidal coils every $1\frac{1}{4}$ miles. In the 93 pairs of No. 14 gauge conductors in the cable the result was a total standard cable equivalent of 11 miles and in the 19 pairs of

No. 16 gauge conductors a standard cable equivalent of 15 miles. Encouraged by these successes the American Telephone and Telegraph Co. proceeded to extend the use of loading cables and in 1912 and 1914 the New York to Washington line (235 miles) and the New York to Boston line (240 miles) were also loaded.

In the long-distance circuits in England open lines carried on poles had been strengthened by increasing the size of the copper conductors. In 1895 the limit was reached on the London–Leeds–Edinburgh line when conductors weighing 800 lb. per mile were erected. It was evident that loading by the addition of inductance was necessary to improve the transmission characteristics both of open and underground lines. In 1915 the Post Office carried out a most important experiment. They looped backwards and forwards the 110-mile circuits on the London–Birmingham cable producing equivalent lengths of 220, 440, 660 and 880 miles. By the insertion of inductance coils at 2½ mile spacing commercial conversation was obtained up to 660 miles, so proving that underground circuits could be provided safely for linking important cities anywhere in Britain. The coils at first had air cores but as the technique developed soft iron cores were adopted and later dust cores were proved to be the most efficient. Phantom circuits were also loaded separately and the coils were assembled in sealed cast-iron pots each holding as many as 64 coils.

The first long-distance telephone circuits from this country to the continent of Europe were carried through a four-core gutta-percha-insulated submarine cable laid across the Channel between Dover and Calais in 1891. The weight of the conductor was 160 lb. per nautical mile[8] and good commercial speech was obtained between London and Paris. The first submarine gutta-percha insulated cable to be loaded was laid from Dover to Calais in 1910. The insertion of the inductance coils involved new problems which are described by O'Meara in the paper quoted and within the next few years other loaded submarine cables followed. These enabled the London telephone service to be extended as far as Basle and Geneva.

These were all coil loaded circuits but another form of loading was also tried out to a limited extent. Known as continuous, or Krarup loading after the inventor, it consisted in adding inductance by wrapping the copper conductors closely with a fine soft iron wire from end to end before applying the dielectric. A four-core gutta-percha-insulated cable of this kind was laid across the Straits of

Georgia in 1913 joining Vancouver and Victoria Island. It had two physical circuits and one phantom.[9]

To round off this phase of the application of new science to telecommunication, reference may be made to the position reached in the development of long-distance telephone circuits in the United States. After the successful pupinization of the New York–Washington and New York–Boston lines, the American Telephone and Telegraph Co naturally turned its attention to transcontinental possibilities and, by loading and reinforcing overhead lines, a New York–Denver circuit of 2,200 miles was achieved as a first stage. On 25 January 1915, a commercial telephone service was inaugurated between New York and San Francisco, a distance of 3,400 miles over open wires weighing 870 lb. per mile. Today the United States has a vast system comprising some 200 million miles of wire and practically any two telephones can be interconnected on demand.[10] Twenty-five years ago the average American inter-city call gave a conversation efficiency represented by a distance of 35 feet apart between talkers while today it may be represented by the two talkers standing only 6 to 12 feet apart. This remarkable achievement has been the result of many other innovations since the First World War in an ever-increasing crescendo of progress—repeaters, coaxial cable, carrier transmission, multi-channel voice frequency telegraphs, radio relay lines—culminating within the last three years in the magnificent transatlantic telephone cable link.

The story of the invention and application of the thermionic triode valve is primarily that of radio communication, to be discussed in a later chapter, but it must be anticipated here because of the valuable contribution which the valve has made over the past forty years in the field of line telecommunication. The valve amplifier was introduced during the First World War and opened up the way to better transmission and reduction in size of conductors. Two-wire amplified circuits were at first developed for inland trunk cables in this country, but by 1939 had been replaced by four-wire circuits with losses reduced to practically zero. Halsey says in his excellent survey given before the British Association,[11] in speaking of the period immediately before the Second World War: 'The heyday of inductive loading based on the work of Heaviside was past, audio frequency transmission was ceasing.' Not only had the repeater reduced the need for capacity correction in the line by the addition of inductance, but carrier technique, introduced in 1920, had made it possible to transmit more than one conversation over one pair of wires.

The adoption of a multiplex system in which the speech transmission was carried by a modulated high frequency current advanced slowly; at first one extra circuit and then three on a single pair of open wires and one extra on a cable pair. The war intensified the effort following the fall of France as so many internal service circuits were required. The important development of 'negative feed-back' had enabled greater stability and control of linearity to be obtained in repeaters, so avoiding interference. In 1937 this new tool had made it possible to carry twelve channels on each pair of a cable in the frequency range 12 to 60 kc/s,* a channel spacing of 4 kc/s. This involved improvements in the type of filter used and the extensive employment of quartz-crystal (piezo electric) resonators.[12]

By 1939 cables with repeater stations every 22 miles formed the normal provision of long-distance circuits in Britain. In trunk cables the chief development was in changing over from the audio frequency range to frequencies up to 60 kc/s for carrier. The quad cable with 20 lb. air-spaced paper-covered conductors and loaded with 88 mH coils at 2,000 yards spacing, were only satisfactory up to 3,400 c/s whereas 24-pair 40 lb. cables used in pairs 'go' and 'return' were suitable for the 12-channel carrier systems. This did, however, involve reducing the 20-mile spacing to 16 miles for subsequent 24-channel working in the frequency band 12 to 108 kc/s adopted later.

By the end of the Second World War the trunk network of this country had been almost completely converted to carrier working with a consequent enormous increase in the number of available trunk circuits. The pre-war 6,800 services over 25 miles in length were increased to over 17,000 and dialling by operators on such circuits had been widely introduced.

A recent statement on the American development of carrier[10] shows a vast increase using frequencies up to 200 kc/s on open wires and up to 8 Mc/s† on cable in common use on most of the long-distance telephone circuit mileage in that country. This advance has involved a great deal of perfection in network theory and many detailed refinements to ensure sharp separation of frequency bands. By the use of two different paths for opposite direction of transmission, balance in repeaters is facilitated and echoes minimized.

One of the greatest contributions to the extension of the carrier system has been the development of the coaxial cable. The first to be laid in this country was the London–Birmingham in 1937, which

* Kilocycles per second. † Megacycles per second.

had four individual coaxial tubes 0·45 in. diameter with repeaters inserted every 8 miles. The effective band width was 0·5–2·1 Mc/s, providing immediately for 280 circuits. From 1945 fewer 12-channel carrier cables were laid but more cables with coaxial pairs. In 1947, by inserting repeaters at 6-mile intervals, the band width was increased to 60–2,850 kc/s accommodating 600 telephone circuits. By 1957[13] the number of telephones in this country had risen to over seven million and by reducing the spacing of repeaters to three miles in order to meet the demand for long-distance conversations it was anticipated that 1,000 telephone circuits plus a 405-line television circuit could be provided by each coaxial tube. The design of the tubes has already passed through many phases from the solid centre wire with a spiral of cotopa string, having an outer tube of 0·45 in. internal diameter consisting of several copper tapes. The string was then replaced by hard rubber discs and one single copper tape made to form a tube. Polythene discs were later introduced. A recent American claim is that two such coaxials can provide either 1,800 telephone channels or 600 telephone channels, plus a 4·2 Mc/s television circuit in each direction.

Mention can only be made of the introduction of voice frequency telegraphs during the past few decades or of the way in which telegraphy and telephony have approached one another. When the telegraph system of this country was reorganized in the 1930's teleprinters replaced all earlier systems and the voice frequency telegraph adopted first 18 channels and then 24 channels, each 120 c/s wide. These were operated on telephone lines, a technique far in advance of anything previously attempted.

One of the most spectacular developments in land telephone transmission has been the employment of the microwave telephone link. Between the two world wars difficulties arose in connection with the ever-increasing number of wavelengths and the 'crowding of the ether', as it was termed. But in 1931 a remarkable proposal was tested which, at one stroke, extended the facilities for transmitting a vastly greater number of simultaneous radio services. This was the microwave link by which, using a parabolic reflector and an aerial less than an inch in length, high quality speech could be transmitted over a distance of forty or fifty miles with an expenditure of less than one watt. The wavelength used was under 20 cm. and in the first demonstration the link was set up successfully between St. Margaret's Bay, Dover, and a point near Calais, a distance of 35 miles. By 1934 the method had been so perfected that the link was taken into commercial use with a mixed telephone and teleprinter service.

With the advent of television the need arose for greatly increased transmission facilities, but the newly developed coaxial cable provided most of what was required. One section of the network, however, gave an opportunity for a microwave link which stretched away north from Manchester to Kirk o' Shotts, between Glasgow and Edinburgh, a distance of over 250 miles, with seven repeater stations on the way, each with its parabolic aerial. A frequency of 4,750 megacycles was used. At the time of the Coronation, 1953, television pictures were beamed from the high tower of the University of London to Dover and Cassel in France with intermediate relays. Passed on to the French broadcast system in Paris, the pictures were converted by suitable adjustment from the British 405-line standard to the French system. Already in many parts of the world multi-channel microwave links are in successful operation on telephone and television systems.

Trans-oceanic telephony has made rapid strides since the middle of the century. Before the Second World War there were a few audio-frequency circuits from this country to the continent with a limited application of superimposed carrier circuits. The improved dielectric paragutta enabled a pair of coaxial cables to Holland to carry twelve circuits. But for real long ocean cables little progress had been made for fifty years. There were no telephone cables and ocean telegraphy continued on a direct current basis with only a moderate increase in speed. Continuous loading of a few had made possible some improvement. For instance, in 1926 a cable laid between Fanning Island and Bamfield, 3,458 nautical miles, was continuously loaded with a new alloy, Mumetal, and gave a speed of working of 250 words a minute against the former usual achievement of 25 words a minute, but the big step forward awaited the combination of the new dielectric polythene and the repeater, both products of intensive scientific research. The first submerged repeater was laid between Holyhead and the Isle of Man in 1943, and this was followed in 1946 by the Anglo-German cable. A repeater had also been inserted in an existing transatlantic telegraph cable and had permitted an increase of 50 per cent in the operating speed. Repeaters were also inserted in various shallow water cables between Britain and the Continent and considerable experience had been obtained in the United States by the Bell system, in particular on the Key West–Havana cable.

After many years of collaboration on the proposal to lay a transatlantic telephone cable, a joint decision was taken in 1953 by the authorities in Britain, the U.S.A. and Canada. An intensive three years of preparation and execution followed until on 14 August 1956 the final splice completed one of the most remarkable feats of electrical engineering ever accomplished. The system, consisting of two separate single-core cables, provided on completion 29 telephone circuits between London and New York, six between London and Montreal, and a number of other circuits on the Canadian land

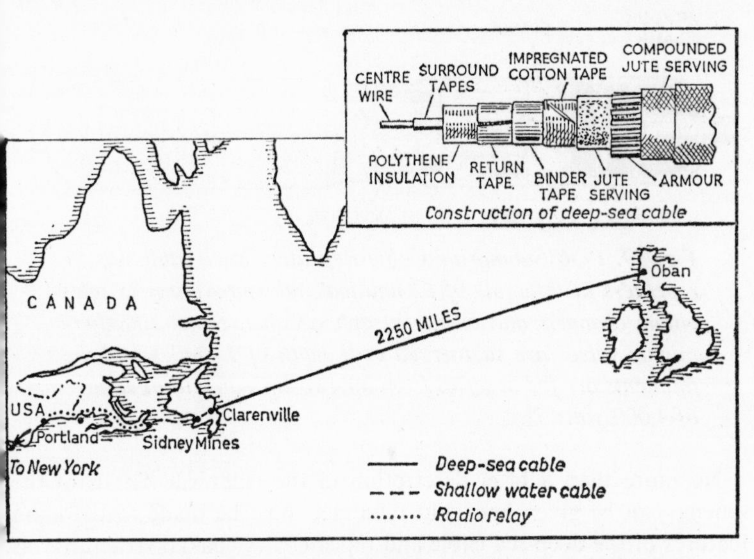

Fig. 38. 1956 Transatlantic Telephone Cable. Two single-core cables operate as 'go' and 'return' for the speech currents. The conductor of 0·1318 inch copper wires is surrounded by three 0·0145 copper tapes and insulated with polythene over which a 0·016 copper tape is laid. The armouring consists of 24 high-tensile steel wires.

section. The maximum length of main circuits is 4,157 miles, London to Montreal, and 4,078 London to New York. Some of the circuits are permanently connected through to European capitals, the longest being the New York to Copenhagen, 4,948 miles. The vast amount of work which went into the preparation and installation of the cable is fully described in a Symposium held in 1957.[14]

Each of the two cables on the long intercontinental section between Newfoundland and Scotland has 51 repeaters and the working frequency range of 144 kc/s provides the 35 telephone channels and one telegraph channel. All the 'go' channels are in one cable and the 'return' channels in the other. Some 300 thermionic valves are thus operating in these submerged repeaters at depths up to more than 2¼ miles. By means of the separate 4 kc/s channels the equivalent number of telegraph channels which could be obtained is 864, a great achievement and a striking advance over the early telegraph cables limited to about three words a minute.

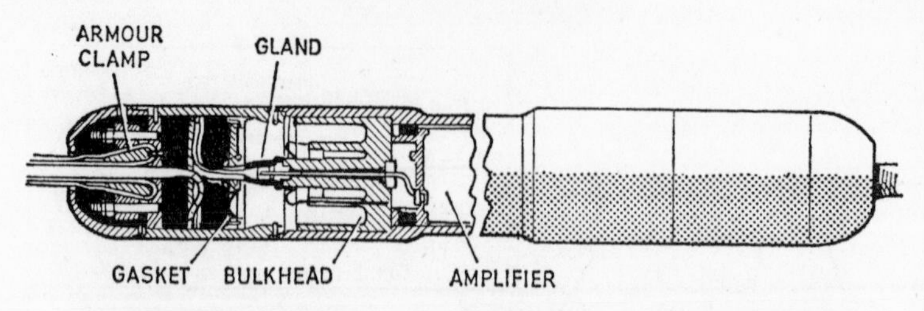

Fig. 39. 1956 Submarine Cable Repeater. Each cable has 51 repeaters at intervals of 37 nautical miles providing 35 telephone channels and one telegraph channel. Some 300 thermionic valves are submerged at a depth of 2½ miles and the filaments are fed in series from the two ends on a total voltage of 4,000 volts D.C.

No more than a brief description of the technical details of the scheme can be given here, but reference must be made to the main features of the deep-sea cable and repeaters. A coaxial structure was adopted for the cable. The central conductor consists of a copper wire 0·1318 in. diameter surrounded by three 0·0145 copper tapes and then a layer of polythene to a diameter of 0·620 in. The 0·016 copper tapes constituting the return conductor, applied with a long lay over the dielectric, are followed by a lapped copper tape as portection against the teredo. Over special beddings are 24 high-tensile steel armour wires and two jute servings, bringing the overall diameter up to 1·21 in.[15]

The repeaters, housed in bulges in the cable, are of intricate and precise construction and are inserted at intervals of 37 nautical miles. Each repeater is divided into its 17 elements, which are enclosed in polymethylmethacrylate cylinders and arranged head and tail so as to be flexible. The string of such units is further protected by steel rings, copper tube, protective coatings and special armouring wires. The structure on completion is 8 ft. long and 2⅞ in. diameter with a taper at each end about 20 ft. in length down to the cable diameter.

Prolonged and detailed consideration was given to the design and construction, in particular the avoidance of water leaks at a depth of over two miles and to ensuring reliability in service without maintenance attention after laying. The D.C. power for the valves is supplied by constant current generators at both terminal stations and the one current, passing through all repeaters in series, requires a total voltage of just under 4,000.

Dramatic as have been many of the stages in the progress of electrical engineering, no event in its previous history could approach the successful completion of the first transatlantic telephone cable. A century ago a small band of enthusiasts overcame great difficulties in completing the first successful transatlantic telegraph cable, an epic which will live long in our annals. By their dogged determination they succeeded in spite of ocean storms and a serious lack of scientific knowledge. But today, by the application of many minds on both sides of the Atlantic and the application of a vast potential of science, electrical engineers have added an even greater contribution to human progress.

REFERENCES

1. *J. Soc. Tel. Engineers*, vol. 15, p. 6, 1886.
2. *J. Soc. Tel. Engineers*, vol. 15, p. 274, 1886.
3. *J. Soc. Tel. Engineers*, vol. 16, p. 42, 1887.
4. Oliver Heaviside, *Electrical Papers*, vol. 2, p. 120, 1892.
5. *Trans. A.I.E.E.*, vol. 16, p. 98, 1899.
6. Michael Pupin, *From Immigrant to Inventor*, New York, 1925.
7. Hayes and Gherardi, *International Electrical Congress*, St. Louis, 1904.
8. O'Meara, *J.I.E.E.*, vol. 46, p. 310, 1910.
9. *Electrician*, 22 Aug. 1913.
10. Green, *Electrical Engineering*, vol. 78, p. 470, May 1959.
11. *Engineering*, vol. 184, p. 432. 1957.
12. Author's Presidential Address, *J.I.E.E*, vol. 93, p. 1, 1946.
13. Radley, Presidential Address, *J.I.E.E.*, vol. 104, Part B, p. 1, 1957.
14. *Proc. I.E.E.*, vol. 104, Part B, Suppl. 4, pp. 1–126, 1957.
15. *Proc. I.E.E.*, vol. 102, Part B, p. 117, 1955.

CHAPTER XVI

Power Cables

We have seen in earlier chapters how the first attempts to carry the electric current underground were made by the pioneers in telegraphy, and it was only towards the end of the 1870's that public distribution of current for lighting purposes created a new field of electrical engineering. After sporadic attempts to prevent access of moisture to underground conductors by coating the wires with tar-impregnated twine, telegraph engineers turned to gutta-percha, recently introduced from Malaya, and this form of underground cable had a vogue for many years. It was natural, therefore, that when the arc lamp installations were carried out on the Embankment, Holborn and elsewhere, this practice should be adopted.

Gutta-percha was quickly found to possess undesirable characteristics: it cracked through ageing on exposure and softened under load, allowing the conductor to become decentralized. Vulcanized rubber was then tried and found to be so successful that it established a position which was maintained unchallenged for many years. Rubber cables were introduced by Hooper in 1859 and very quickly demonstrated their superiority over gutta-percha. Not only were they used for underground work, but also extensively for the overhead distribution systems in London from the time of the Grosvenor Gallery station onwards. This system, as we saw in Chapter IX, operated at a voltage of 2,400 and the rubber-covered conductors of 19 wires, each No. 15 gauge, were some of the first made by the India Rubber and Gutta-Percha Co. at their new Silvertown factory in Woolwich.

In his early work on developing the system for 'subdividing the electric light', described in Chapter IX, Edison wrapped copper rods with fibrous spacing materials and drew them into iron pipes, which he then filled with oil or compound. In one design which became very popular, the rods were drawn to a segmental section and placed in the tube with the flat sides facing one another and kept apart by stiff paper spacers.

In his Kensington Court system Crompton distributed the current through a system of existing subways by means of bare copper rod carried on porcelain insulators fixed to the walls of the subway. When, however, in extending his system he reached the limits of the available accommodation, he constructed 2-ft. square conduits immediately below the surface of the public footways and carried the three copper strips for his three-wire system on insulators fixed to wooden balks stretched from side to side of the trench.

Other conduit systems followed that of Crompton including one devised by Kennedy for the Westminster Co. in 1889, in which the conductors were supported on porcelain bridges and another by Latimer Clark and Muirhead for the Pall Mall and St. James's system. In the latter system short lengths of iron troughing were substituted for the brick trench and copper strips 2 in. wide were carried, on edge, in porcelain bridges. Hunter and Hazell report[1] that a section of this system which remained in service for nearly 40 years, still exists beneath Sackville Street.

Modifications of various kinds were introduced into these original conduit designs both in London and abroad, but the risks of creepage over the insulators, sagging of the strips and ingress of water or gas were found to be serious disadvantages. Instead of insulating an exposed conductor only at points it became clear that a continuous coating of insulating material from end to end was desirable and the electric cable was the solution.

In an intermediate stage between the bare conductor assembled on site on insulators, and the insulated cable made in a factory and laid as a complete entity, there appeared a system which, though not extensively used, has always attracted attention because of its original features. In 1887 David Brooks introduced in the United States a system in which stranded conductors were lapped with jute yarn, then covered with hessian tape, braided overall and drawn into iron pipes previously laid in the ground. The jute coverings were impregnated on site as the lengths were drawn in and the pipe was finally filled with a viscous oil. From our present-day knowledge of the features required to produce a good dielectric for high-voltage cables we can see that Brooks was working along the right lines. Notwithstanding extensive experiments by various companies in the United States, however, and the completion by Johnson and Phillips of a number of installations in this country one of which, at Worcester, remained in service at 2,400 volts for 40 years, the system never established itself.

While rubber cables were gaining popularity and the Brooks system and various duct proposals were under trial a development appeared from a non-electrical source which had a successful career for many years though destined not to survive eventually. This was

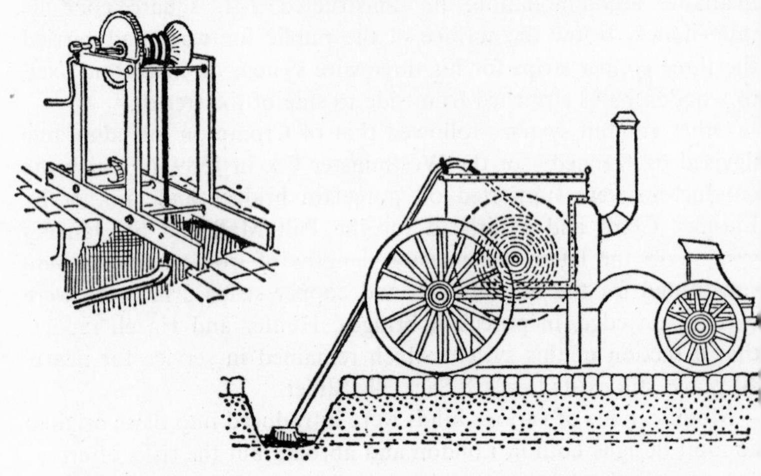

Fig. 40. Brooks' High-voltage Oil-filled Cable System, 1887. An early oil-filled system in which hessian-taped jute-lapped cores were drawn into iron pipes laid in the ground direct from an impregnating tank. The pipes were subsequently filled with oil.

the vulcanized bitumen cable. In 1881 W. O. Callender, who used Trinidad bitumen in his road-surfacing business, conceived the idea of treating it in some way to make it suitable for insulating electrical conductors. In the experiments which followed the suggestion it was found that, by heating cotton seed oil with an admixture of sulphur and adding a proportion of bitumen, a substance could be produced which was suitable for cable making. This 'vulcanized bitumen', as it was called, was patented and put to use by a new company at Erith under the management of W. O. Callender's eldest son, Thomas Callender, whose name has remained famous in the cable industry ever since.

The vulcanized bitumen cable made a great contribution to the development of some of the early supply systems. Eleven pairs were laid across Waterloo Bridge to supply the Jablochkoff installation

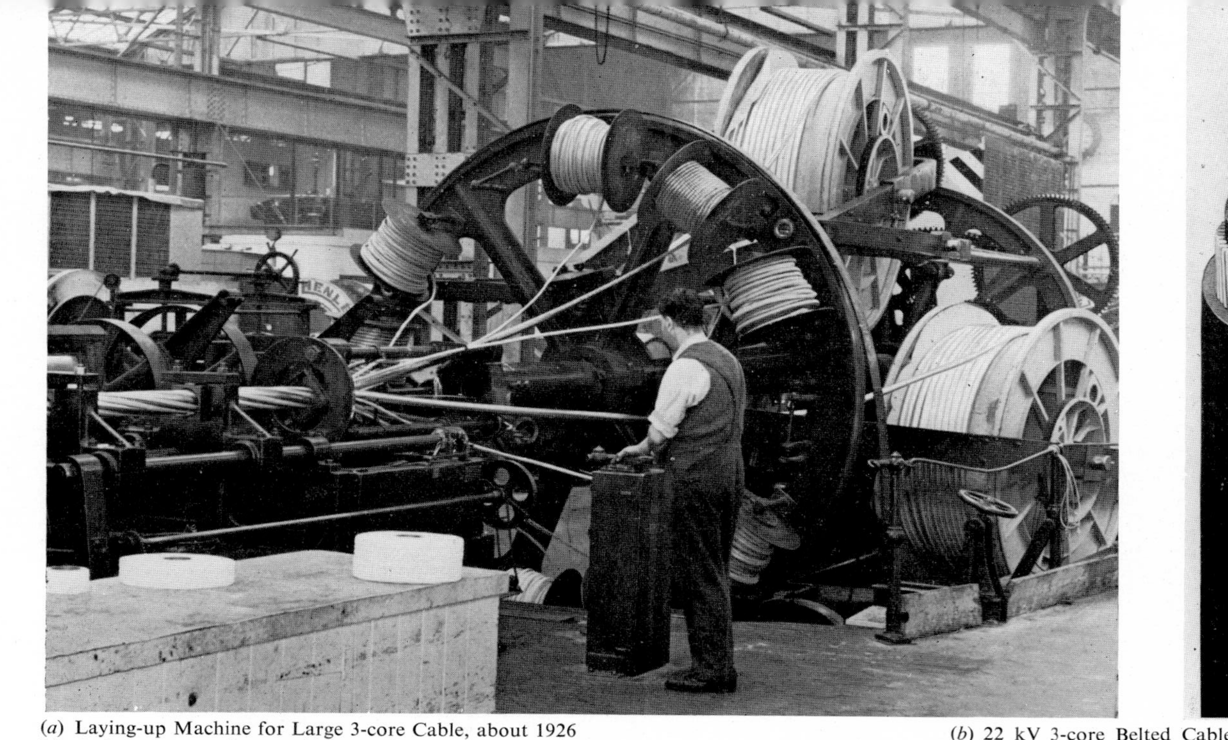

(*a*) Laying-up Machine for Large 3-core Cable, about 1926
Much thought was given to this operation with a view to securing adequate filling of interstices with jute or paper packings—shown in centre of picture—and maintaining the correct angle of the cores by using floating carriages for the main bobbins—shown on right.
(*Photo: Associated Electrical Industries Ltd.*)

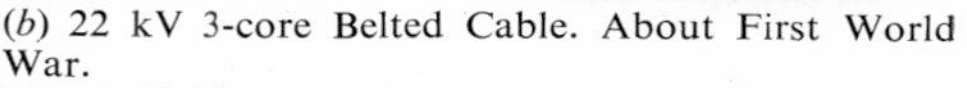

(*b*) 22 kV 3-core Belted Cable. About First World War.
The individual conductors were separately paper lapped and then a further lapping of paper applied over the three after laying up. After impregnation with rosin oil compounds a lead sheath was applied followed by protective servings of jute and armouring wires. (*Photo: Associated Electrical Industries*)

PLATE XXXVIII

(*a*) 33 kV 3-core Screened Cable, 1924
In view of mysterious failures of belted cables at these high voltages, Hochstadter proposed encasing each core in its own metallic screen. This simple precaution was widely adopted. It completely eliminated slow destructive discharges which had caused the trouble and brought in a new era in high voltage cable technique.
(*Photo: Associated Electrical Industries Ltd.*)

(*b*) 110 kV 3-core Oil-filled Cable, about 1950
Emanueli's proposal, made about 1920, for filling high-voltage cables with a fluid oil maintained under continuous pressure from both ends, was a revolutionary idea and proved to be most effective. With cables of this kind higher and higher voltages have become possible. The illustration shows the construction of an oil-filled cable with the oil ducts in the form of spiral metal spacers.
(*Photo: Associated Electrical Industries Ltd.*)

(*c*) 380 kV Oil-filled Cable, about 1955
In recent years oil-filled cable technique has reached a point where any transmission line voltage can be carried underground.
(*Photo: Associated Electrical Industries*

at the Covent Garden Opera House and provincial supply companies began to use the type extensively. 'V.B.' cables provided a complete answer to the fear of moisture ingress underground and, as its use extended, a method of laying in troughs filled up with bitumen was adopted. Troughs of iron or wood were provided with wooden bridge pieces shaped to carry the cables and after individual lengths had been jointed together, and the joints insulated with bitumen tape, the troughs were filled with molten bitumen and the flanged covers placed in position. Hundreds of miles of this 'solid system', as it was called, were laid in its various forms well on into the present century.

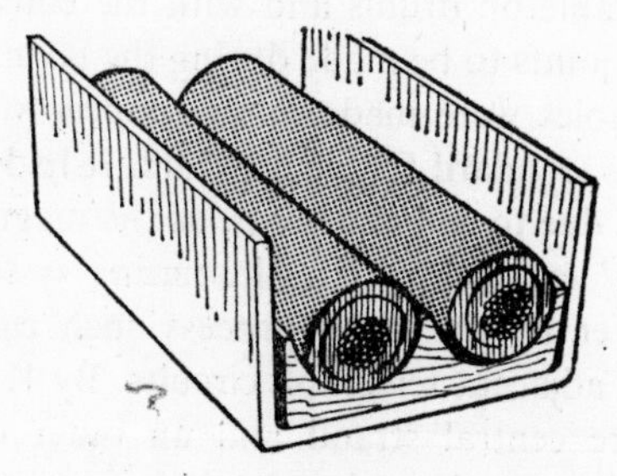

Fig. 41. Callender Solid V.B. System. From the early days of public lighting in London, vulcanized bitumen insulated cables were laid in troughs filled with hot bitumen underground. They gave good service for many years.

A type of power cable which had a vogue for some time because of its simplicity in manufacture was the varnished cambric cable with a lead sheath. The insulating material was a cotton tape impregnated and coated with varnish composed of linseed oil, gum resins and 'driers'. The cloth was little more than a mechanical carrier for the highly insulating varnish. To facilitate sliding of the layers during bending of the cable a 'slipper' compound was applied during manufacture, but this amount of compound was not sufficient to cause trouble through draining when the cable was suspended in a vertical position. Today non-draining cables are mostly made of the usual impregnated paper design, but with specially viscous compounds.

The widely used impregnated paper cable of the present day in which the dielectric is lapped spirally with paper strip, impregnated with an insulating oil, and protected against moisture by a continuous lead sheath was first introduced in the United States in 1886 when the Norwich Wire Company was formed to make such cables. Three years later a Mr. Atherton, of Liverpool, became interested in the process while on a visit to America and on his return secured the co-operation of electrical men including Ferranti, in the formation of the British Insulated Wire Company. A factory was established at Prescot and by the end of 1891 the manufacture of both low- and high-voltage cable had been commenced.

As already described in Chapter X, Ferranti had made and laid his rigid 10,000-volt cable to link the new Deptford Station to the Grosvenor Gallery. In this he had used impregnated paper but the application of paper in strip form, now proposed, enabled a flexible cable to be produced, a cable which could be manufactured in long lengths transportable on drums and with the consequent reduction in the number of joints to be made during the laying operation.

Early paper cables were made in the concentric form and this type continued as standard for many years. It had the advantage of low sheath losses due to eddy currents, as the magnetic effects of the 'go' and 'return' currents on an alternating system cancelled one another; moreover by the same process such cables caused little interference with adjacent telegraph circuits. By 1893 a construction with a seven-wire central strand and an outer conductor of flat copper strips had been adopted and other paper cable factories had been set up both in this country, in the United States and on the continent of Europe.

The use of a lead sheathing to prevent ingress of moisture to a fibrous dielectric was not new at this period. In 1882 the Berthoud-Borel Co. had been formed on the Continent with the object of manufacturing jute insulated cables. These were constructed of layers of jute twine laid over the central conductor with consecutive layers having opposite directions of lay. The covered core was then impregnated with a hot compound consisting of boiled linseed oil and resin and lead sheathed. The jute cable was used extensively on the continent but never competed seriously in this country with the bitumen cable, although to a limited extent it was used up to recent times.

With the development of the systems and particularly with the introduction of three-phase working, the practice was established of enclosing within one common lead sheath two or three separately insulated conductors. The conductors often continued in the form of circular strands, but the economic advantages of shaping the strands in oval or clover-leaf form were soon recognized. As early as 1900 the cables from the new power station at Wood Lane to supply the Kensington and Notting Hill Companies at 5,250 volts were made on the clover-leaf construction.

The voltage for which underground paper cables were required rose from the beginning of the century until by the time of the First World War 22,000 volts was considered to be a commercial proposition. Various refinements were introduced. For instance as the thickness of dielectric between two cores was twice that between core and earth, a disparity considered undesirable, the thickness of the core insulation was reduced and the core-to-earth thickness increased again by applying a belt paper over the cores after laying up. This construction became standard and was widely adopted for voltages up to 33,000. Much attention was given to such refinements as methods of paper lapping, construction of wormings and details of the impregnating process and there was general confidence in a policy of raising the voltage and meeting the requirements by increasing the dielectric thickness, but warnings came in the form of slow failures. Cables were soon breaking down after many months, in some cases years, of service. It became evident that in some way the step-up from 20 kV to 30 kV had introduced dangerous conditions.

Much concern was caused among both cable-makers and users by the new situation. Dielectric science received a boost which it had never had before and intense investigations were carried out in laboratories set up on both sides of the Atlantic. During the war Martin Hochstadter suggested a type of three-core cable in which each core was provided with an external lapped metallized tape. The tape was perforated to permit the flow of oil during the impregnating process and the effect was to provide an equipotential surface over each core, so ensuring a radial stress in place of the previous uncertain combination of radial and tangential stresses. Although the mechanism of failure was only elucidated within the few years following the war, it was the Hochstadter cable which immediately provided the practical solution to the mysterious breakdowns even before the cause was properly understood. It did so by eliminating the slow burning of the papers at the surface of the cores due to the imperfect contact of the cores with one another. No longer was separation of the cores, due to mechanical deformation, dangerous. There was now no stress in the space. In an alternative three-core cable known as the S.L. type, the same result was achieved by constructing the three-core cable of three single lead-covered cores twisted together under one common armouring. The 'H' type became very popular and opened

the way to still further advance in the voltage which could be carrie
on underground cables.

Many improvements resulted from the intensive research an
development work set in motion in the cable industry between th
two wars[2] Much was learned about the power factor of a dielectri
and its change with voltage and temperature. Different oils an
compounds were tried out to improve the breakdown strength an
stability and the vacuum-impregnating process was made mor
scientific after many decades based on rule of thumb.[3] The fetis
for manilla paper gave way to the informed use of wood-pul
paper of high purity and specified porosity. One of severa
major mechanical improvements introduced into the build-u
of the cable was the pre-spiralling of stranded shaped conductors
Formerly the conductor was first paper-lapped and then twisted i
the laying-up process. This resulted in serious distortion of th
papers but at the end of the war an American company adopted th
method of first twisting the conductor and then lapping on th
papers, no further twisting being necessary on laying up. Within a
few years the practice spread to cable factories all over the world.

One of the processes in the manufacture of lead-sheathed cable
which had considerable influence on progress was the application o
the sheath. About the year 1800 the first attempts were made t
extrude a continuous lead covering on to a cable. A tube larger tha
the cable was adopted and subsequently this was drawn down i
order to fit the core. Within a few years various types of hydraulic
press were invented for the purpose and to avoid the subsequent
drawing process. At an important stage in cable history one of its
interesting features was the lack of confidence which arose in connection with the lead sheath. Risks of water percolation were so great
that some cables were made with a double sheath, the inner one being
provided with ribs so that a space between the two could be filled
with a waterproofing compound. Many inventors, led by Robertson
of the United States, studied the flow of the metal in a separate steel
die block through which the cable passed as the heated lead flowed
through carefully designed ports on its way to the annular space
between the point and die where the tube emerged in contact with the
lead. In this process of improvement in lead covering the years
following the First World War saw many advances. Devices such as
the Judge straight-through lead press and the Pirelli continuous extruder survived other original proposals by Dunsheath designed to
produce a seamless tube, and are in wide use today.[4]

Mention must also be made of the use of aluminium for cable sheaths. The weight of lead as a sheathing metal has always been considered a disadvantage and attempts have been made from time to time to substitute the light metal aluminium with its stronger mechanical characteristics. The ductility of aluminium being much less than that of lead the problems of application to the cable are, of course, greater and different processes have been adopted in attempts to overcome this disadvantage. In one method which has been used to a considerable extent since about 1947 the cable is drawn into a tube with a larger diameter and the tube subsequently reduced in diameter by being drawn through a steel die. The direct extrusion of aluminium has been adopted to some extent in spite of the serious technical difficulties, but the metal shows no sign of replacing lead at present.

Although these many improvements increased the reliability and extended the voltage range of normal cable design a very fundamental change which was introduced in 1920 calls for special comment. Luigi Emanueli, an Italian, designed a cable in which the viscous compound, formerly standard practice, was replaced by a thin mineral oil.[5] The conductor was built up on a spiral tape which provided a duct along which the oil could flow longitudinally to compensate for volume changes due to heating and cooling on load and at intervals along the route reservoirs to accommodate the surplus oil were provided. From the very start this cable was a great success and has been installed all over the world for the highest voltages. The design went straight to the root of the trouble as the stresses were increased—the ionization of voids within the dielectric. Emanueli eliminated the voids by keeping the cable continuously full of oil under a small pressure.

A few years after the introduction of the oil-filled cable Bennett in the United States invented the Oilostatic system in which screened single cores are drawn into a steel pipe which is afterwards filled with insulating oil maintained under pressure. This type has been entirely successful but inventors have been no more content to accept finality in this field than in any other.

A gas under pressure is also capable of reducing the risk of ionic discharge within a dielectric, a fact appreciated by Fisher and Atkinson in the United States in 1920. Only during the past few decades have British manufacturers taken this idea seriously and a number of modified types of gas cables have been introduced with various advantages, viz., the compression cable, the gas-cushion cable, the flat-pressure cable, the gas-filled cable, and the impregnated gas-pressure cable. The relative merits of these different types are fairly assessed in Hunter and Hazell's book and while it is difficult to summarize the position today it may be indicated by the fact that in Vancouver an installation of 14 miles of oil-filled cable was completed in 1957 to carry 700 amperes A.C. at 230 kV,[6] and that a scheme is under way (1960) for connecting the French and British power systems by submarine power cables across the English Channel. This will be a D.C. link transmitting 160,000 kW, equivalent to 800 amperes D.C. at 200 kV, and will connect the 275 kV British super-grid with the French 225 kV system. An excellent summary of 25 years' development has recently been given by Herman Halberin.[7]

It is evident that the weakest part of a power cable system is in the joints. The cable is made under cover and carefully controlled factory conditions which ensure uniform quality. The joints, however, have to be made in the open, or at best with temporary shelter from the weather. Cable-makers have recognized the risks due to faulty joints and from the earliest times have devoted much thought to the problem. Special joint boxes were designed for the Edison system in the 1880's and, as the voltages carried by underground cables rose, so did the attention paid to detail increase. From about 1920 the cable-makers intensified their development work in this field and today the subject is covered by detailed designs and instructions on a wide scale. A good illustration is to be found in the instructions issued in such publications as that put out by Henleys,[8] which has reached several editions.

Retracing our steps to the introduction of rubber-insulated cable in 1859 it is interesting to note that the design of this type of cable did not change to any material extent over a period of fifty or sixty years. A layer of natural rubber was applied directly over the conductor and two layers of vulcanized rubber compound were superimposed. The design no doubt originated from the fact that pure rubber has excellent electrical properties but is liable to take up moisture and swell when exposed whereas vulcanized rubber is much more moisture resisting. Moreover the sulphur used in the compound of the vulcanized rubber attacked the copper conductor, and the pure separating layer helped to prevent this action which was also combated by tinning the wire. The various theories on this question were discussed by Beaver in his excellent paper in 1914[9] and his book in 1926.[10]

As more became known about the ageing properties of rubber and reliability of rubber cables in service so the science of rubber cable manufacture advanced and, between the wars, in parallel with the application of science to the impregnated paper cable, rubber cables were greatly improved. Certain ingredients in the complex rubber compounds were traced as the cause of long time failure by a special committee of manufacturers set up by the Admiralty. New chemical substances known as inorganic accelerators and de-oxidants became available and during the Second World War, owing to the loss of

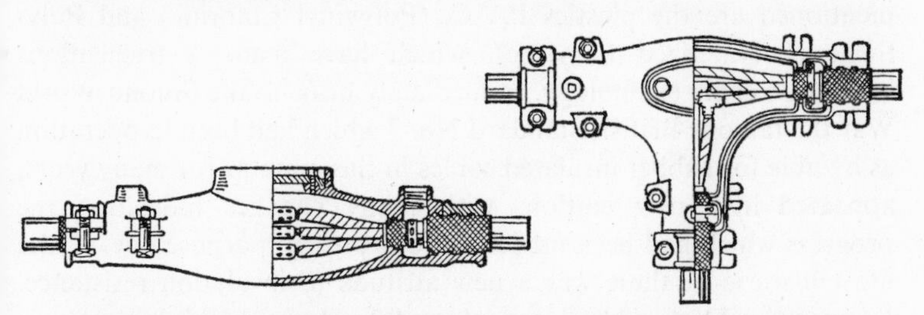

Fig. 42. Cable Joint Box from c. 1920. The extensive use of impregnated paper cable led to the design of a wide range of convenient iron joint boxes suitable for compound filling.

Malaya, frantic efforts had to be made to replace natural with artificial rubber for cable-making. Neoprene (Polychloroprene), a man-made substitute, was found to have special features, one of which was its fire-resisting properties, and attained wide recognition for cable work.

Other rubbers which became famous dielectrics were butyl rubber (isobutylene and isoprene), which was found to be capable of operating at much higher temperatures than natural rubber, up to 85° C., and was more resistant to deterioration by ozone, and silicone rubber, made by an entirely new process which resulted in most dramatic heat-resisting properties—indefinite use at 150° C. When destroyed by fire silicone-insulated conductors were found to retain sufficient of the pure silica to leave a covering of the wire and keep it insulated. Further, silicone wires withstood exceptionally low temperatures satisfactorily.[11]

A major modification of rubber cable manufacturing methods is that known as continuous vulcanization which has been introduced during the past twenty years. The method replaces former methods in which strips of rubber were rolled on to the conductors longitudinally by grooved rollers. The extrusion process is employed for covering and the covered conductor passes direct from the extruder into a long metal tube filled with high-pressure steam. Here the vulcanizing process is carried out as the wire passes along before it emerges into the air at the far end of the machine. The process is continuous and there is no stoppage even for connecting on new lengths of wire in front of the extruder. As machines are replaced no doubt the continuous process will supersede all former methods.

The complete list of substitutes for natural rubber which the cable industry has evolved is a long one. In a different class from those mentioned are the plastics P.V.C. (Polyvinyl Chloride) and Polythene (Polymerized Ethylene), which have made a tremendous impact on cable technology. Immediately prior to the Second World War the famous British Standard No. 7 which had been in operation as a guide for rubber insulated cables in this country for many years, appeared in a new edition, with many changes indicating the progress which had been made. For our present purpose one of the most historic of these was a new attitude to insulation resistance. Formerly rubber cables were classed by 'megohms' but voltage requirements were substituted. A Commonwealth Conference in 1953 formulated new standards for rubber cable compounds which abandoned the former high insistence on electrical insulation resistance and a new British Standard which allowed a rubber content of 50 to 60 per cent was adopted.

Thus, in this apparently humdrum branch of electrical engineering history, continuous and fundamental progress has been made. The modern successor to the crude rubber or bitumen cable of half a century ago, as exemplified by the plastic insulated and sheathed cable with its wide variety of pleasing colours for identification purposes, has certainly added both precision and beauty to utility, and in the field of paper insulated cables recent spectacular advances to higher and higher voltages show no signs of abating.

REFERENCES

1. Hunter and Hazell, *Development of Power Cables*, p. 51. London, 1956.
2. Dunsheath, 'Science in the Cable Industry', *J.R.S.A.*, vol. 54, p. 506, 1926.
3. Dunsheath, 'Dielectric Problems in High Voltage Cables', *J.I.E.E.*, vol. 64, p. 97, 1926.

4. Dunsheath, 'The Continuous Extrusion of Cable Sheaths', *J.I.E.E.*, vol. 80, p. 353.
5. L. Emanueli, *High Voltage Cables*. London, 1929.
6. *Electrical Engineering*, vol. 77, p. 816, 1958.
7. 'Insulated Power Cables', *Electrical Engineering*, vol. 78, p. 604, 1959.
8. Henley, *Practical Cable Jointing*. London.
9. Beaver, *J.I.E.E.*, vol. 53, p. 69, 1914.
10. Beaver, *Insulated Electric Cables*. London, 1926.
11. B. B. Evans, *Distribution*, vol. 31, pp. 257, 283, 1959.

CHAPTER XVII

The Electron in Engineering

The application of the electron forms the basis of a vast branch of modern electrical engineering and in some of the preceding chapters, particularly Nos. VIII and XV, the subject has already received consideration in the fields of illumination and telecommunication. There remains, however, an important part of electrical history to recount in tracing the development from J. J. Thomson's remarkable discoveries which in 1897 established the free electron as a vital factor in modern engineering practice.

In the year 1883 Edison had observed that the bulb of a carbon filament lamp became coated with a black deposit. He also discovered that when a metal plate was suspended inside the bulb and connected externally through a galvanometer to the positive end of the filament, a steady current flowed into the plate as long as the filament was heated.

In 1904 Professor J. A. Fleming, of University College, London, after studying both the Edison effect and the work on discharge through gases being carried out by Thomson and his collaborators at Cambridge, saw an application in the new field of wireless telegraphy in which Marconi was then making such rapid strides. But to appreciate fully the significance of Fleming's idea it is necessary to trace first the course of another tributary to our main stream, this time the development of wireless telegraphy itself.

The idea of transmitting intelligence electrically without wires may be said to have originated in the experiments of Morse, who in 1842 succeeded in sending signals across a canal by means of four sunken earth plates, two in the transmitting circuit on one bank and two in the receiving circuit on the other bank.[1] The system was applied practically by Sir William Preece twenty-five years later and again by field units in France in the 1914–18 War. In this case the enemy discovered that by sinking earth plates he also could pick up the signals, although the employment of Aberdonians to speak and receive the messages introduced a simple but effective code which overcame the objection.

Earth conduction systems had no great practical use for long-distance signalling, and it was only when electric or Hertzian waves were employed that any significant advance took place. In 1864 Maxwell had predicted that electromagnetic waves could travel through space with the velocity of light and it was known through the work of Henry, Kelvin, Kirchoff, Heaviside, Helmholtz and others that the frequency of the oscillatory discharge of a charged condenser was controlled by the capacity and inductance of the circuit including the spark gap.

With this knowledge in mind, Heinrich Hertz, a professor at Karlsruhe, carried out in 1886 a series of experiments of great fundamental importance to the future of radio-telegraphy.[2]

A predecessor at Karlsruhe, for some reason unknown, had made some flat coils of wire insulated with sealing wax mounted coaxially and with the ends brought out to open terminals. Hertz observed that when a Leyden jar was discharged through one of these coils a spark passed across the gap between the ends of the other coil. The idea of 'tuning' one circuit to another by adjusting the capacity and inductance occurred to Hertz and he quickly realized that, in this way, resonance could be obtained and more powerful results achieved. The Leyden jar became unnecessary and the effect could be obtained by simply connecting an induction coil direct to one of the coils.

Hertz extended his investigations over a comprehensive range of experiments with powerful results on our knowledge of high-frequency oscillatory currents and spark discharge, but for our present purpose we need only record the two main achievements. He found in the first place that the primary circuit, instead of being a coil, could comprise a straight conductor with a spark gap at the centre of its length, and terminating at each end in a large metal ball. Secondly, and most fundamentally, he established the fact that the transmission mechanism between the primary circuit and the secondary detector coil was a straightforward wave motion in space. He measured the length of these electromagnetic waves, related the wavelength to the frequency of the oscillatory current in the transmitter, and demonstrated such phenomena as reflection, interference, standing waves, and so on. He showed that the waves were transmitted with the velocity of light, 300,000 kilometres a second and, what was so important, the electromagnetic action did not fall off as the square of the distance.[3] He soon found it possible to detect the wave with his resonator—a small coil with a tiny spark gap—at a distance of many metres with original frequencies of the order of a hundred million a second producing waves of only a few metres in length.

Hertz carried out over many years a wide range of investigations in the fields of high-frequency oscillatory circuits and the propagation of electromagnetic waves, as evidenced by his writings in the *Annalen der Physik und Chemie*, collected as a Treatise in 1891[2] and by the comprehensive modern display of his original apparatus in the Deutsches Museum in Munich. It is clear, however, that he did not see the possibilities of employing Hertzian waves for radio-communication. The question was raised by a correspondent, Huber, and Appleyard, in his book quoted, reproduces a reply sent by Hertz on 3 December 1889, in which he says 'If you could construct a mirror as large as a Continent, you might succeed with such experiments, but it is impracticable to do anything with ordinary mirrors, as there would not be the least effect observable.'

Other students of the work of Hertz were more optimistic. In the year following this letter Professor Branly of Paris noticed that an electric spark occurring near an ebonite tube of metal filings increased their conductivity appreciably.[4] Such a tube of metallic powder does not normally pass the current when placed in circuit with a battery and galvanometer, but after being subjected to the rapid oscillatory current from a condenser discharge the conductivity is increased to such an extent that appreciable deflections are obtained on the galvanometer. This device at once offered a much more sensitive detector of electromagnetic waves than Hertz's spark gap.

On the death of Hertz in 1894 Sir Oliver Lodge delivered a memorial lecture at the Royal Institution,[5] at which he repeated many of Hertz's experiments, gave the name 'coherer' to Branly's detector, and showed that two metallic spheres in light contact with one another displayed the same phenomenon. He also demonstrated a de-cohering device which automatically restored the high resistance of the coherer through mechanical tapping by an electric bell mechanism in the battery circuit, a proposal also made by the Russian, Popoff.

In the year 1899 it became known[6] that, some years before Hertz's fundamental discovery of electromagnetic waves in 1886, Professor Hughes, the inventor of the microphone, which he had used in telephony, had observed 'invisible electric waves which evidently permeated great distances'. These came from a coil or frictional electric machine and were detected on sensitive microphone contacts. From 1879 to 1886 he had carried out an extensive investigation in this field and had demonstrated aerial transmission over distances of several hundred yards to a number of well-known scientists. The results created great interest. Professor Stokes, one of the two honorary secretaries of the Royal Society, however, after a successful three-hour demonstration, considered that there was no evidence for electromagnetic waves, but that all the results could be explained by known electromagnetic induction effects. This so discouraged Hughes that he refused to publish his results and so left the field clear for Branly and Marconi, to whom he later paid a generous tribute.

There is no doubt that the credit for the application of Hertz's discoveries to practical radio telegraphy must go to Guglielmo Marconi, an Italian who, within a year of Lodge's Royal Institution lecture, transmitted signals in Bologna, a distance of over a kilometre. Marconi first of all improved the coherer as a detector of electromagnetic waves by reducing its dimensions and employing nickel and silver filings between two silver terminal blocks in a sealed glass tube from which he evacuated the air. For the transmitter and receiver he adopted long vertical wire aerials suspended from masts or towers, the transmitting and receiving devices being inserted between the aerial and earth plates.

In his early transmitter Marconi connected the secondary terminals of an induction coil across the sphere spark gap in the aerial circuit and inserted a sending key in the primary battery circuit. The coherer in the receiving aerial circuit had a battery and telegraph sounder connected across it. With this elementary but effective form of wireless telegraph Marconi applied for a British patent and approached Sir William Preece, the Engineer in Chief of the Post Office.

From 1896 there commenced a series of successful experiments in which the British Post Office gave Marconi every facility. The question of tuned circuits, known at first as 'syntony', assumed importance in the work of Sir Oliver Lodge and others and trials were carried out on Salisbury Plain and elsewhere. In 1897 an eight-mile circuit was set up across the Bristol Channel, to be followed by an eighteen-mile link from Poole to the Isle of Wight. The first commercial use of tuning was in providing two simultaneous transmissions between Poole and St. Catherine's Point, a distance of 30 miles, and by March 1899 equipment had been improved to such an extent that communication was established across the Channel between the South Foreland Lighthouse and Wimereux, near

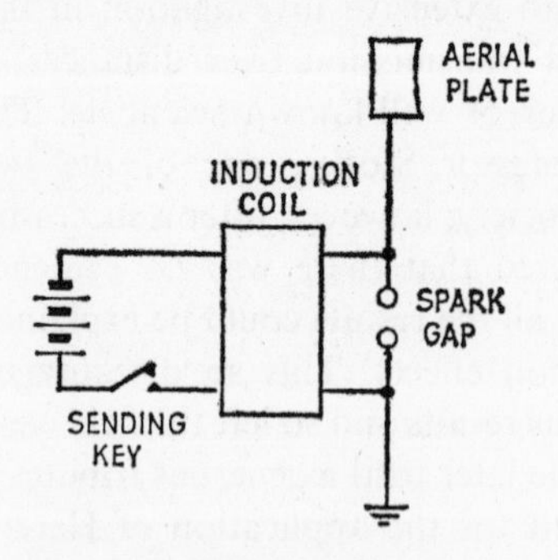

Fig. 43. Marconi's First Transmitting Circuit. In 1894 Marconi succeeded in transmitting feeble signals by means of a spark gap in the aerial circuit.

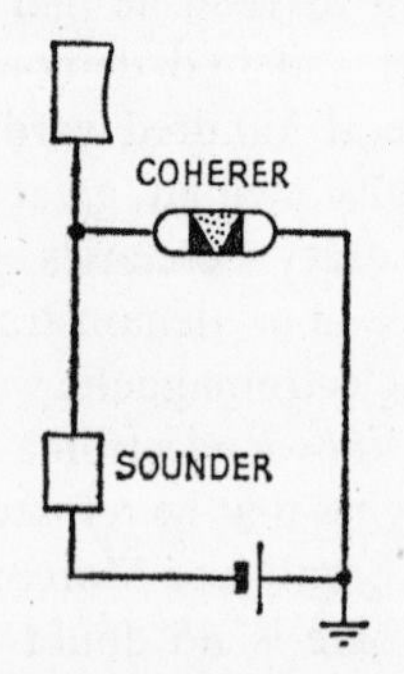

Fig. 44. Marconi's First Receiving Circuit. This employed the coherer and there was no tuning.

Boulogne. Later the same year, during a lecture before the British Association meeting in Dover, radio communication was established with the French Association meeting simultaneously in Boulogne.

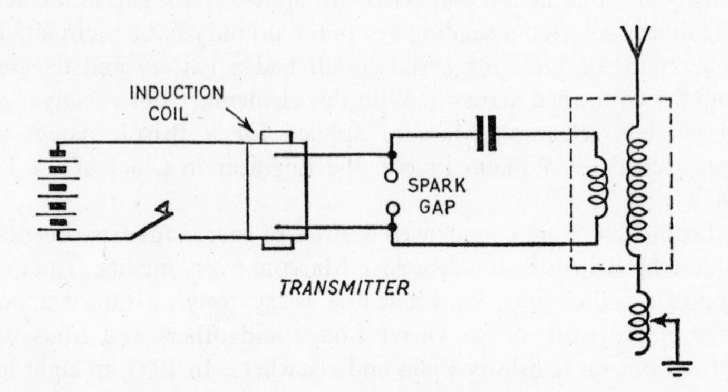

Fig. 45. Marconi's Major Circuit Modification. Removal of the spark gap from the aerial circuit in the transmitter to an oscillatory circuit provided long trains of waves.

In 1900 Marconi made a major modification of the transmitting circuit. The spark gap was removed from the main aerial and placed in an oscillatory circuit consisting of a condenser of several Leyden jars in parallel and the single turn primary of a transformer, the secondary winding of which had many turns and was in the aerial-to-earth circuit. The energy was fed in from the induction coil across the spark gap and an adjustable tuning inductance in the aerial enabled long trains of waves to be produced. Similar tuning facilities in the receiving aerial increased the effectiveness of transmission to such an extent that very soon distances of several hundred miles could be covered satisfactorily.

The spark system of Marconi having made such strides, he was encouraged to contemplate bridging the Atlantic and in 1900 the first high-power radio station was constructed at Poldhu in Cornwall. In place of the induction coil fed by a voltaic battery an alternator driven by a 25 h.p. oil engine worked through transformers to give a voltage of 20,000. The aerial system consisted of 50 wires supported on masts 200 feet high. Towards the end of 1901 Marconi left for St. John's, Newfoundland, and set up temporary aerials by means of kites and balloons, with which, on 13 December 1901, he received the agreed three-dot signals which were being transmitted from Poldhu. The wavelength used was from 2,000 to 3,000 feet.

In 1902 Marconi invented a new type of detector to replace the coherer which practice had shown to have certain defects. In the magnetic detector a loop of fine iron wires was made to travel slowly down the centre of two concentric coils which thus formed a transformer. One of these coils was connected in the receiving aerial circuit and the other to a telephone receiver. Near to the moving wire were disposed two horseshoe permanent magnets which thus magnetized the wire as it travelled along. When the high-frequency current from the receiving aerial passed through the primary coil it interfered with the magnetism in the tape and produced a 'click' in the telephone. As the wave trains arrived at a frequency of some hundreds a second, the result was a continuous musical note in the telephone as long as the waves were arriving. By subdividing these on the Morse code system through the transmitting key, telegraph messages were received by the operator wearing headphones. The magnetic detector soon completely replaced the coherer.

To limit the disturbance due to atmospherics and to raise the pitch of the received signal higher than was possible with the make and break mechanism of the induction coil, different forms of interruptor were sought. Rotary contactors, in the form of discs with external spokes passing near to fixed studs, were driven by the alternator shaft and became common practice.

The main technical problems in providing long-distance radio telegraphy having been overcome, the Marconi's Wireless Telegraph Company quickly established radio stations around the coast of Britain and other countries and equipped large numbers of ships. In 1910 those in this country were taken over by the Post Office and a regular ship-to-shore service established as part of the National telegraph system. Dramatic rescues were achieved in a number of accidents at sea and by the beginning of the First World War this new phase of electrical engineering was firmly established both technically and commercially.

It was soon recognized that transmission by bursts of damped vibrations was not satisfactory and considerable thought was given to replacing spark discharge by a continuous wave system. In 1900 W. Duddell had suggested using the electric arc to produce persistent electric oscillations by coupling it to a resonant circuit, but it was only when Poulsen, a Danish engineer, in 1903 enclosed the arc in hydrogen gas and placed the arc in a powerful magnetic field that the system was entirely successful and produced undamped oscillation at a frequency of a million a second. In another system very high-frequency alternators were developed by Alexanderson in the United States, Latour in France and Goldschmidt in Germany producing frequencies up to 100,000 cycles a second, suitable for generating powerful undamped waves of over 10,000 feet wavelength. Until the application of the high-power thermionic valve these generators were widely used in long-distance radio stations.

At the beginning of this chapter we saw that although the development of the thermionic valve originated in the Edison effect, it had to await the completion of the early stages of 'wireless'. Having traced the latter through its first few decades built on spark telegraphy we now see the two streams come together in Ambrose Fleming's invention of the diode in 1904. In his early associations with Marconi and studies of the work of Hertz, Fleming had seen the need for a rectifier which would convert the received trains of high-frequency oscillations (a million per second) into trains of intermittent but unidirectional current. The former would have no effect on the most sensitive telephone receiver whereas the rectified current would produce a detectable musical note in the receiver with the frequency of the trains of sparks (about 100 per second). J. J. Thomson

Judge Straight-through Cable Press, about 1930

Appreciation of the risks of defects in extruded lead cable sheaths resulted in extensive study and development of new designs of press. In the Judge press the problems were successfully overcome by passing the mass of molten lead around the cable to make the "joint." The slug was subsequently extruded with less risk of "split-lead." (*Photo: Associated Electrical Industries Ltd.*)

(*a*) Electrolytic Refining of Aluminium
Electrolytic processes in industry have developed extensively from Davy's original discoverie made 150 years ago. Based on Heroult and Halls discovery in 1886, aluminium is extracted i large quantities by electrolysis. Heavy currents through carbon rods separate the pure meta from the oxide. The metal falls to the bottom and is siphoned off as shown.
(*Photo: British Aluminium Co. Ltd.*)

(*b*) 3-electrode Steel Furnaces, about 1930
A tilting refractory-lined vessel containing the ore is heated by heavy 3-phase currents passing through massive carbon rods and the refined metal is poured into crucibles. (*Photo: British Iron & Steel Federation*)

had elucidated the Edison effect by showing that electrons or negatively charged particles were emitted from the hot filament in a high vacuum electric bulb and Fleming recognized the possibilities of a solution of his problem. He had some lamps constructed in which a connection to a metal cylinder surrounding the filament inside the bulb was brought out as a third terminal. By connecting the negative end of the filament and the 'plate', as the cylinder was called, in the circuit carrying the received high-frequency current he found that, as anticipated, the current was rectified and could be detected by the telephone receiver.

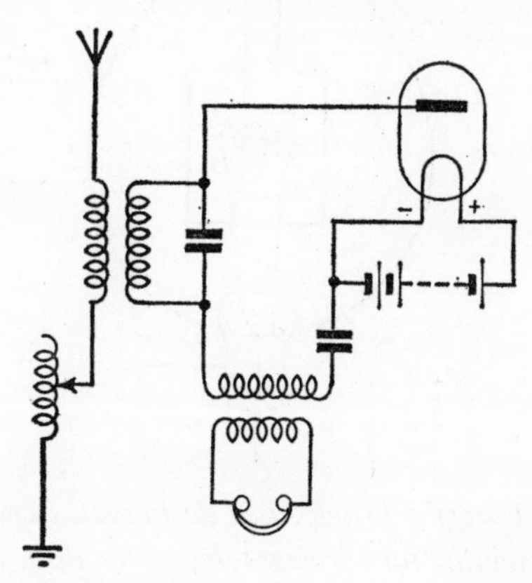

Fig. 46. Early Radio Receiving Circuit with Diode. The diode, invented by Fleming in 1904, provided a valuable device for rectifying weak radio signals, making them detectable by a telephone receiver.

In view of its wide popularity among amateurs in the early days of wireless mention must be made of another and very simple type of rectifier. This was the crystal or 'cat's whisker'. It was found that certain crystals in contact with one another possessed a better conductivity in one direction than the other. Consequently they converted an alternating current into a unidirectional current. In the popular form a metallic wire was held in light contact with carborundum and could be moved about to obtain the most sensitive spot. As we shall see later the great transistor development arose out of this simple early device.

Two years after Fleming announced the diode, that is in 1906, Dr. Lee de Forest in America made a fundamental improvement to Fleming's valve by adding a third electrode between the filament or hot cathode and the plate which has become known as the grid. The introduction of the grid by de Forest gave the thermionic valve its most valuable feature, the ability to amplify a signal, and it was this feature which formed the basis of the subsequent phenomenal application of the valve in radio communication and in other fields.

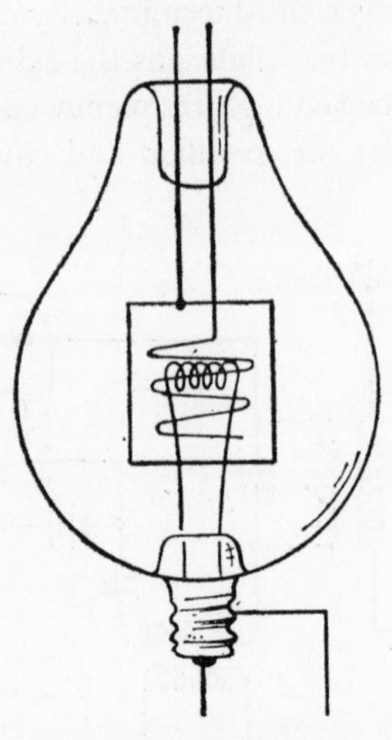

Fig. 47. De Forest's Triode. Lee de Forest, experimenting in 1906 with detection for wireless telegraphy, discovered the possibilities of a thermionic valve in which a grid of wires was inserted between the filament and a nearby plate. His patent specification is not very clear but seems to embody the construction shown above.

Lee de Forest had been working for some years on detectors for wireless telegraphy, using a gas flame between platinum electrodes in a battery circuit.[7] One day he had a filament lamp made with a plate inside but still employing the battery circuit with a telephone to detect changes in conductivity caused by the tiny aerial currents. He appears to have had no thought of rectifying the received high-frequency alternating current, but among his various arrangements was one in which the aerial was connected to a strip of tinfoil attached to the outside of the glass valve. He then had a valve made in which two plates were inserted, one on each side of the hot filament. This led him to place one of the plates between the filament and the other plate and subsequently to the important stage of making this in the form of a wire grid.[8] A patent application in

October 1906 covered simply the amplification of feeble electric currents, but three months later he filed another relating it specifically to wireless telegraphy. The amplifying properties of the triode are illustrated by the characteristic curves now so well known.

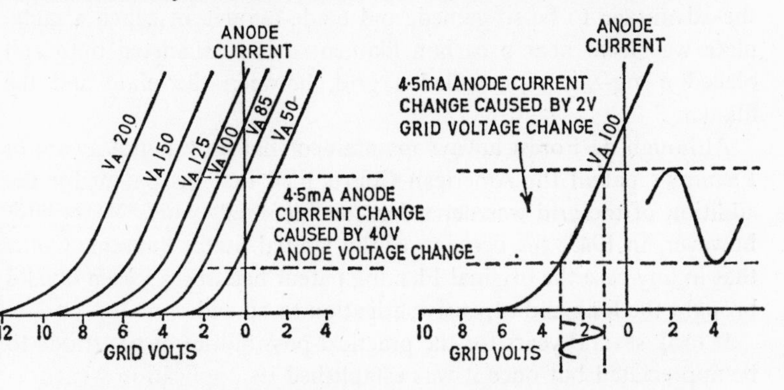

Fig. 48. Graph of Thermionic Valve Characteristics. The characteristic curves of the Triode, now so well known, illustrate the feature of amplification which de Forest employed without at the time fully appreciating its significance.

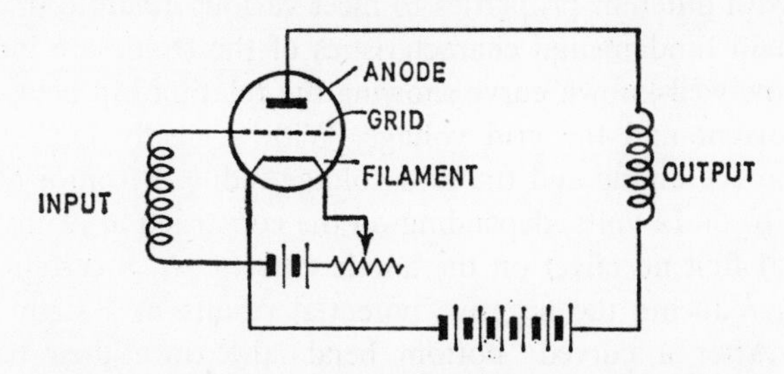

Fig. 49. Simple Amplifying Circuit. The simple application of the triode for radio reception which developed from de Forest's discovery.

As was shown by Professor G. W. O. Howe in a lecture on the genesis of the thermionic valve,[9] the subsequent history of the relations between Fleming and de Forest was a very unfortunate one. A confused situation continued for years during which time they were involved in endless litigation and very much bitterness. That Fleming fully recognized the fundamental difference between his own and de Forest's contribution, however, is clear from the following extract from his autobiography.[10]

'Sad to say, it did not occur to me to place the metal plate and the zig-zag wire in the same bulb and use an electron charge, positive or negative, on the wire to control the electron current to the plate. Lee de Forest, who had been following my work very closely, appreciated the advantage to be so gained, and made a valve in which a metal plate was fixed near a carbon filament in an exhausted bulb and placed a zig-zag wire, called a grid, between the plate and the filament.'

Although de Forest always maintained that he was not aware of Fleming's patent the American Courts held that his patent for the addition of the grid was dependent on Fleming's work. Years later however, in 1943, the decision of the United States Supreme Court that in any case the original Fleming patent had always been invalid brought the long and expensive litigation to an end.

It took several years for the practical possibilities of the triode to be appreciated but once it was established its application went forward in leaps and bounds, encouraged to a great extent by the service requirements in the First World War. The possibility of employing amplifying valves in cascade provided a device of fantastic sensitivity and in parallel with this extension came a vast range of circuits with different properties to meet various conditions.

The main fundamental characteristics of the triode are indicated by the now well-known curve showing the relationship between the anode current and the grid voltage. With a steady D.C. voltage applied to the anode and the grid voltage reduced from a negative value of 10 or 12 volts, depending on the construction of the valve, there is at first no effect on the anode current. At a certain value, however, reducing the negative potential results in a slight anode current. After a curved 'bottom bend' the trace then becomes straight and remains so for a considerable range of grid voltage until it again flattens out at the 'top bend'. There is a family of such characteristic curves for different values of the anode voltage, the curve moving to the left as the anode voltage is increased.

Considering one of these curves, say the one for 100 volts on the anode, a change of 1 volt on the grid produces 2 mA change in the anode current. To produce the same change in the current by altering the anode voltage would require something like 40 volts. Thus the amplification of this particular valve would be 40 and the apparent or mutual conductance of the valve under these conditions would be 0·002 amperes per volt, a value obviously much greater than that calculated for the load circuit based on the anode voltage alone.

The usefulness of the triode, however, is in the amplification of alternating currents and the same curve can be used to illustrate how this takes place. With a steady positive voltage applied to the anode, say 100 volts, a small alternating voltage of say 1 volt applied to the grid produces an alternating voltage in the anode circuit and an alternating current of amplified magnitude. But it was discovered in the early days of radio that not only could the triode amplify the tiny received signals but it could also rectify them, as was done by the diode, and so make them detectable by the telephone receiver. This

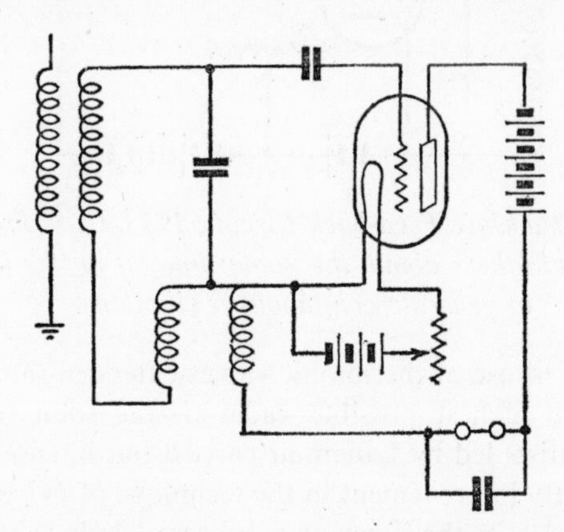

Fig. 50. Armstrong's Circuit for Simultaneous Rectification and Amplification, 1915. By running the triode with the grid potential maintained in the 'bottom bend', Armstrong produced a unidirectional amplified current.

was done by Armstrong[11] by maintaining the steady potential of the grid at such a point in the 'bottom bend' as just to stop the anode current. When the alternating voltage was received only half the cycle was thus effective and the amplified current was unidirectional. A simple way of achieving the required grid potential without using a special battery was by the employment of a high resistance grid leak which permitted a charge to accumulate from the action of the anode voltage and so achieve the result without interfering with the effectiveness of the incoming signal.

Just prior to the First World War the phenomenon of feed-back was discovered by C. S. Franklin and several inventors hit on the use of the triode for generating oscillations. In a simple oscillating circuit with reaction coupling, or feed back, the condenser of the primary oscillator is connected to the grid of the triode and the anode circuit is led back through a reaction coil linking with the inductance of the primary circuit. In this way the oscillations are reinforced and instead of dying away are maintained at the level where the energy fed back equals the resistance losses.

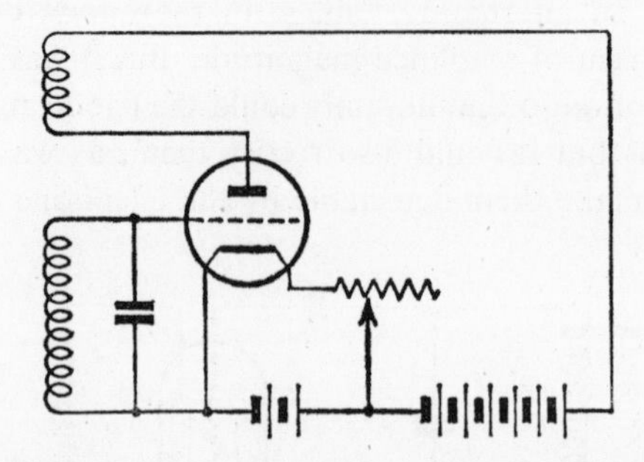

Fig. 51. Franklin's Feed-back Circuit, 1913. C. S. Franklin and several others about the same time hit on the idea of 'feed-back' to generate continuous oscillations.

As the range of use of thermionic valves extended the importance of securing the highest possible vacuum was soon realized and American scientists led by Langmuir carried out intense research in this field.[12, 13] By improvement in the technique of valve production the 'hard' triodes in the form of a spherical bulb with a straight tungsten filament were soon being turned out in large numbers. The grids were of helically wound molybdenum or nickel and nickel cylindrical plates were used.[14] The filaments were oxide coated. As Sir Edward Appleton points out in his excellent summary, the chief result of the war from the electronics view point was that people began to understand how a valve worked and how important was its internal physical dimensions and structure. He concludes with an interesting picture of the way in which the high-speed electrons are shot through the holes in the grid without being caught so that 'the grid does the work but the anode gets the benefit'.

The form and constitution of cathode received much attention in the United States and Langmuir's discovery, that mixing only a small percentage of thorium with tungsten would increase the emission at a given voltage enormously, enabled the temperatures to be dropped. Instead of running the filament at about 2,000° C., satisfactory working could be obtained at a dull heat, so increasing the life of the valve. The Germans also began to coat filaments with barium and strontium with similar advantages. For certain purposes, too, it was found desirable to separate the emissive part of the cathode from the heater, which provided the necessary heating by thermal radiation.

One of the major difficulties experienced with the triode during the years when its use was being so rapidly extended was the existence of an electrostatic capacity between the grid and the anode which introduced an undesired coupling between the two circuits outside the valve. Several workers attacked the problem by the addition of a screen between the anode and the grid, starting with Scholtkey in 1916, but it was H. J. Round who in 1926 brought the idea of a

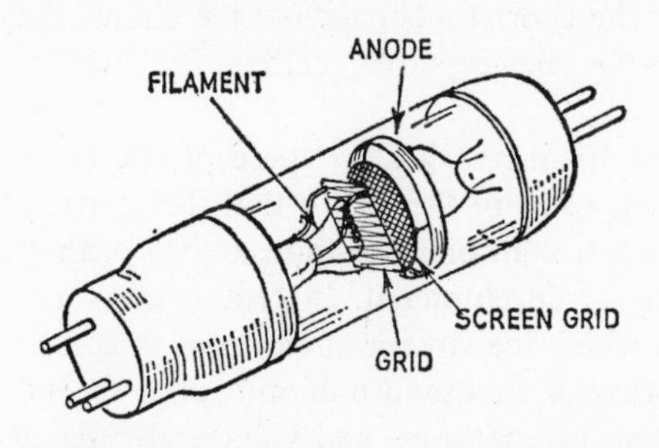

Fig. 52. Round's Tetrode, 1926. To overcome difficulties due to the electrostatic capacity between grid and anode, H. J. Round introduced a second grid known as the screen-grid. The valve became known as the Tetrode.

screen-grid into practical use in this country by the introduction of his famous S625 valve, nowadays known as a Tetrode. As the screen is in such a position as to form part of a circuit—filament, grid, screen and the greater part of the load battery—it removes from the anode load circuit the undesirable capacity effect.

The next stage in valve design arose from the observed secondary emission from the anode due to its bombardment. These low-velocity electrons migrated to the screen in the tetrode instead of returning to the anode and so interfered with the linear relation between anode current and voltage. In 1926 the Dutch engineers, Tellegen and Holst, added therefore a further grid between the screen and the anode of the tetrode and produced the pentode. This suppressor grid, as it was called, maintained at a low potential, usually that of the filament, prevented the trapping of the secondary emission and at the same time had no appreciable effect on the main stream of high-velocity electrons passing through on their way to the anode. Because of their improved efficiency pentodes have assumed great importance as amplifiers in radio receivers. Up to the beginning of the Second World War their use had become very general.

It is quite impossible here to cover the vast number of modifications which have been introduced into the construction of thermionic valve design. Fourth, fifth and even sixth grids have been used and to meet limited space requirements a multiplicity of diodes and triodes with their associated connections have been accommodated in one and the same enclosing envelope. Early in the Second World War, to meet the demand for ever shorter wavelengths, even the inductance of the short leads carrying the circuit through the glass envelope became unacceptable, calling for drastic alterations in design.

In the search for increasing the speed of operation of thermionic valves it became clear in the 1930's that the inertia of the electron, that tiny mass less than one two-thousandths of that of a hydrogen atom, might limit development. In centimetric wave technique the cycle through which the stream of electrons in a valve must operate may be less than a thousand-millionth of a second. This time is insufficient even if dimensions and voltage are adjusted to practical limits and attention was turned entirely to new methods depending on velocity modulation. Instead of the space charge from the cathode being modulated in quantity on its way to the anode a pulse technique was introduced. Some progress in the system was made in Germany in 1934 but the real development took place in America several years later.

In the velocity modulator amplifier a high velocity beam of electrons is subjected to acceleration and retardation, without change of direction, by a high-frequency cavity resonator controlled by the incoming signal. Thus the constant stream is caused to bunch up as it floats along in a drift space where the accelerated electrons catch up on those which have been slowed down. These bunches then pass through a second resonator, inducing currents which can be amplified. Feed-back can be arranged between the second and first resonators to set up oscillations but in a more usual form the bunched beams from the first resonator are reflected by a negative electrode and so retrace their steps to effect the same result. This is the principle of operation of the Klystron.

Another electron device which has assumed outstanding significance during the past few decades is the magnetron, an oscillator for producing frequencies of hundreds of megacycles per second suitable for centimetric wave radar and other applications requiring outputs of hundreds of kilowatts. The development of the magnetron sprang from the early work of Zacek in 1924, who produced oscillations with cylindrical diodes having a magnetic field superimposed. Later, Yagi and Okabe split the anode into two halves and were able to use the properties of the curved electron paths produced by the deflection of a magnetic field.

The cavity magnetron which was produced in 1939 by a team in the Physics Department of the University of Birmingham under Professor Oliphant, which included Professor J. T. Randall and Dr. H. A. H. Boot,[15] made great strides during the Second World War. It had a straight electron-emitting cathode passing concentrically down the centre of a copper block anode with inward projecting 'poles' eight or so in number, separated by circular cavities in the metal. A transverse magnetic field caused the electrons on leaving the cathode to follow a curved path in the space between the cathode and the anode. Certain of them were retarded and finally added power to the circuit on reaching the anode. Others being accelerated missed the anode and returned to the cathode, where they added to the emission and started off again to contribute a further portion of the energy via the anode. For radar purposes the high powers were generated in pulses of only a few microseconds, but even then artificial cooling was needed.

On the borderland where electronic physics, originating in the discoveries of Cockcroft and Walton at Cambridge in the nineteen-thirties, becomes electronic engineering stands another already established major device, the cyclotron invented by Lawrence. The pioneer of a family of particle accelerators, both linear and circular, the cyclotron consists primarily of two 'Dee' boxes of the general shape of a flat pill-box, complete with lid, cut through across a diameter situated horizontally between the pole pieces of a massive electromagnet which produces a vertical field passing through the box. A high-frequency high-voltage potential is applied between the two insulated halves of the box, the voltage being of the order of 100,000 and the frequency so adjusted that particles within the box, rotated in circular orbits by the magnetic field, are caught at the gaps at the correct frequency to accelerate them twice in a revolution. The particles are produced initially by a filament cathode and after being accelerated to the desired energy are drawn out at the periphery by a charged deflector plate and are caught in a target-chamber for use in many different ways.

By the development of the Lawrence cyclotron to larger sizes during the Second World War atomic particles were produced with energies of tens of millions of volts but it was soon found that a major change in the mechanical construction was necessary. Better and more economical results could be obtained by substituting a hollow ring for the drum-shaped container. Electrical engineers and physicists in the leading countries combined in a mass effort which resulted in accelerators of astonishing size and performance. An outstanding example is the machine now in operation at the Geneva European Centre for Nuclear Research (CERN), an international organization, inaugurated in 1960. This machine produces particles with energies up to twenty-five thousand million volts and cost ten million pounds to construct and install. The ring, which is 656 feet in diameter, has one hundred electromagnets disposed around the circumference, each of which weighs 30 tons.

When it is considered that the satisfactory transport of electrical particles in an accelerator requires a vacuum of a very high order it will be appreciated that the construction of the containing envelope alone has been a major engineering achievement. It must resist collapse and be furnished with joints between components of unusual soundness. The magnet system too has involved the solution of exceptional mechanical and metallurgical problems.

An application of the electron which has assumed great scientific and industrial importance is the electron microscope. This remarkable instrument has made possible the visual examination of small objects far beyond the range of the optical microscope which, as is well-known, is limited by the resolution associated with the wavelength of light. By using wavelengths only one hundred thousandth those of visible light the electron microscope carries down the limiting dimensions of an object which can be examined from something in excess of one hundred thousandth of a centimetre to about one ten-millionth o a centimetre, or, in optical terminology, to one millimicron (10 Ångstrom). The best modern electron microscopes achieve a magnification of over 150,000.

The fundamental feature of the electron microscope is the use of electrostatic or electromagnetic 'lenses' for focusing a beam of electrons emitted from a hot cathode. These were proposed by Aston as early as 1919[16] but Busch is usually given the credit for showing that magnetic coils could have the properties of lenses and in 1926 showed the way to the electron microscope.[17] In 1932 two Germans, Knoll and Ruska,[18] constructed a practical instrument in which a powerful magnetic field was produced by a circular shielded electro-magnet surrounding the electron beam with a small gap facing inwards.

In a modern instrument the electron beam is generated by an electron gun situated at the upper end about 6 feet from the floor, the beam passing downwards through the tubular construction containing magnetic lenses, object stage and many refinements to the final viewing screen of fluorescent material under which is a camera for making permanent records. These two items are situated at table level for convenience in operation. A subsidiary cabinet contains the 100,000-volt transformer, rectifiers for providing the D.C. potential on the cathode, various transformers for heating the filament and other auxiliary equipment. The pumps for maintaining the high vacuum of one ten-thousandth of a millimetre of mercury are situated in the lower part of the microscope.

As will be apparent from even this brief description the electron microscope is a complex and expensive piece of apparatus, but during the past two decades it has been developed to such an extent as to become a most valuable tool in the investigation of the properties of materials. In spite of the high cost of these instruments several hundreds are already in daily use in university, industrial and research establishments throughout the world.

Passing on to a very different application of electrons making electrical history we must refer briefly to the early steps which have ultimately and within only a few decades given us that vital component of modern everyday life, television.

The history of television starts with the development of the cathode ray tube. Once it was known that in a vacuum a continuous stream of electrons was emitted by a heated filament—the cathode—many practical applications suggested themselves. The electrons could be accelerated electrostatically and could be detected on reaching a screen terminating the end of the tube by their property of causing a coating of some substance such as zinc sulphide to emit visible light. Once started off in a straight line the electrons could be directed owing to their electrostatic and electromagnetic properties at any desired angle. Thus the position of the spot on the screen could be moved about at will by a signal applied to controlling coils or plates near to the beam.

In 1897 Braun made a cathode ray tube for experimental purposes, but it was only in 1905 when Wehnelt suggested coating a platinum filament with lime that the necessary emission for practical purposes was achieved at a reasonable voltage and many experimenters became interested. Within a few years an arrangement of components developed in which the electrons were accelerated by a tubular anode a few inches in front of the cathode maintained at about 1,000 volts or more D.C. positive. This charge was produced by the output from the secondary of a high voltage transformer passed through a valve-rectifier. The beam of electrons was focused on to the viewing screen at the enlarged end of the tube by an electron lens consisting of a

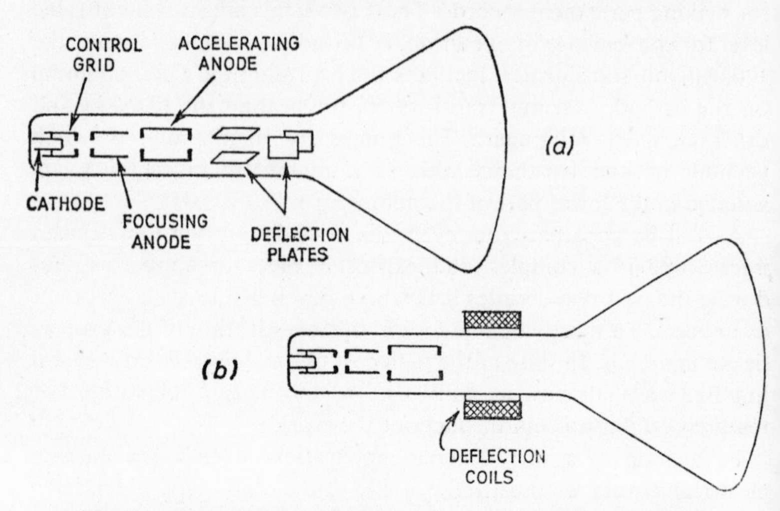

Fig. 53. Cathode Ray Oscillograph, c. 1911. About this date Ryan brought together a number of ideas in which it had been proposed to control the deflection of a stream of electrons forming a visible spot on a screen. Many developments have taken place in the use of both electrostatic (*a*) *and electromagnetic* (*b*) *deflecting plates and coils.*

small negatively charged metal tube which crowds the stream of electrons together and thus acts in the same way as a convex lens on a ray of light. Magnetic focusing was also introduced as already referred to in the description of the electron microscope.

The brightness of the spot on the screen was controlled by passing the electron stream on its way from the cathode through a preliminary grid to which was applied a voltage negative relative to the cathode and adjustable so as to control at will the number of electrons passing.

In its original use as an oscillograph the cathode ray tube was provided with two metal plates to which the fluctuating voltage to be indicated was connected. These were above and below the electron pencil and so caused the spot to rise and fall as the voltage changed. Two other plates at right angles to those were used for the X-movement of the spot and were supplied by what was termed the time-base circuit. The principle was employed by which a condenser charged from a steady voltage through a resistance provides a measure of time. The condenser charged at a definite rate to a voltage at which it was made to discharge and the operation repeated, the spot flying back to zero in the meantime. Many different circuits have been evolved to give time bases of linear form.[19]

The idea of television came long before the cathode ray tube on which modern television depends. In 1875 an American, G. R. Carey,[20] conceived the idea of a mosaic similar to that of the human retina but consisting of a large number of tiny selenium cells which would be electrically sensitive to the influence of light, each cell being connected by a wire and battery to one of a similar series of small electric lamps. Professors Ayrton and Perry took up the same idea in 1877 and later simple working models were constructed by Senlecq, a Frenchman. An Englishman, Shelford Bidwell, carried out experiments in 1908 but gave up the idea on discovering that for a picture two inches square 150,000 circuits would be required between transmitter and receiver. The next idea was the conception of scanning the picture to be transmitted and the use of revolving discs, rapidly vibrating mirrors, and so on, the consecutive signals being sent along one circuit. It is evident from the excellent account of proposals already made before 1928 for seeing electrically at a distance, which appeared under the article already referred to by A. Dauvillier, a Parisian physicist, that the main principle of modern television had been enunciated and only awaited the use of the electron in the cathode ray tube to provide a sound engineering system.

In 1911 a rather elementary form of cathode ray tube was used with some success by a German, Rosing, who transmitted pictures over a wire.[21] But it was not until 1924 that A. Campbell Swinton published[22] definite proposals for scanning both at the transmitting and receiving ends by the use of cathode rays.

At this stage the transmission end of the system became the vital element and in 1932 McGhee and Tedham, inspired by Campbell Swinton made a photo-sensitive mosaic in the form of a disc of aluminium with an oxidised surface on to which they deposited silver by evaporation through a fine mesh stencil. The silver was then rendered photo-sensitive by oxidation and baking in the presence of caesium. The result was not very sensitive, but the idea became later the basis of the Emitron television camera.

In the meantime Zworykin, after tentative efforts in 1924 and working in the R.C.A. Victor laboratory in New Jersey, made a dramatic advance in the production of the Iconoscope. In November, 1931, he set up in a cathode ray tube a screen of mica about four inches square carrying a mosaic of minute photo-electric cells, the back face of the mica having a metal signal plate connected to an amplifier.

The picture or view to be transmitted was focused optically on to the front of the screen and at the same time a beam of electrons from an electron gun in an oblique side tube was made to scan the same surface. The effect of the light on the photo-sensitive elements was to cause an emission of electrons and to change the potential of the element. The arrival later of the scanning beam restored the potential to an equilibrium value. The ultimate result of this operation was to impose on the backing or signal plate, and through it to amplifying circuits, a continuous record interpreting the intensity of light on every picture element in turn.[23]

In this country investigations by McGee and Lubszynski continued on the Iconoscope or Emitron in the laboratories of the E.M.I. and between 1935 and 1937 the Super-Emitron was produced with a tenfold increase in sensitivity.[24] In this form the optical image is focused onto a transparent photo-electric cathode and the stream of electrons is accelerated by a high voltage on the anode. The mosaic itself is not photo-sensitive. Continual improvements are being made in television cameras but so far the principles just described are the basis of modern practice.

The successful production of a camera which would convert a scene into a series of electrical impulses as just described was the major contribution to the advance of high definition television, but there were also problems to be overcome in designing suitable receivers notwithstanding the position reached by the cathode ray tube as an oscillograph.

The process of scanning consists of tracing lines across the picture starting, say, at the left-hand top corner and taking each tiny element in turn along the line to the right-hand limit, then starting a second line and so on to the bottom of the picture. Tracing then recommences at the top and repeats the process. Both successful cinematography and television reception depend on the well-known phenomenon, persistence of vision, so that the scanning of all the lines on the receiving tube must occupy such a short interval of time that the eye retains the image until the next picture element arrives. Without this limitation there would be flicker. In practice it was found that the scan must repeat itself forty times a second for good viewing. The number of picture elements necessary was found to be 160,000, which meant that the electrical circuits must transmit 6,400,000 signals a second. Moreover, the receiving scanner must be synchronized with the transmitter and this is done by transmitting special synchronizing signals at the end of the line scan. By suitable devices in the electronic circuit, the spot flashes back from the end of one line to begin the next line relatively slowly and, by the process known as 'interlacing', the odd lines starting from the top are first traversed and then the even lines, which has the advantage of halving the necessary impulse rate. Various numbers of lines have been adopted by different countries. In Britain the 405-line system was standardized with only 28 lines used for synchronizing and great care has been taken to balance horizontal and vertical definition.

No attempt can possibly be made here to cover such matters as the development in detail of the multitude of valves which were called for by the Services during the war in 1939–45. The story was reviewed generally in the author's Presidential Address to the Institution of Electrical Engineers in October 1945,[25] and in detail by many subsequent conferences commencing with one of Radiolocation early in 1946.[26]

To complete this chapter a word must be said on the Transistor, which during the past few decades has assumed such an important position in electronic engineering as a rival to the thermionic valve.

In many branches of engineering history the first seeds sown often remain dormant for many years after which rediscovery occurs and in the new climate of knowledge fructify rapidly. So it has been in the field of semi-conductors, on which transistors depend. Well over a century ago Roschenschold,[27] a German, discovered that under certain conditions solids known as semi-conductors had a rectifying action on an alternating current passed through them. Apparently the next application was the cat's whisker radio detector used by amateurs in the early days of wireless. In 1924 Lossev[28] returned to the cat's whisker and experimented with many complicated substances. His greatest achievement was in using zincite with carbon points or steel wires in contact. About the same time selenium rectifiers became available commercially and a few years later the copper oxide rectifier. Knowledge was accumulating in the new field of semi-conductors, which are neither ordinary metallic conductors like copper nor insulators like porcelain but something in between. Starting with a classical exposition by A. H. Wilson in 1931,[29] a vast literature has grown up on the theory of operation of semi-conductors culminating in an international convention held in London in 1959[30] to the proceedings of which the reader must be referred.

The important discovery that a triode could be constructed by placing two cat's whiskers near to one another on a piece of germanium with only a few thousandths of an inch between them was discovered in the Bell Telephone Laboratories by Barden and Brattain in 1948.[31]

In the application of the transistor both rectifiers and amplifiers are now being produced commercially in vast numbers for radio, telecommunication, many industrial uses, hearing aids where lightness in weight is so important, and for the fighting services. The output of units manufactured in different countries of the world per annum must already be approaching the thousand million mark. The most frequently used materials are germanium and silicon, both of which have to be refined to a high degree of purity for the purpose involving extensive and complicated techniques.

The length of this chapter, notwithstanding the cursory treatment which has been necessary, is a good indication of the great and growing influence of the application of the electron to electrical engineering—much has been omitted—from X-rays to direction finding and aircraft altimeters. A reference book on this subject is today a massive tome, and even lists of items such as valves fill many pages of a catalogue. The seeker after information can now only be referred to the many excellent works already available on different aspects of the subject.

REFERENCES

1. Fleming, *Fifty Years of Electricity*. London, 1921.
2. Heinrich Hertz, *Electric Waves, being Researches on the Propagation of Electric Action with Finite Velocity through Space*. English translation by D. E. Jones, London. Macmillan, 1893.
3. Rollo Appleyard, *Pioneers of Electrical Communication*. Chapter on Hertz. London, 1930.

(*a*) High-frequency Induction Heating
An example of many processes made possible by the application of high-frequency currents. Current in the square coil induces eddy currents in the tip of the lathe tool resulting in rapid local heating.
(*Photo: Messrs. Jessop, Sheffield*)

(*b*) Progress in Street Lighting, 1949
Photograph shows the uniform lighting of Old Bond Street, which set a new standard for public lighting.
(*Photo: Associated Electrical Industries Ltd.*)

(*a*) Eversheds First "Megge
Insulation Tester, 1905
This well-known instrument h
passed through a number
improved designs over the pa
half-century but still retains t
two main features of Evershe
original portable box—a cra
handle for driving a D.C. gene
ator and a scale from which t
insulation resistance is read
in megohms.
(*Photo: Evershed Vignoles Lt*

(*b*) The Avometer
A pioneer universal measuri
instrument in wide use.
(*Photo: Avo Ltd.*)

4. *The Electrician*, vol. 27, p. 221. June 26, 1891.
5. *Proc. Royal Institution*, vol. xiv, p. 321, 1894.
6. *The Electrician*, vol. 43, p. 40, 5 May 1899.
7. Lee de Forest, *Father of Radio, an Autobiography*. Chicago 1950.
8. *Trans. A.I.E.E.*, vol. 25, p. 735, 1906.
9. *Thermionic Valves*. Jubilee number, J.I.E.E., 1955.
10. Fleming, *Memories of a Scientific Life. An Autobiography*, p. 144. London, 1934.
11. Armstrong, *Proc. Inst. of Radio Engineers*, vol. 3, p. 215, 1915.
12. *Phys. Rev.*, vol. 2, p. 402, 1913.
13. *Proc. Inst. of Radio Engineers*, vol. 3, p. 261, 1915.
14. Appleton, *Thermionic Valves*. Jubilee number, J.I.E.E., 1955.
15. I.E.E., *Proceedings Radio Location Convention*. March 1946. p. 182.
16. *Phil. Mag.*, vol. 38, p. 709, 1919.
17. *Ann. Phys.*, vol. 81, p. 974, 1926.
18. *Zeit f. Phys.*, vol. 78, p. 318, 1932.
19. V. E. Cosslett, *Electron Optics*. London, 1946.
20. *Revue Gen. d'Electricité*, vol. xxiii, p. 6, 1928.
21. *Zeit f. Schwachstromtechnik*, vol. 5, p. 172, 1911.
22. *Wireless World and Radio Review*, vol. 14, p. 51, 1924.
23. *Proc. I.R.E.*, vol. 32, p. 16, 1934.
24. *J.I.E.E.*, vol. 84, p. 468, 1939.
25. *J.I.E.E.*, vol. 93, Part I, p. 17, 1946.
26. *J.I.E.E.*, vol. 93, Part IIIA, No. 1, 1946.
27. *Ann. Pogg.*, vol. 34, p. 437, 1835.
28. T. R. Scott, *Transistors and other Crystal Valves*. London, 1955.
29. *Proc. Royal Society*, A. vol. 133, p. 458, 1931.
30. *J.I.E.E.*, Special Report on Semiconductors. 1959.
31. *Phys. Rev.*, Series II, vol. 74, p. 230, 1948.

CHAPTER XVIII

Measurement Instruments and Standards

From the earliest days of electrical engineering there has been a continuous and growing need in every branch for methods and apparatus to determine the magnitude of different electrical and magnetic quantities. At all stages the many scientific, mathematical and mechanical problems involved have exercised the best brains in science and in industry. Standards have been developed internationally and great ingenuity displayed in the production of a vast range of instruments and circuits of all kinds. At one end of the scale are instruments of astonishing accuracy for scientific work and calibration while, at the other, a great industry has grown up for the production of instruments which, while being robust in character, cheap and easily manipulated by the unskilled, possess a reasonable accuracy suitable for the purpose for which they are designed.

Generally speaking, measuring devices have developed in step with the progress of discovery of phenomena and it is not surprising to note, therefore, that long before there was any such thing as a steady electric current, that is before 1800, attempts were being made to estimate the strength of a magnet and the value of a static potential. A crude measure of the strength of a lodestone or of a permanent steel magnet was its lifting power, while the measurement of potential started with a suggestion by John Ellicott, of Chester,[1] who noted the mechanical force created by the attraction of a charged body. The idea was taken up by Gralath, who described a simple electrometer based on the same principle,[2] Cavallo in 1775 and, more significantly, Rev. Abraham Bennet in 1787,[3] invented the electroscope. The latter is described by Mottelay[4] as 'the most sensitive and the most important of all known instruments for detecting the pressure of electricity'. It consisted of a glass cylinder which was covered with a projecting brass cap, made flat in order to receive upon it whatever article or substance is to be electrified and having an opening for the insertion of wires and a metallic point to collect the electricity of the atmosphere. The interior of the cap held a tube which carried two strips of gold leaf in lieu of the customary wires or threads (evidently other simple electroscopes were already in use), and upon two opposite sides of the interior of the cylinder were pasted two pieces of tinfoil directly facing the gold-leaf strips. The cap was turned around until the strips hung parallel to the pieces of tinfoil so that any electricity would cause the strips to diverge and make them strike the tinfoil which would carry the electricity through the support of the

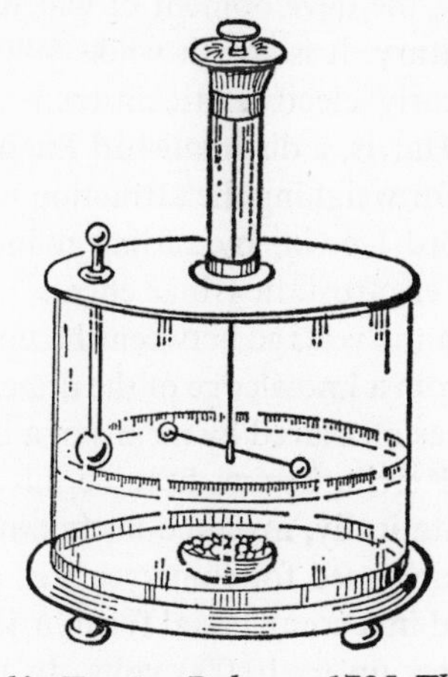

Fig. 54. Coulomb's Torsion Balance, 1785. This historic device, consisting of a fibre-suspended arm carrying a balanced charged ball within a glass case and a fixed second ball, may be considered the father of electrical measuring instruments. The attraction or repulsion between the balls was measured by the angle of torsion in the suspension wire.

cylinder to the ground. In 1802 movable coatings were embodied by Pepys (not the Diarist, 1633–1703) to increase the sensitivity of the instrument and, in 1805, Weiss, a German, improved the insulation by adding an internal glass tube. In this form the gold leaf electroscope was employed for many years and proved a powerful tool in the hands of experimenters.

In 1785 the man who may be claimed as the founder of quantitative measurement in the field of electrostatics, Charles Augustin de Coulomb, invented his famous torsion balance electrometer. This consisted of a long silken thread carrying at its lower end a balanced straw covered with sealing wax. The straw rotated inside a large diameter glass tube which carried degree markings on the cylindrical surface and the silken thread was fixed to a rotatable torsion head at the upper end of a small diameter glass tube extension of the cylinder.[5] In these memoirs it is stated that the balance was so delicate that each degree of the circle expressed a force of only one hundred thousandth of an English grain and that with a single fibre of silk 4 inches long the straw was turned through 90° by bringing a rubbed sealing wax to within a distance of one yard.

Before turning to the larger and more practical field of electromagnetic instruments, the development of which came in at the end of the nineteenth century, it is worth while following up the rather limited number of early electrostatic instruments to more recent times. In 1834 Snow Harris, a distinguished English scientist, devised a simple mechanism for weighing the attraction between two charged discs and in 1855 Lord Kelvin, by adding refinements, including a guard ring to prevent electrostatic fringe effects, produced a precision voltmeter from which the voltage between the upper and lower plate could be calculated from a knowledge of the dimensions and the force of attraction which was measured by means of a balanced lever.[6]

A few years later Kelvin, having become interested in measuring high voltages electrostatically, invented his famous quadrant electrometer which, after being used for some years as a laboratory instrument, was constructed in a commercial form in 1887 as an indicating voltmeter for pressures up to 10,000 volts. In this instrument two vertical butterfly sheets of metal, forming a cellular space between them, were fixed in a protecting case with their axes at an angle of 45° to the vertical. Between them was a thin aluminium sheet which was pivoted horizontally to swing in between the fixed plates. This had a pointer moving over a scale at the upper end and weights were fixed to the lower end so that when no difference of potential existed between the fixed and moving plates the pointer rested at zero on the scale and the blade was poised just outside the cellular space. On the application of the voltage the moving plate was drawn into the space to an extent determined by the magnitude of the difference of potential which was registered on the scale. In 1890 Kelvin introduced a more sensitive instrument consisting of a series of multiple horizontal vanes on a vertical suspension and a multi-cellular box into which the blades moved on attraction. A horizontal needle moved over a circular front scale. This voltmeter, sometimes known as the 'carriage lamp' type because of its shape, was widely used for voltages up to 200.

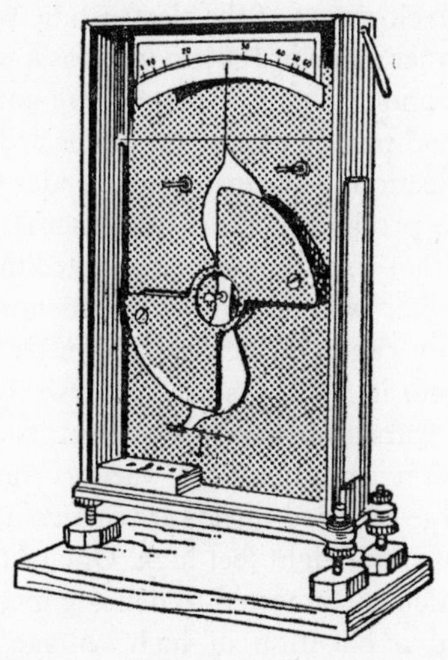

Fig. 55. Kelvin's Electrostatic Voltmeter, 1887. Kelvin produced a commercial quadrant voltmeter for pressures up to 10,000 volts on the principle of attraction between oppositely charged plates. Gravity control was employed.

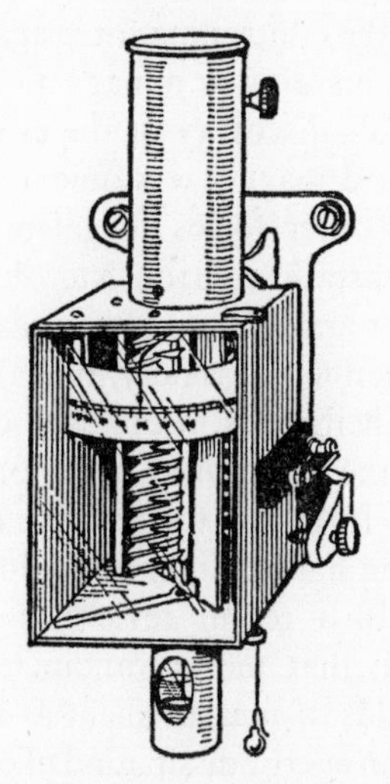

Fig. 56. Kelvin's Carriage Lamp Voltmeter, 1890. By increasing the number of quadrants and using a torsion wire control, Kelvin added considerably to the sensitivity of his quadrant voltmeter.

An interesting development of the electrostatic voltmeter arose out of a proposal by Potier made in 1881 to use it as a wattmeter. For this purpose the needle and one pair of plates were connected across the load while the second pair of plates was connected to a resistance in the circuit. The deflection of the instrument under these conditions is proportional to the product of the voltage across the load and the current through it, that is, to the watts dissipated in the load. For this purpose the wattmeter was adopted by Addenbrooke in 1900 for measuring losses in cable dielectrics and after improvement by Patterson and Rayner in 1913 was used extensively in industry.

Between the two wars a further type of electrostatic voltmeter was devised by Abraham and made in various ranges by Everett–Edgcumbe. Two large rounded shields are carried at the upper ends of two insulating rods about eight feet high. One of these has a central moving plunger which is connected to delicate mechanism inside the screened cover and application of high voltage between the two shields causes the plunger to be attracted electrostatically and to move a pointer over a scale. The instrument was designed for ranges up to 500 kV.

The great majority of electrical measuring instruments use the electromagnetic force in a coil of wire carrying a current to move a pointer across a scale, the controlling force against which the current operates being gravity, a spring or a magnetic field. As we have seen in Chapter IV, Oersted's discovery of the effect of the current in a wire on a nearby pivoted magnet was quickly followed by Schweigger's 'multiplier' and other forms of galvanometer. In one form Nobili introduced the astatic control, later developed by Kelvin by the addition of a mirror and a beam of light as pointer, in which two magnetic needles are connected rigidly, one above the other, on the suspension wire with their poles in opposite directions. This astatic construction reduced the controlling force of the earth's magnetic field and increased the force due to the current-carrying coil so that when in addition a long light pointer was added this became a very sensitive instrument. In a recent review of instrument progress,[7] D. C. Gall has shown that the galvanometer still holds a strong position for certain fields of measurement as a good one will give a usable indication with an energy dissipation of only 10^{-16} watt without any of the noise problems which often accompany the more recent use of thermionic valve circuits for measuring very small currents.

Precision measurement of current was established by Pouillet, when, in 1837, he devised the tangent galvanometer. In earlier forms of galvanometer it was recognized that the deflection, for various reasons, was not proportional to the current. In Nobili's, for instance, when the deflection was large the poles of the magnet emerged from the coil. Pouillet substituted a large circular coil for the small flat one and reduced the length of the swinging magnet. Instead of employing a torsion fibre suspension he employed a pivot suspension with control by the earth's magnetic field on the magnet itself. From

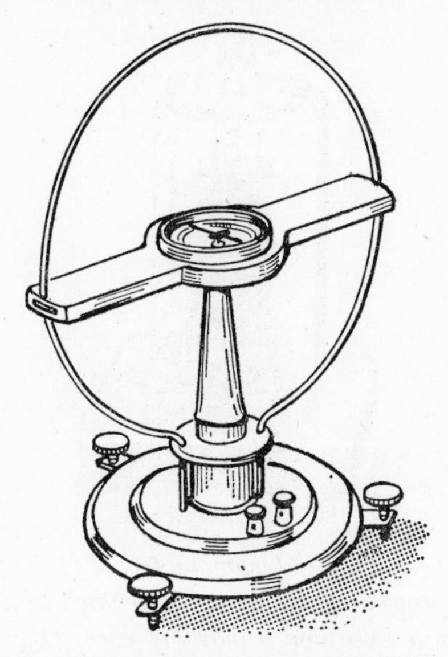

Fig. 57. Pouillet's Tangent Galvanometer, 1837. Pouillet devised the tangent galvanometer, an original form of practical precision current-measuring instrument. By using the earth's field for control and keeping the magnet small in comparison with the deflecting coil, the current was directly proportional to the tangent of the angle of deflection.

the geometry of the instrument the current was accurately proportional to the tangent of the angle of deflection, the angle being measured by means of a long light pointer moving over a flat circular scale. The tangent galvanometer long remained a favourite instrument for laboratory use.

In 1841, largely as the result of Ampère's analysis of the action of current-carrying conductors on one another, Weber, a German, invented the electro-dynamometer. This was taken up actively by Frölich, of Siemens and Halske, forty years later and became a valuable instrument for the measurement not only of current but also of the power in a circuit. Fundamentally the torsion dynamometer, as it was also called, consisted of a fixed vertical coil inside which, or embracing it, was pivoted or suspended a second coil which could rotate against a controlling force consisting of a bifilar wire or a spiral spring. The torque on the suspension was measured by means of a torsion head which was rotated over a scale of angles until the

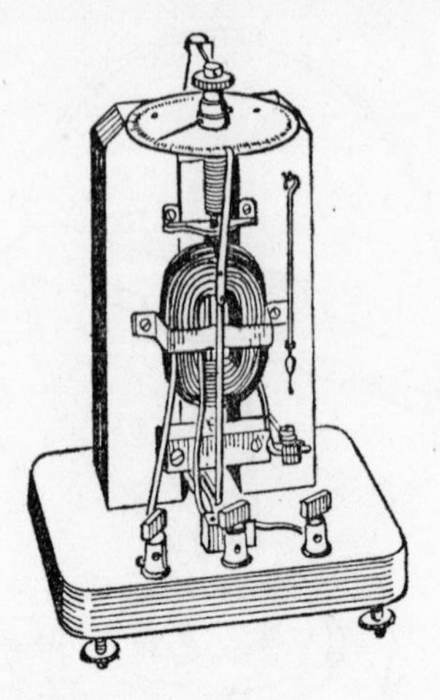

Fig. 58. Siemens' Electrodynamometer, 1841. The power in a circuit is measured by the torque developed in a suspending wire between a fixed and a movable core. One coil with a high-resistance winding in series accounts for the voltage component and the other of stouter wire the current component.

electromagnetic force between the fixed and swinging coil was balanced and the pointer on the latter brought back to zero. The connections to the moving coil were made through mercury cups.

The dynamometer type of instrument came to be used to measure current, in which case the current flowed through the two coils in series and the deflection, or balancing torque, was proportional to the square of the current. From the nature of the interaction between the two coils an interesting feature of the dynamometer type of instrument is that it can be used equally well for alternating and for direct current. Moreover, when used for power measurement, in which case one coil carries the current feeding the load to be tested and the other the voltage or current through a high resistance connected across the load, the instrument takes account of the power factor in the circuit and registers true watts.

Dynamometer types of electrical measuring instruments have been more widely used for laboratory than for industrial purposes and as substandard instruments for checking portable types of instruments. In the early days of high-voltage cable testing, about 1920, the Duddell Mather wattmeter was used for measuring dielectric losses

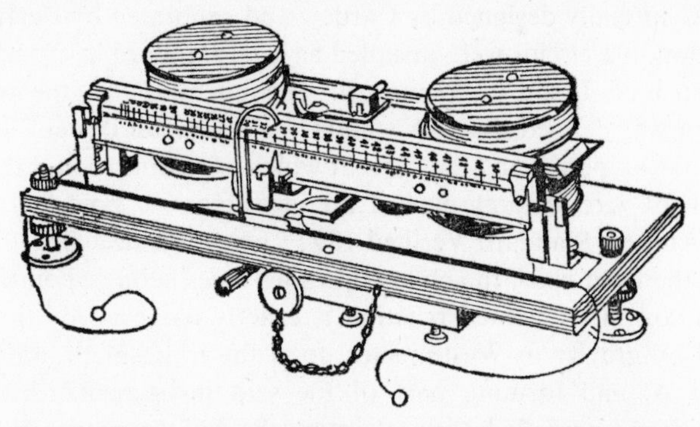

Fig. 59. Kelvin Ampere Balance. The force of attraction between a series of fixed coils and a complementary series of coils fixed to a balance arm is balanced by an adjustable weight on the steelyard principle. The instrument has been much used for standardizing purposes.

and power factor at pressures up to 100,000 volts and frequencies of 50 cycles. Employing the same fundamental principles and by the introduction of suitable modifications a number of switchboard type instruments were introduced early this century.

Towards the end of the nineteenth century a current-measuring device operating on the dynamometer principle which took a prominent part in the progress of precision measurement was the Kelvin ampere balance. Following up the researches of Weber, Kelvin invented the current balance which bears his name. He substituted an arrangement whereby the forces in two separate sets of coils carrying the same current and at opposite ends of a balance arm assisted one another. Two flat movable coils were fixed to the ends of a light metal beam supported by a metal ligament of fine wires at its centre to carry the current in and out of the coils. Above and below each of these coils were fixed coils in close proximity and all six were connected in series. The direction of current in the three coils at one end of the beam is such that the end tends to rise, the current in the three coils at the other end tends to make that end fall, and the torque so generated was balanced by a sliding weight on a steelyard arm carried by the beam. As the mechanical force was proportional both to the current in the fixed coils and that in the movable coils, i.e. on the product of the two currents as in all dynamometer arrangements, the current was proportional to the square root of the torque. Many years afterwards, in 1894, by an Order in Council, two such current balances specially designed by Cardew and calibrated by electrolytic deposition in a circuit were installed as legal standard instruments at the Board of Trade in Whitehall. The operation of the ampere standard was defined thus: 'A standard of electrical current denominated one ampere being the current which is passing in and through the coils of wire forming part of the instrument marked "Board of Trade Ampere Standard Verified 1894" when, on reversing the current in the fixed coils, the changes in the forces acting upon the suspended coil in its righted position is exactly balanced by the force exerted by gravity in Westminster upon the iridio-platinum weight marked A, and forming part of the said instrument.' A similar specification controlled the volt standard by the movement of a sighting wire in relation to two fixed points. In his book published in 1894,[8] Sir David Salomons stated with apparent satisfaction that the accuracy was within one-fifth of one per cent!

It is interesting at this stage to jump forward sixty years and to note that values of standards maintained at the National Physical Laboratory now go to the sixth decimal place.

Possibly the most fundamental of all electrical measuring instruments is the Wheatstone bridge which, historically, takes precedence in the vast array of devices which sprang from Oersted's experiment of 1820. Within thirteen years Samuel Hunter Christie[9] described a differential arrangement of conductors which formed the basis of Wheatstone's application in his *Differential Resistance Measurer*, published in his 1843 Bakerian lecture.[10] In this lecture Wheatstone unreservedly gave the credit for the idea to Christie but he made so many practical additions that the bridge became widely assigned to him and is now always known by his name.

At this stage of development there was no accepted standard of electrical resistance and Wheatstone adopted one of his own which was the resistance of a copper wire 1 foot in length having a diameter of 0·071 in. and weighing 100 grains. He made up a number of resistance boxes which today are on display in the museum at King's College. On one of these the value is marked in 'miles'.

The four elements of the Wheatstone bridge were set up practically in two forms. In one, the slide wire bridge, the place of the ratio arms was taken by a straight wire stretched along a metre scale divided at any point at will by a sliding contact connected to the galvanometer. In parallel with the slide wire were stout metal bars with gaps and terminals for the insertion of the resistance to be tested and of a standard resistance with which it was to be compared.

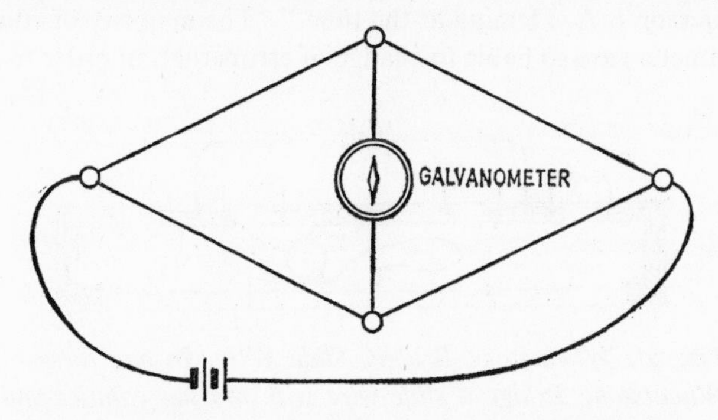

Fig. 60. Wheatstone Bridge. Diagrammatic. In 1843 Wheatstone announced under the title 'Differential Resistance Measurer' the assembly of resistances which have become famous as the bridge.

In the form which has been employed extensively, first for telegraph testing and later in all fields of electrical measurement, the standard resistance was divided into sub-divisions in the form of bobbins assembled in a portable box and connected to heavy brass plug terminals on a sheet of ebonite forming the cover. Decade arrangements in which the plug blocks were arranged in circles or replaced by a rotary switch enabled any value to be obtained quickly from 1 up to, say, 10,000.

Another early major contribution, closely related to the Wheatstone Bridge in form, was the potentiometer devised by Poggendorff in 1841,[11] in which a steady fall of potential was generated along a wire by a controlled current through it and the wire calibrated in units of potential by connecting a standard cell in opposition so adjusted that no current was drawn from the cell. Thus a standard of e.m.f. was obtained in terms of linear dimensions on the wire which could be used in various ways as, for instance, in determining the drop across a known resistance in a circuit and consequently the current flowing in that circuit.

Special forms of potentiometer were constructed by Crompton, Nalder and others in which the slide wire was replaced by coils of wire connected in series within a container and on the lid was a multi-contact switch by which contact was made through a rotating switch arm to the junctions between coils. The relative position of the potentiometer and other electrical measuring instruments at the end of the nineteenth century can be appreciated by a statement published by Professor J. A. Fleming at the time.[12] 'The majority of commercial ammeters are so liable to change of errors that, in order to avoid

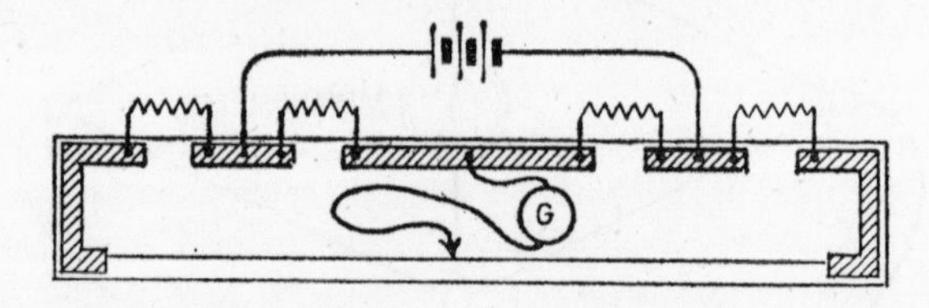

Fig. 61. Wheatstone Bridge. Slide Wire. In one form of Wheatstone Bridge a slide wire with moving contact and key provided a convenient variable arm.

the expenditure of time in constantly calibrating them, it is better in many cases to rely on the potentiometer as a means of measuring current directly.'

In a branch of electrical engineering where the struggle for ever-increasing accuracy continues, it is interesting to record here the stage reached in this field. A recent (1960) modern precision potentiometer designed by the Cambridge Instrument Company gives readings in steps of one-tenth of a microvolt up to 2 volts on a single range with an accuracy of 0·001 per cent. It is constructed in the form of a double potentiometer, a primary one with two dials, one of which controls a 20-way switch and the other a 100-way switch, giving steps of 0·1 volt and 0·001 volt respectively. The secondary potentiometer is provided with two dials controlling 100-way switches, one of which gives steps of 0·00001 volt and the other steps of 0·0000001 volt. With this extraordinarily high order of accuracy, temperature effects on the voltage of the standard cell used become important and compensation is incorporated to cover temperature changes from 10° C. to 35° C. for a cell with a nominal e.m.f. of 1·0186 absolute volts at 20° C.

While so much interest was being aroused from the middle of the nineteenth century in the measurement of resistance a movement of great significance to the whole future of electrical engineering was started by Professor W. Thomson (later Lord Kelvin) to establish and co-ordinate electrical standards. Until about 1850 all units of resistance were based on arbitrary size and weight of a wire and in 1851 Professor Weber had suggested a fundamental system of electrical and magnetic measurement. In 1861 at a meeting of the British Association, on the proposal of Professor Thomson a committee was set up to determine the most convenient unit of resistance and the best form and material for the standard. With Professor Wheatstone, Dr. Matthiessen, colleagues on the Atlantic cable inquiry described in Chapter XIII, and others, Thomson went into the inquiry with a will. During the course of its meetings over seven years he devised instruments of fundamental importance, revolving coil methods for the absolute measurement of a resistance, apparatus for the absolute measurement of current and special electrometers for the measurement of electromotive force.

The annual reports of the committee collected together by Fleeming Jenkin and published in 1913,[13] provide a remarkable record of the evolution of the C.G.S. system of units and the experimental methods adopted for evaluating different units. In its first report, published in 1862 the Committee decided to adopt, for the first time, a coherent system of units for resistance, current and electromotive force and, having considered the relative merits of the electrostatic and electromagnetic systems, decided on the latter. It also adopted the continental measure for length (centimetre) and weight (gramme). The use of the tangent galvanometer was prominent in determining the value of a current first in terms of the earth's magnetic field and then eliminating this factor and obtaining an absolute unit by the application of the Gauss and other devices by then well known.

The Committee decided that a material standard of resistance should be prepared in a permanent form specified as 'the unit of 1862' and being equivalent in C.G.S. terms to 1,000 million centimetres per second. Latimer Clark suggested the name OHMAD for this unit, which was ultimately adopted as OHM. In 1867 Kelvin produced a long report for the Committee on the determination and co-ordination of units and in 1870 the Committee was dissolved, its work for the time being completed and a number of coils of wire of different alloys were deposited at the Cavendish Laboratory as standards.

In the second and subsequent reports, some of which were masterpieces of exposition on the subject of fundamental electrical measurement, the methods by which more and more accuracy in determination was achieved were described. The heat generated by a known current in a wire unit of resistance was measured to calibrate the wire. Spinning coils of carefully measured dimensions were used to determine the value of resistance by calculation of the e.m.f. generated. Standard physical 'ohms' were made up which showed discrepancies between one another and those of earlier times of less than 2 per cent and were deposited at Kew.

Much interest was displayed in these coils during subsequent years and in 1881 Lord Rayleigh, using the revolving coil method in the Cavendish Laboratory, carried out check tests on them to see to what extent they represented the theoretical ohm. His results led him to consider that the B.A. unit was only 0·9893 of the real value. Other investigators, including Dr. Glazebrook and Professor Ayrton, dissatisfied with the standard, carried out tests which confirmed the discrepancy and Lorenz invented his classical method of test which was employed in different standard laboratories on both sides of the Atlantic.

At the International Electrical Exposition held in Paris in 1881 much dissatisfaction with the position was expressed by practical electrical engineers, who were still using widely 'Weber' to denote the unit of current, measuring electrical pressure in terms of the equivalent number of Daniell cells and resistance in terms of miles of telegraph wire[14] and the new names ampere, volt and ohm were adopted.

In 1891 the Board of Trade set up a Standards Committee 'for the measurement of electricity for use in trade" which adopted the C.G.S. system of electrical units with the values of units determined by the original British Association Committee and adopted the names which that Committee had proposed, ohm, ampere, volt for the units of electrical resistance, electric current and electric pressure. Two years later an important International Electrical Congress was held in Chicago under the presidency of Helmholtz with Britain represented by Preece, Ayrton, Silvanus Thompson and Alexander Siemens.

This Congress recommended the following as legal units:

Resistance. The International Ohm based upon the ohm equal to 10^9 units of resistance of the C.G.S. system of electromagnetic units represented by the resistance to an unvarying current by a column of mercury at the temperature of melting ice 14·4521 grammes in mass of a constant cross section and of a length of 106·3 cm.

Current. The International Ampere, one-tenth of the unit of current of the C.G.S. unit represented by the current which when passed through a solution of nitrate of silver in water deposits silver at the rate of 0·001118 of a gramme per second.

Electromotive Force. The International Volt, which is the e.m.f. that steadily applied will produce a current of one International Ampere represented by $\frac{1000}{1434}$ of the e.m.f. of a Clark's cell at 15° C.

The unit of quantity, the International Coulomb, was that quantity transferred by 1 ampere flowing for 1 second.

The unit of capacity, the International Farad, was the capacity of a condenser which a charge of 1 coulomb would raise 1 volt.

The unit of work, the International Joule, was 10^7 units of work in the C.G.S. system or the energy expended when 1 ampere flows through a resistance of 1 ohm for 1 second.

The unit of power, the International Watt, 10^7 units in the C.G.S. system or the power when the expenditure of energy is at the rate of 1 joule per second.

The unit of induction, the International Henry, or the induction in a circuit when a pressure of 1 volt is induced by a change of 1 ampere per second.

On 23 August 1894 an Order in Council legalized the Chicago recommendations as regards resistance, current and pressure and on 10 January 1910 a further Order in Council brought the standards up-to-date in certain respects based on the recommendations of a further important International Conference of Electricians held in London in 1908. By this time the same units of resistance, current and e.m.f. had been adopted in all the principal countries of the world.

Progress in improving the accuracy of electrical standards has gone on through the activities of the National Physical Laboratory, which was founded in 1891. In 1942 Hartshorn published an excellent summary of the legal and international position of the various electrical units reached by that time.[15] To bring this aspect of our story up to date an official statement was published in 1947[16] by Sir Charles Darwin, the Director of the N.P.L., adopting an important decision of the International Conference on weights and measures held in Paris in October 1946. In this it was stated that:

From 1 January 1948 the International system of Electrical Units so far used in the Laboratory based on certain material standards will be superseded by units derived from the centimetre, gramme and second, i.e. the so-called 'absolute units'.

1 International Ohm = 1·00049 absolute ohms.
1 International Volt = 1·00034 absolute volts.
1 International Ampere = 0·99985 absolute amperes.

also that as regards apparatus manufactured after 1 January 1948 acceptance will be within errors expressed in absolute units.

One final word may be said here on fundamental units. Over the years a great deal of discussion and experiment has taken place on the electrostatic and the electromagnetic systems, both based on the centimetre, gramme and second. Practical physical standards have also been established which have come more and more into agreement with the best theoretical results. The practical units, ampere, volt, ohm, etc., have departed by powers of 10 from the accepted electromagnetic units for the same quantities. In the M.K.S. system first proposed by Giorgi in 1901 metres are substituted for centimetres and kilogrammes for grammes, which results in a consistent system in which the ampere becomes the fundamental unit of current, the volt, the unit of potential, and the ohm, the unit of resistance. Thus in the M.K.S. system we no longer have to employ the factor 10^{-1} for converting absolute units to amps or 10^{8} for converting absolute units to volts.

In 1940 one standard system was suggested to replace the former three-absolute electrostatic units, absolute electromagnetic units and international standard units, and as we have already seen was promulgated in this country in 1948.

With all its advantages, this M.K.S. system still possessed certain disadvantages when calculating electric and magnetic flux, the factor 4π still appeared. A solution proposed by Oliver Heaviside was to rationalize the system by defining a unit magnetic pole as one which produced one line of force instead of 4π lines as previously and one unit electric pole as the charge originating one tube of electric flux. By these changes, electric and magnetic unit systems have been simplified and the rationalized M.K.S. system has now become the agreed practical system on the recommendation of the I.E.C. made in July 1950. Full details are given in modern books on electricity and magnetism.[17, 18]

About the year 1880 electrical measurement took a strong leap forward in the production of instruments. For a century measurement of potential had advanced steadily from Coulomb's balance

Plate XLIII

(*a*) Campbell Galvanometer, 1910–12
An early outstanding example of instrument design aimed at increased sensitivity. (*Photo: Cambridge Instrument Co.*)

(*b*) Unipivot Galvanometer, 1903
A major development in the design of portable instruments was made by R. W. Paul. This removes the need for accurate levelling, is highly sensitive yet resistant to shock, and thus has formed the basis of a wide range of electrical measuring instruments. (*Photo: Cambridge Instrument Co.*)

PLATE XLIV

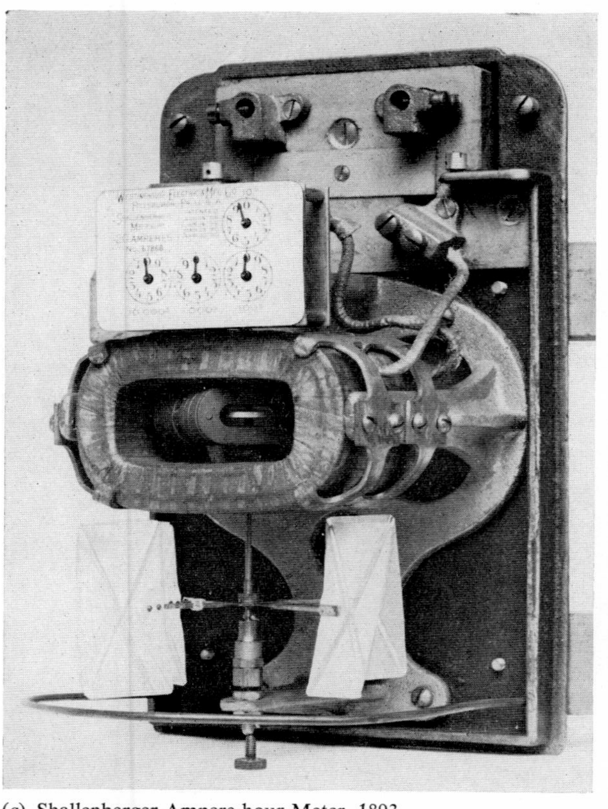

(*a*) Shallenberger Ampere-hour Meter, 1893

(*b*) Modern 12-element Oscillograph

to Kelvin's electrostatic voltmeter. Measurement of resistance by bridge methods had been in use for forty years and was established as a routine in telegraphy and experimental laboratories. The B.A. units were fast replacing arbitrary lengths of wire for resistance and the potential of one Daniell cell as a unit of electromotive force had been superseded by the volt. The ampere had ousted the Weber as the unit of current. But currents were still measured by non-direct methods—deposition of silver in an electrolytic bath, by 'drop'

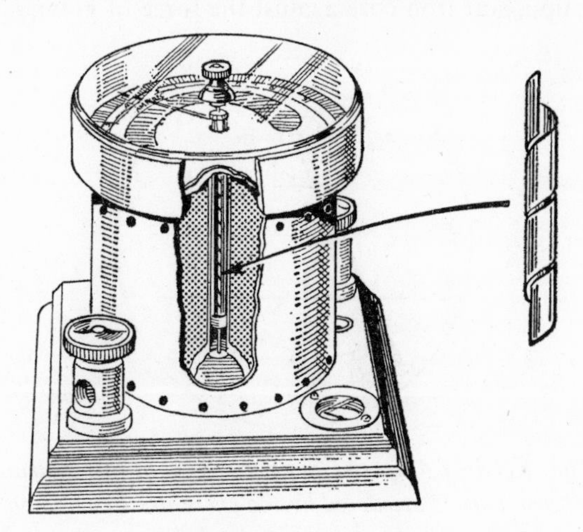

Fig. 62. Ayrton and Perry's Spring Ammeter. The attraction of an iron core within a tubular coil compressed a flat strip spiral which resulted in a twist at one end to which was attracted an indicating pointer.

across a resistance or by non-portable galvanometers. Direct reading instruments of great accuracy as we know them today were not available. It was the establishment of a bulk supply of the electric current which suddenly established the necessity and brought the invention.

Deprez and others constructed modified galvanometers to make them direct reading and in 1876 Kohlrausch invented a moving iron type of instrument in which an iron core was drawn into a solenoid, operating a lever moving over a vertical scale. But the first practicable portable ammeter was that invented by Ayrton and Perry in 1879. A light magnetic needle was pivoted between the poles of a powerful horseshoe magnet inside a coil. The controlling magnet maintained the needle and a pointer attached to it in the zero position until current was passed through the coil when the deflection against the control of the magnet gave a measure of the current in amperes. Other forms of moving iron instruments quickly followed. Ayrton and Perry produced the magnifying spring ammeter, an ingenious arrangement whereby a soft-iron core in the centre of a vertical axis coil compressed a long spiral of flat metal, so producing a rotary motion of the pointer on a horizontal circular scale. Kelvin made his ampere gauge in which a coil at the upper end of a glass-fronted case drew up a soft iron core against the force of gravity. A simple

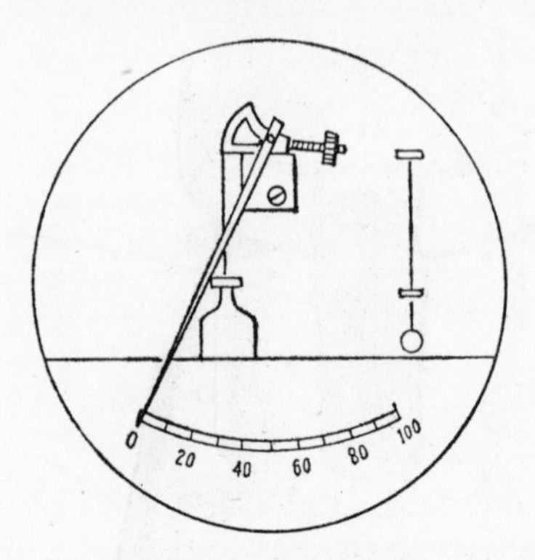

Fig. 63. Kelvin's Ampere Gauge, c. 1895. In this instrument a soft iron core is attracted by the current-carrying coil and the suspending cord causes rotation of a spindle carrying the indicating pointer.

suspension carried the upper end of a pointer, the lower end of which moved across a vertical scale.

Several types of moving iron ammeters with gravity control were designed in which the iron core, instead of being directly drawn into a solenoid, was pivoted to move eccentrically within the core. In one of these, constructed by Schuckert, a horizontal coil with its axis at right angles to the front face of the instrument surrounded a curved iron plate so pivoted that as it rotated it moved away from the axis of the coil. With no current flowing the weight of the iron plate brought the needle to the zero position and the application of current caused it to rotate towards the stronger part of the field.

Another gravity-controlled ammeter very similar to the Schuckert was the Nalder instrument which was widely used. In this instrument an eccentric bunch of iron wires replaced the curved iron sheet. S. F. Evershed also made a useful gravity ammeter in which the horizontal

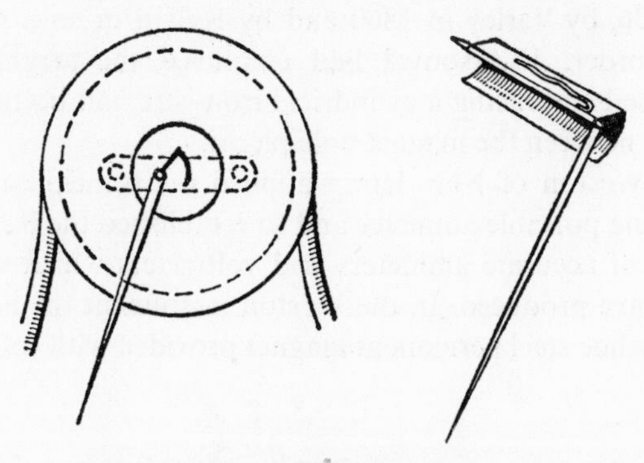

Fig. 64. Schuckert's Moving Iron Ammeter. The movement carries a curved strip of soft iron and is so mounted eccentrically within a horizontal coil at right angles to the face of the instrument that a twisting torque is produced by the field due to the current in the coil.

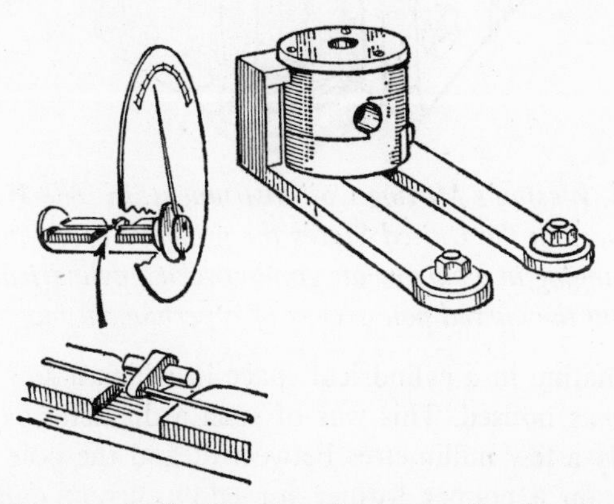

Fig. 65. Evershed's Gravity Ammeter. In this design of ammeter, a small piece of soft iron was drawn into the gap between soft iron bars in which the current induced a field.

coil encircled two small slabs of soft iron with a gap between them wide enough to admit a pivoted iron rod. The current in the coil magnetized the two pieces of iron and drew the pivoted iron into the gap, carrying with it the coupled pointer.

A modification of major importance in instrument design was the substitution of the moving coil system for the moving iron. The suspended coil had been used in galvanometers by Sturgeon as far back as 1836, by Varley in 1860 and by Kelvin in constructing his siphon recorder. D'Arsonval had employed the arrangement in which the coil embracing a cylindrical iron core had been housed in a short gap between the magnet pole pieces.

In 1888 Weston of New Jersey applied the principle to the first direct reading portable ammeter and so established the basis for that long series of accurate ammeters and voltmeters which subsequent designers have produced. In the Weston instrument the main frame was a horseshoe steel permanent magnet provided with soft iron pole

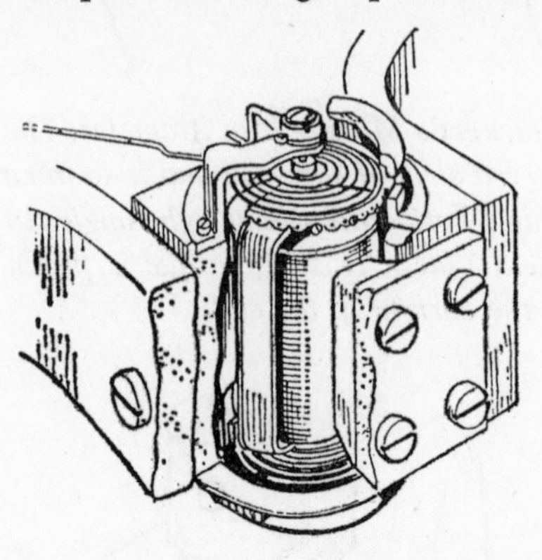

Fig. 66. Weston's Moving Coil Instrument. In 1888 Weston introduced in the United States the principle of the moving coil swinging in a narrow air gap between a cylindrical iron core and the curved pole pieces of a permanent magnet.

pieces terminating in a cylindrical space in which a soft iron cylindrical core was housed. This was of such a diameter as to leave a space of only a few millimetres between it and the pole faces. The coil, wound on a copper former for eddy current damping, was situated in the gap, supported above and below in jewelled bearings. The control was provided by light spiral springs between the coil and the pivots which also served to carry the current to the coil. The success of the Weston ammeter soon led to the adoption of the moving coil principle by many manufacturers and instruments for many purposes including switchboard use followed.

Among the major developments in portable instrument design must be mentioned the unipivot movement developed by Paul, and adopted by the Cambridge Instrument Company for a wide range of instruments where accurate level is difficult and high sensitivity must be secured at the same time as resistance to shock.

The same company has recently (1960) introduced another major fundamental improvement in moving coil galvanometer movements. Instead of carrying the coil either by the ordinary suspension or on pivots, a taut suspension consisting of short metal bands is employed. These bands, besides supporting the coil, provide control and electrical connections. There are thus no jewels or pivots to wear or delicate hair springs to get out of adjustment. Combined with this unique suspension the former external permanent magnet and centre soft-iron core of the moving coil instrument are replaced by a central

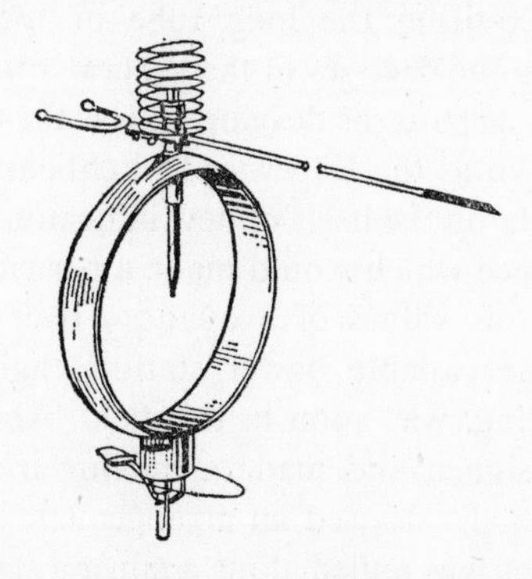

Fig. 67. Paul Unipivot Movement. This outstanding and ingenious movement was invented by R. W. Paul in 1903.

permanent magnet of high-quality steel with a cylindrical soft-iron core surrounding the moving coil.

Another type of current measuring instrument which appeared during the earlier period though primarily for the measurement of pressure was the hot wire voltmeter devised by Cardew in 1883. This instrument depended on the fact that when a wire is heated by an electric current it increases in length. In one form taken by the Cardew voltmeter the two ends of a fine platinum-silver wire about 2½ mils. in diameter and about 10 ft. long were fixed to terminal blocks behind an indicating dial. A double loop was formed by passing the wire down inside a metal encasing tube to two fixed pulleys and back again over an upper central jockey pulley which was spring supported to keep the wire taut. This movable pulley was connected by simple gearing to a pointer in front of a scale. As the current increased, the temperature of the wire rose, its length increased and the pointer moved across the scale calibrated in volts. From the early photographs of power-station switchboards it is clear that these Cardew voltmeters were widely used.

Towards the end of the century Hartmann and Braun introduced an improvement on the Cardew hot wire voltmeter by employing, instead of the direct increase in wire length, the sag of a taut wire. A wire attached to the centre of the current-carrying wire was kept taut by a spring and was passed round the arbor of an indicating pointer, so moving the pointer as the main wire heated or cooled. This arrangement permitted the housing of all the equipment in a circular case instead of requiring the long tube of the original Cardew instrument. Among the changes in the general appearance of switchboard instruments, as patterns developed, was the type which became very popular, known as the Edgewise Switchboard Instrument.

Following quickly on the heels of new indicating instruments came the recorder, a device which would make a permanent record showing the instantaneous values of a quantity over a period of time. Arthur Wright, the capable power station engineer of the early Brighton undertaking, was soon in this field. About the year 1886, having already designed and made a moving iron ammeter and a house service meter, he devised an arrangement in which a strip of carbon coated paper was pulled along against a spring while a needle operated by a plunger and solenoid marked the paper. The solenoid was connected across the mains in series with a 10,000-ohm resistance. At one time four such recorders were in use recording the station voltage at a speed of 2 cm. an hour. Subsequently many uses for recording instruments were found and up to the present time improvements have been continuous. Today they are employed in a vast range of industrial processes.

During the first few years of the electric supply industry a demand arose for energy meters to determine the quantity of electrical energy supplied to a consumer and both electrolytic and electromagnetic systems were employed. In the former, used by Edison, the current passed through a solution of zinc sulphate between zinc plates. The customer was charged on the weight of zinc deposited and where considerable currents were involved heavy shunts were connected across the cells and a suitable adjustment made.

An early form of supply meter of the electromagnetic type was that first used by Ayrton and Perry in 1882 and later improved by Dr. Aron. This had two pendulum clocks operating on one train of gears. One of the pendulums had a bob in the form of a coil of fine wire swinging inside a fixed coil of heavy wire. The inner coil was connected across the mains and the outer in series with the consumer's supply so that the joint effect of the two currents on the rate of swing of the pendulum was proportional to the watts in the circuit. Thus the total error between the two pendulums as registered on the dials was a measure of the energy consumed.

The first of the large class of motor meters was that devised by Elihu Thomson. Two large fixed air-cored coils acted as field magnets and carried the main current. The armature, mounted on a vertical spindle, was wound with fine wire and rotated between the coils connected in series with a resistance across the mains. The lower

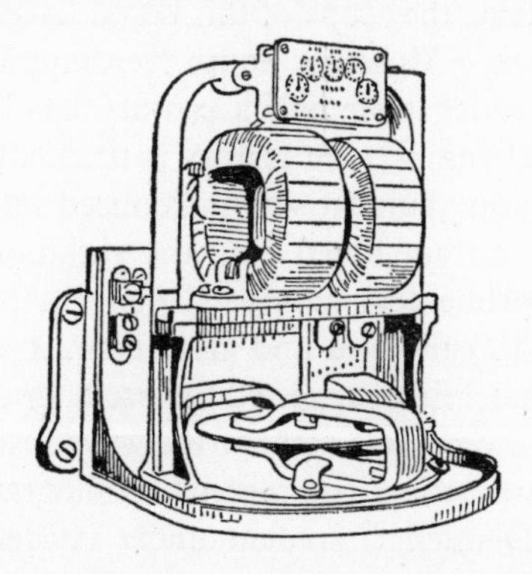

Fig. 68. Elihu Thomson's Energy Meter, 1882. Electromagnetic energy meter. Air-cored motor type with eddy current restraining disc.

end of the armature spindle carried a horizontal copper disc which rotated freely between the adjacent poles of several strong permanent magnets. The drag due to eddy currents in the disc produced a retardation proportional to the speed so that, as the turning moment was proportional to the product of volts and amperes the speed was proportional to the watts, and the total number of revolutions recorded on the dials the number of units of energy consumed. The dial readings were brought into legal Board of Trade units by adjusting the position of the magnets.

With the introduction of alternating current supply it was found that special precautions had to be taken in adapting the earlier types of energy meters, such as the prevention of eddy currents in the main circuit components, etc. Meters specially designed for A.C. circuits were usually of the induction type. The Shallenberger, for instance, produced in 1898, had a light horizontal iron disc rotor. A large vertical field coil in two parts carried the main current and between it and the rotor, disposed at an angle of 45° to the main coil, was a short-circuited coil of copper loops. The effect of the current in the main coil and that induced in the short-circuited coil was to produce a rotating field and a torque on the disc proportional to the square of the main current. A retarding fan of aluminium vanes offered an air resistance proportional to the square of the speed so that the actual speed of rotation was proportional to the current flowing. This meter naturally recorded only ampere-hours and not watt-hours.

Many other supply meters were designed at this time but a description of these is outside the scope of this book. Excellent examples are on view at the Science Museum, South Kensington.

An instrument which over half a century has contributed handsomely to electrical engineering history is the Duddell oscillograph. As soon as alternating current was introduced interest arose in the wave-form of the current illustrating the variation of current with time. Various experimenters developed point-by-point methods but they were tedious to carry out and great interest was aroused when in 1892 M. A. Blondel first suggested the oscillograph, an instrument which would give a complete view of the wave instantaneously. The man to produce a practical commercial oscillograph, however, was William du Bois Duddell, a student under Ayrton at the City and Guilds College, London, and later a collaborator of Professor Marchant at Liverpool University.

Duddell first met Horace Darwin, the head of the Cambridge Instrument Company, in 1897, when he brought to him his ideas for a galvanometer with an extremely small periodic time and a viewing device which would spread out the deflections on a time scale in such a way as to display the shape of a current wave repeating itself 50 or 100 times a second. An announcement of the instrument was made at the British Association held in Toronto in the same year and the first description was published at a meeting of the British Association.[19, 20] The original instrument made by Duddell is in the Science Museum, South Kensington, London.

By the use of a pair of tight flat bronze strips, carrying a tiny mirror submerged in oil for damping purposes, Duddell produced an instrument with a free vibration period one ten thousandth of a second, the shortest periodic time of any galvanometer so far made. A synchronous motor drove an oscillating mirror and Duddell proudly claimed 'with this new instrument one can investigate with accuracy cycles of changes which take place in so small a time as one three-hundredth of a second'. The oscillograph has proved a valuable tool in many different spheres and has been employed in large numbers all over the world. Although valuable improvements have been effected

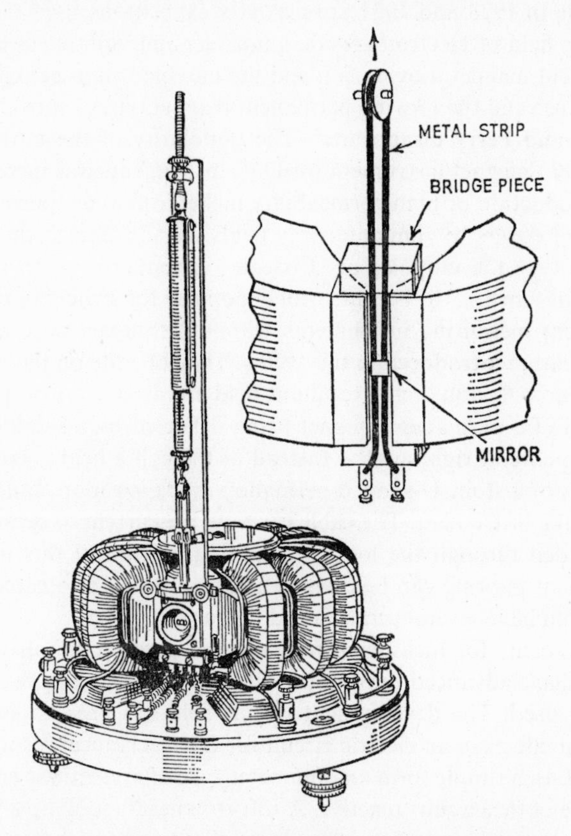

Fig. 69. Duddell Oscillograph, 1897. A galvanometer with a taut strip, suspending in the air gap of a powerful electromagnet a coil carrying a mirror, had an extremely small periodic time. A mechanically operated viewing device spread out the deflections on a time scale.

on the original design the fundamental idea today remains identically that first worked out by Duddell sixty years ago.

From 1900 the story of electrical measurement has advanced at a rapid and ever-increasing rate with a vast array of instruments and methods made available for the electrical engineer. Nothing more can be done here than to indicate a few of the more outstanding features and to refer the reader to such reviews of the subject as appear from time to time in the *Journal* of the I.E.E. Two by C. V. Drysdale in 1928 and 1931 are masterly expositions.[21, 22]

In the field of electromagnetic ammeters and voltmeters both the permanent magnet moving coil and the moving iron types came into general use and the moving permanent magnet type as introduced by Ayrton and Perry disappeared. The popularity of the moving coil permanent magnet instrument for D.C. instruments was increased by the introduction of high permeability nickel iron alloy, permalloy or Mumetal covers to eliminate stray fields. In addition to the bipolar Weston type the unipolar or 'Cirscale' introduced by Record with long scales up to 210° became widely popular for switchboard use.

Current measuring instruments without terminals were an interesting feature introduced in the 1920's. They operate on the principle of the early Ayrton–Perry resultant field ammeter, an iron needle in the field of a permanent magnet being deflected by a small current-carrying core at right angles. Instead of the coil a field is created by the ends of a stout U-shaped permalloy or stalloy loop standing out behind the instrument. The cable carrying the current to be measured is threaded through the loop. In the first version of this iron ring instrument the ring can be divided and clamped over a cable or bus bar or can have several turns passed through the loop.

Instruments for measuring frequency of alternating current electrically have advanced beyond the vibrating reed type though these are still used. The development in this field has been to substitute resonant effects in an electric circuit for the mechanical resonance of the reed. In a simple form an ohmmeter type of system was employed with one of the circuits reactive. A soft-iron needle took up a position dependent on the relative strengths of the current with reactive and non-reactive circuits and so gave a measure of the frequency. In another form the moving coil of a dynamometer was split at the centre and the two halves received current from two resonant circuits, one having a period higher than the normal for the instrument and one lower.

Many portable instruments were invented to meet the increasing demands for conductor and insulator testing such as the world-famous Megger of Evershed, and miniature instruments of various kinds have become very popular. The application of thermo-junctions has made permanent magnet moving coil instruments suitable for use on alternating current up to the highest frequencies, great care having been taken with the design of junction as in the case of the one produced by the Cambridge Instrument Company which has housed the junction in a vacuum container.

Between 1930 and 1945 considerable attention was paid to the improvement of the electric supply meter and prepayment meters were designed and introduced on a wide scale. Moulded cases replaced the iron ones. Much attention was paid to the bearings, which had always given trouble, and researches by Mr. Shotter in

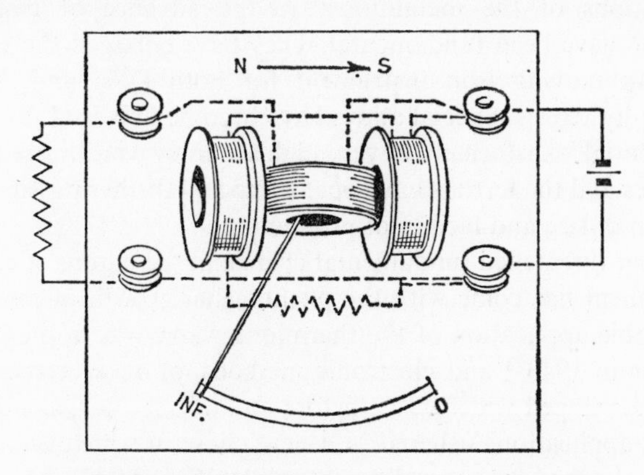

Fig. 70. Evershed's Megger. Copy of a contemporary drawing which illustrates the principle of the original Ohmmeter designed by Evershed. A small permanent magnet was suspended at the intersection of the centre lines of two fixed coils at right angles. The scale read directly in ohms.

sapphire cups, steel points and special oils assumed classic importance. Improvement in the steel used for braking magnets has resulted in greater stability so that today a meter should not fail for any of the former reasons in less than eight to ten years' life. Previous errors have been eliminated and now, through improvement in testing departments run by the supply authorities, there is no difficulty in complying with the strict legal requirements imposed by the Board of Trade.

With all forms of electromagnetic instruments a critical factor in development over the past few decades has been the quality of the magnetic component. Just prior to the First World War the importance of this factor became recognized and aluminium-nickel-cobalt alloys made their appearance. Before this date practically all the special steels for permanent magnets used in instrument manufacture were imported, but the position changed rapidly and now magnets produced in this country are as good as any. Much credit for the progress must be given to Evershed, whose two papers on the subject in 1920 and 1925 created a great impression.[23, 24] The measure of improvement effected over the years is to be seen in the figures: whereas tungsten steel gave a coercive force of 90 oersteds, 15 per cent cobalt gave 190 and recently Alnico has raised the figure to 600. The contributions of the metallurgist to the advance of instrument technique have been fundamental. They have enlarged the scope of the cheap moving iron instrument for both D.C. and A.C. by reducing hysteresis. Special alloys have become universal for current and potential transformers of very high accuracy while more recently dust cores and the ferrites have opened up still further possibilities in reduction of size and facility of production.

Another threatened fundamental change in the nature of electrical measurement has come with the phenomenal growth of radio. The first notable application of the thermionic valve was in the Moullin voltmeter in 1943[25] and electronic methods of measurement have already developed widely. Just before and during the Second World War the applications ushered in a new phase in which the possible limits of the feed-back amplifier are still unforeseeable.

This chapter ought clearly never to be closed; two vast fields of electrical measurement—industrial process control and medical research and diagnosis have not yet been mentioned—but reference to one outstanding example must suffice.

For a century and a half it has been known that contraction of the muscles of the heart is accompanied by the generation of an electric potential and medical men have recognized the diagnostic possibilities of examining this potential. Professor Einthoven, of Leyden University, a distinguished physiologist, looking for a very sensitive form of galvanometer with which to investigate the question, tried various constructions and ultimately adopted the single-wire d'Arsonval galvanometer already used—unknown to him—for submarine telegraphy by Ader. One by one he overcame the problems of securing the required sensitivity and stability and in 1903 published his paper on the subject 'A New Galvanometer'[26] which became known as the String Galvanometer.

With the paper were published cardiograms which compare very favourably with those obtained today on modern equipment, but the equipment was large and unwieldy. About the year 1908 Einthoven invited the co-operation of the Cambridge Instrument Company in the production of an improved instrument and after three years a set was installed in University College, London. Other firms on the Continent and in the United States produced their own designs and electro-cardiography became an established medical science.

The improvement of instrumental details, the nature of contacts applied to the patient and the provision of convenient photographic attachments occupied many years, but by the end of the First World War cardiographs were being made for hospitals all over the world. The demand for a portable instrument became insistent and no sooner had the Einthoven principle been modified for this purpose about 1928, than the arrival of the thermionic valve amplifier changed the whole picture. A cathode ray screen was first adopted and then in 1948 a direct-writing cardiograph was produced. Today the existence of the portable instrument known as the 'Transrite', weighing about 20 lb., is a tribute to the electrical engineers who have combined widely divergent sciences into a valuable diagnostic aid. By the addition of a cathode ray tube the operating surgeon and his anaesthetist may also keep before them during an operation a live tell-tale on the patient's condition.

REFERENCES

1. *Phil. Trans*, vol. 45, p. 195, 1748, and vol. 45, p. 195, 1749.
2. Mottelay, p. 166.
3. *Phil. Trans.*, vol. 77, p. 26, 1787.
4. Mottelay, p. 289.
5. *Mem. de l'Académie Royale des Sciences*, pp. 569–611. 1785.
6. *British Assoc. Report*, 1855, pt. ii, p. 22.
7. *J.I.E.E.*, vol. 95, Part I, p. 27, 1948.
8. Whittaker, *Electric Light Installation*. p. 138. London 1894.
9. *Phil. Trans.*, vol. 123, p. 95, 1833.
10. *Phil. Trans.*, vol. 133, p. 303, 1843.
11. Poggendorff's *Annalen*, vol. 54, p. 161, 1841.
12. *Handbook for the Electrical Laboratory and Testing Room*, p. 376. London 1901.
13. *Reports of the B.A. Committee*. Camb. Univ. Press, 1913.
14. R. E. Crompton, *Reminiscences*. London 1928.

15. *J.I.E.E.*, vol. 89, pt. 1, p. 526, 1942.
16. *J.I.E.E.*, vol. 94, pt. 1, p. 342, 1947.
17. H. G. Mitchell, *Fundamentals of Technical Electricity*. Methuen, 1952.
18. G. R. Noakes, *Text Book of Electricity and Magnetism*. Macmillan, 1954.
19. *The Electrician*, vol. 39, p. 637, 1897.
20. *B.A. Report*. 1897.
21. *J.I.E.E.*, vol. 66, p. 596, 1928.
22. *J.I.E.E.*, vol. 69, p. 170, 1931.
23. *J.I.E.E.*, vol. 58, p. 780, 1920.
24. *J.I.E.E.*, vol. 63, p. 725, 1925.
25. *J.I.E.E.*, vol. 58, p. 1039, 1930
26. *Annalen der Physik*. IV Series, vol. 12, p. 1059, 1903.

CHAPTER XIX

Professional Organization

In a field where discovery and the application of science have advanced side by side on an ever-broadening front, as in electrical engineering, exchange of ideas between those interested has always been a vital factor in progress. For centuries past groups of scientific men have formed for the dissemination and discussion of the available knowledge and for hearing about new discoveries and the working of new inventions. Today the Institution of Electrical Engineers stands pre-eminent as the professional organization and learned society for all those who are engaged in elucidating electrical science and applying it to the uses of man.

For many years before the formation of a professional body for electrical engineering, the pioneers who were laying the basis of the science found a suitable forum in the Royal Society. Many fundamental communications on electrical engineering by Sturgeon, Kelvin, Wheatstone, Hopkinson and others have appeared in its *Transactions*.

During the second half of the eighteenth century small societies were founded by groups of men interested more in the development of engineering and technology than in pure science. The Lunar Society, so called because the meetings were arranged at the time of full moon in order that the members could find their way home easily, included many well-known pioneers whose names became famous—Watt, Boulton, Murdock, Priestley and others.

In this connection it is interesting to quote if only as an indictment of the slow progress of civilization the fact that during the First World War the Council of the Institution of Electrical Engineers, in a discussion on the dates and times of meetings in the special circumstances then obtaining, decided that they 'should be held during hours of good moonlight', but here the reason was more the deterrent effect of moonlight on enemy air raiders than its convenience to the members.

The subjects discussed by the Lunar Society covered a wide range including, besides science and engineering, the great political issues of the day, which at times brought upon them the wrath of the mob.

Another society, known as the Smeatonians, which was formed about 1771, comprised many well-known civil engineers, John Smeaton, Robert Stevenson, George Rennie among others: and from its early rules appears to have been a model for subsequent societies. One of the objects was to 'pass the evening in that species of conversation which provokes the communication of knowledge more readily and rapidly than it can be obtained from private study or books alone.' The society was for 'real engineers, actually employed as such in public or private service', and those who could 'have been real engineers had it not been their good fortune to have it in their power to employ others in this profession'!

The Smeatonians paved the way for The Institution of Civil Engineers which was established in 1818 and incorporated by Royal Charter ten years later, a date when engineering may be claimed to have become a recognized profession.

At this time, electrostatics and magnetism had been the subject of study by many scientists over several centuries. Lightning conductors were in use and the compass was established for navigational purposes. But when the Institution of Civil Engineers was founded the steady electric current had only been known for eighteen years, that is since the voltaic pile had been announced through the Royal Society in 1800. Two further years were to elapse before the broad path to electrical engineering was opened up by the recognition of electromagnetism through the discovery of Oersted. The Admiralty had, however, told Ronalds (in 1816) that telegraphs were wholly unnecessary but, eleven years after the 'Civils' received their charter, the Great Western Railway Company were operating a commercial line (1837). By this date Faraday had made his important discoveries and both dynamos and motors were being designed culminating in the provision of mechanically generated electric light for lighthouses in 1858. In the same year as the great Exhibition, 1851, a public telegraph service was opened between England and France through a cross-channel submarine cable and this was quickly followed by the phenomenal growth of submarine telegraphs throughout the world.

During this period the forum for electrical engineering discussions was divided between the Royal Society, The Institution of Civil Engineers and the British Association for the Advancement of Science, but the costly failures of the 1857 and 1858 transatlantic

Plate XLV

Alternating Current Bridge, 1935–50
modification of the Wheatstone Bridge to make it suitable for use with alternating current been the subject of much research over the past forty years. Many refinements have today o a convenient accurate instrument.
to: Cambridge Instrument Co. Ltd.)

Modern High Precision Potentiometer
Potentiometer of Poggendorff (1841) has been constantly made in improved form. Crompton others replaced the slide wire by coils and by 1900 it was looked upon as a fundamental od of measuring voltage and current accurately. The achievement to date can be appreciated ference to the instrument depicted with which steps in voltage of 0·000001 can be obtained rately.
to: Cambridge Instrument Co. Ltd.)

PLATE XLVI

(*a*) Picture Transmission by Wire
This historic picture transmitted in 1924 indicates the standard

(*b*) Picture Transmission by Wire
In comparison with the illustration above, this recent result

cables and the important government inquiry set up in 1859 began to direct attention to the specialist nature of electrical engineering. The telegraph had been extensively developed in the United States over the previous twenty years and Great Britain, along with other European countries, was becoming alive to the need for speeding up its more modest efforts. There was an increasing demand for more telegraph engineers with a scientific knowledge so that it was a logical step when on 17 May 1871 a group of men closely associated with the development of telegraph communications met at No. 2 Westminster Chambers, Victoria Street, London, 'To consider the expediency of forming a Society of Telegraph Engineers having for its object the general advancement of Electrical and Telegraphic Science, and more particularly for facilitating the exchange of information and ideas among its members.'

Among the eight names recorded in the minutes of the meeting were those of men who were already distinguished for their contributions to telegraph engineering. Whitehouse, who occupied the chair, was known for his experiments on the speed of signalling in submarine cables and his association with the laying of the Atlantic cable. Sabine, son-in-law of Wheatstone, became famous as the author of *History and Progress of the Electric Telegraph*, which had been published in 1867 and had already reached its third edition and for the *Electrical Tables and Formulae*, published in 1871, which he compiled in conjunction with Latimer Clark. For different reasons Loeffler was a noteworthy member of the committee: after his submarine cable activities in association with Siemens Brothers, he died leaving a sum of a million and a half! There were also representatives of signalling departments in the Services including Webber, who rendered outstanding service to telegraph administration and who appears to have been the prime mover in organizing the meeting although in a subsequent public pronouncement he gave full credit to Major Bolton (later Sir Francis Bolton) for the success of the venture.

At the meeting on 17 May 1871 Bolton, who was appointed Honorary Secretary, submitted a list of 'gentlemen connected with the Telegraphic profession who were desirous of becoming Members or Associates of the proposed Society' and in the list of 66 who were elected there appeared many whose names became well-known in telegraph history; in addition to those already mentioned, Latimer Clark, Carey Foster, W. T. Henley, W. H. Preece, S. E. Phillips, Willoughby Smith, Spagnoletti, Charles W. Siemens, E. Graves, R. S. Culley and others.

At a second meeting, on 31 May 1871, Siemens was elected first president, Major Bolton, hon. secretary, Robert Sabine as treasurer and librarian, along with a council of eleven. A committee was appointed to draft rules.

The new and thriving Society of Telegraph Engineers was given a warm welcome by the Institution of Civil Engineers, who from 1872 granted the use of its rooms on the famous site at 25 Great George Street free of expense for its meetings, an arrangement which continued for 24 years. At the opening meeting on the 28th February 1872, C. F. Varley made a prophetic speech which included the words: 'This Society will gradually, by natural selection, develop more into an electrical society than into a society of telegraphy proper: and the moment it is understood that all papers on electricity or bearing directly upon the development of electrical science are admitted, it at once takes the science out of the narrow groove in which it seems to be drifting, into the most extensive of all grooves, because it will be found later to embrace every operation in nature.' The membership rapidly grew and by 1 January 1873 there were 155 members, 170 associates, 25 foreign members, but only 2 students, a total of 352, and during the following year 100 more were added. The constitution of the membership at this stage appears to have been roughly one-quarter from the British Post Office, somewhat less from the Services, a further similar fraction equally divided between railway, foreign telegraph departments and scientists, with a sprinkling of manufacturers.

From 1872 to 1874 the Society had offices at 2 and 5 Westminster Chambers, and then moved to 4 Broad Sanctuary. In March 1903 the offices were moved to 92 Victoria Street, where they remained until the present building in Savoy Place was occupied in 1910. Fortunately for posterity the Society immediately founded a *Journal*, the publication of which has continued in an unbroken series to the present day, but for the first eight years the costs of this feature introduced grave financial difficulties. Great difficulty was experienced in getting in the subscriptions from members and at one time the costs of printing were in arrears with a bank balance of only £44. By the generosity of one or two members and the introduction of improved business methods, however, the Society weathered the storm.

Administrative problems did not detract from progress in the main objects of the Society and the first paper read on 13 March 1872 by R. S. Culley, Engineer-in-Chief of the Post Office Telegraphs, was soon followed by others which were published in Vol. 1 of the *Journal*.

The subsequent interest displayed in the proceedings by University professors throughout the history of The Institution and its predecessor was introduced when Carey Foster discussed the Wheatstone bridge in May 1872. The following month the first of the series of conversaziones was held and proved such a great success that a number of electrical exhibits were later transferred to the Albert Hall, where a public lecture given by W. H. Preece was accompanied by the transmission of a message to India and an immediate reply, a subject naturally creating wide public and press interest.

In 1878 Professor W. E. Ayrton undertook the duty of editing the *Journal* and as chairman of the editing committee received an honorarium of fifty guineas. Interest in the papers and discussions became enthusiastic and at times even acrimonious as for instance when two English cable engineers attacked a statement that the submarine telegraph was of German origin. Dr. C. W. Siemens poured oil on the troubled waters by making the frequently repeated statement: 'Submarine telegraphs are specifically English enterprises. I might go further and say every submarine cable which is now (1876) working is almost without exception, the produce of this country, and has been shipped from the Thames.'

The Society naturally became the sounding board for announcing new discoveries and inventions. In 1873 it heard for the first time of the variation of the resistance of selenium with exposure to light and in 1877 received the announcement of Bell's invention of the telephone. A few months afterwards Bell attended a meeting of the Society of Telegraph Engineers and described the results of his own researches. There were mixed feelings among the telegraph engineers, who saw a possible rival in communication, but when Hughes made his announcement of the microphone to the Society a few months later there could be no doubt about the future of this new method of transmitting and receiving the spoken word.

Towards the end of the 1870's the interest displayed in the field of electric lighting led to a modification of the title of the Society. In 1879, the year after the gas scare, in which there had been a serious slump in gas shares owing to the prospects of electric lighting, a paper was read on 'Historical Notes on the Electric Light' by the honorary secretary, Bolton, and the same year Colonel Bateman-Champain of the Royal Engineers, not primarily a telegraph engineer, was elected to the presidential chair. Public opinion was agitated by the progress made in France in the development of the electric light and our backwardness in this field. The papers read before the Society were moreover extending more broadly over the whole field of electricity and magnetism. The result of these various disturbing factors was to lead the Council to propose the addition to the title of the Society the words 'and of Electricians', a proposal which was accepted by the members on 22 December 1880.[1] The 'of' disappeared later. Significantly on 24 November 1880 Swan read his famous paper on the electric light in which he introduced the incandescent filament lamp.

A feature which has always assumed importance in the progress of professional electrical engineering since the early days of the Society of Telegraph Engineers and even for some years previously has been the contribution made by the engineering press. The author is indebted for much of the contents of this chapter to the well-known history of the Institution of Electrical Engineers by Rollo Appleyard,[2] in which the origin of the leading periodicals is reviewed in a convenient form. *The Engineer* and *Engineering* were founded in 1856 and 1866 respectively and both gave space to electrical subjects, but it was *The Electrician*, first produced in November 1861, which aimed at specializing on electrical engineering. It ran, however, for only a brief period and after three years ceased publication. The name *The Electrician* was taken for a new periodical in 1878 and this time with more success. The journal, under Sir John Pender and Sir James Anderson, became a valuable medium as a repository for every phase of electrical progress and has had a distinguished history over the subsequent years. In 1872 *The Telegraphic Journal and Electrical Review* was launched and continued for twenty years when, under Alabaster Gatehouse and Kempe, it became *The Electrical Review*, aiming rather more at the commercial than the academic side of electrical engineering. Later, in 1891, *Lightning* appeared under the control of Robert Hammond, who also served as honorary treasurer of the Institution for many years. One of the interesting features of *Lightning* was the publication of statistics of the industry; subsequently, under the direction of R. W. Hughman, it changed its title to *The Electrical Times* and has made a considerable impact.

During the period from about 1880 to 1890 the Society of Telegraph Engineers and Electricians, through the influence of Professor Ayrton and his colleague Perry, began to take a special interest in the provision of improved facilities for specialized education for electrical engineers and on 1 November 1879, an event of outstanding importance was the establishment of the City and Guilds of London

Institute at Finsbury. The post of Professor was given to Ayrton, a pupil of Kelvin's, who had completed a successful five years as Professor of Electrical Engineering at the Imperial College of Engineering in Japan. Ayrton threw himself into the work with great energy and, by improvisation in inadequate accommodation and an enthusiasm for teaching and instrument design, quickly built up a reputation which has survived over the years. Within two years Finsbury Technical College began to grow. His success can be measured by a statement made by Perry later to the effect that on looking around a large meeting of the Society in 1886 he estimated that nearly three-quarters of those present were Ayrton's students. The Finsbury and City and Guilds tradition migrated to Exhibition Road, South Kensington, where Ayrton's early but robust plant has grown to the fullness of an important and flourishing component in the University of London.

During the year 1887 agitation arose in favour of modifying the title of the Society of Telegraph Engineers and Electricians, so as to represent more adequately all kinds of electrical engineers and on 24 November of that year a Council resolution, proposed by W. H. Preece and seconded by Alexander Siemens, approving the change to the Institution of Electrical Engineers was carried unanimously. This was confirmed by the body of members and on 1 January 1889 a Certificate of Incorporation was obtained from the Registrar of Joint Stock Companies. At this time the membership was about 1,500 and in the same year the first annual dinner was held with Sir William Thomson in the chair as president and Lord Salisbury representing the Government.

From about this time the meetings of the Institution, still held at the Institution of Civil Engineers, were frequently enlivened by the demonstration of newly discovered phenomena which attracted large audiences. In his presidential address in 1891 Crooks showed unfamiliar electrical effects in vacuo while the following year Tesla delighted the members by his experiments at very high frequency. This was one of the early occasions when the meeting had to be held outside—at the Royal Institution—in view of the anticipated large attendance. Here shocks were received safely to everyone's astonishment at 50,000 volts.

Before the end of the century the Institution had admitted women to its meetings for on 13 December 1893 the Students' Committee requested permission for one of them to bring a lady to their next meeting and the Council agreed. A precedent was clearly established as, within a short time, presidents were addressing the meetings with 'Ladies and Gentlemen' and in 1899, Mrs. W. E. Ayrton was elected a member in recognition of her work on the arc. During this same period the practice of the Institution in sending delegates to meetings of other national bodies was established: Ayrton and Silvanus Thompson attended the Volta centenary celebrations in Como. Many of its well-known members were also involved in the national controversies over the sub-division of the electric light, the A.C./D.C. battle, operation of alternators in parallel, and others, and the opposing views were thrashed out at length in the formal discussions of the Institution.

At this time the Institution contributed other matters of interest to the rapidly growing industry through the activity of its Standardizing Committee. The establishment of a National Standardizing Laboratory for electrical instruments had been considered by the committee for some years as a necessary concomitant to the carrying out of the Provisional Orders issued under the Electric Lighting Acts. Along with representatives of the London Chamber of Commerce, Institution delegates had waited on the Prime Minister to press the proposals but without result. On 13 April 1899, however, the Royal Society invited the Institution to nominate two members to serve on the General Board of the National Physical Laboratory, a privilege which it has enjoyed ever since. It was also in July 1897 that the Wiring Rules first compiled in 1882 and issued from the offices of the Society at 4, the Sanctuary, Westminster, were brought up-to-date and secured wider and more authoritative acceptance. During the same year joint discussions with the Physical Society resulted in the establishment of *Science Abstracts*. The Royal Engineer Corps of Electrical Engineers was also set up at the instigation of the Council about this time and a small detachment under Colonel Crompton rendered service in the South African War. Thus started the friendly relations between the Institution and the fighting services which has continued and grown in usefulness throughout the years.

The establishment of local sections of the Institution became a subject of debate as early as 1880 on the request of Walter T. Glover, the Manchester cable-maker, but it was not until new Articles of Association were adopted on 1 January 1899 that the idea was accepted. At the end of the year local centres, at first known as sections, were formed in Dublin, Glasgow and Newcastle, and a few months later a Manchester centre arose out of an amalgamation of the Northern Society of Electrical Engineers with the Institution.

Today there are twenty-one local centres and sub-centres in these Islands.

By the same changes in the Articles, the class of Associate Member was established to rank between those of Member and Associate, and the letters M.I.E.E. and A.M.I.E.E. were adopted.

During the first decade of the present century one of the outstanding activities of the Institution of Electrical Engineers was the organization of summer visits. Before 1900 such visits had taken place to the Paris Electrical Exhibition in 1881 and 1889 and Switzerland in 1899. Successful visits continued to take place both at home and overseas and in 1904 one of the most important from the practical results which followed was that to the St. Louis Electrical Congress. In co-operation with members of the American Institute of Electrical Engineers, important discussions took place on electromagnetic units and International standardization, one outcome of which was the founding of the International Electrotechnical Commission, which celebrated its jubilee in Philadelphia a few years ago and is today the recognized organization for setting up international standards in the electrical field. Only two years before this event, the Institution had begun to support a Standardizing Committee appointed by the Institutions of Civil and Mechanical Engineers. This Committee became the British Engineering Standards Association in 1918 and later, in 1931, the British Standards Institution. Throughout its growth, the Institution of Electrical Engineers has always maintained the closest touch with this body both in giving financial support and in providing for representation on many of its committees.

During the past half-century the need to subdivide the ever-increasing field of electrical engineering for professional purposes has been the subject of review in the Institution. In 1907 the first step was taken when four Sectional Committees were formed to foster communications dealing respectively with (1) Traction, Light and Power Distribution, (2) Telegraphs and Telephones, (3) Manufacturing, and (4) Electro-chemistry and Electro-metallurgy.

By the year 1900 the facilities available at The Institution of Civil Engineers for meetings of the Electricals were strained to the utmost and the Council turned its attention to finding a building, or a site, on which a building could be erected expressly for the use of the Institution. Various alternatives were under contemplation around Victoria for several years without result. In 1908 a building fund which had been established on a small scale in 1894 and so wisely built up over the years, stood at well over £10,000, with the total assets of the Institution approaching £50,000. At this juncture the present building in the precincts of the Savoy Embankment became available. It was used as the Examination Halls of the Royal College of Physicians and Surgeons, but was no longer required by them as their examinations were now held in their own lecture theatres. The remaining 76 years of a 99-years' lease at a ground rent of £2,201 per annum was secured in June 1909 for the sum of £50,000. To bring the building story right up to date it should be reported that the offices were transferred to the new building in 1910 and have remained there to this day.

One of the earliest uses to which the new building was put was for the first meeting of the International Electrotechnical Commission in England, under the chairmanship of Mr. A. J. Balfour, M.P., a happy augury of the many occasions on which it has since been used for international gatherings. In 1913 the International Radio Conference was held in Paris with the Société des Electriciens. More recently the lecture hall in London has been linked by radio and later by submarine telephone cable with other capitals, including New York for joint meetings of electrical engineers physically separated by thousands of miles yet exchanging views in one common debate. Interchange of visits of presidents and vice-presidents with overseas centres has become commonplace and the Institution building is a mecca for Commonwealth and other visitors to London.

A different phase of professional organization is the way in which other bodies having kindred objects keep contact with the premier Institution. Even before the First World War the practice of placing the lecture theatre at the disposal of such bodies was established and, in his *History*, Mr. Appleyard gives a list of fifteen who had taken advantage of this privilege, including the important Institution of Post Office Electrical Engineers.

Further, in education and training and in many other fields of common professional interest, there has been for many years the closest relation between the three major Institutions, the Civil, Mechanicals and the Electricals, and with universities and technical colleges. The Association of Principals of Technical Colleges and the Association of Technical Colleges have, over the years, co-operated closely in the development of the National Certificate in Electrical Engineering.

In both World Wars the influence of the Institution of Electrical Engineers has been noteworthy in recruiting suitable members for the Forces and in technical co-operation in the provision of electrical equipment and services. But one of the outstanding professional results of the first war was the establishment of the principle of sectional meetings devoted to one particular branch of electrical engineering. A beginning was made with a wireless section, now the Electronics and Communications Section, the first meeting of which took place on 19 November 1919. There was then in existence a body of radio amateurs, the Wireless Society of London, later to become the Radio Society of Great Britain, to which the Institution provided and continues to provide hospitality for its meetings.

Nine years later, that is in 1928, the need for further sectionalization was established. There had grown up the Meter Engineers Technical Association, and this was absorbed into the Meter and Instrument Section. The Overhead Lines Association formed in 1927 flourished independently of the Institution until 1934, when at its own request its members joined the newly instituted Transmission Section to embrace both overhead and underground cable problems.

An event of great significance in the history of electrical engineering took place in 1921 in the granting of a Royal Charter to the Institution of Electrical Engineers, an ambition which had existed since before the war, and new bye-laws were allowed by the Privy Council on 1 June 1922. Two years later a new bye-law conveyed the right of members to use the title Chartered Electrical Engineer.

Another feature of the Institution which must be considered a major factor in the progress of electrical engineering since the profession became organized is the library which today contains something over 12,000 scientific and technical volumes, with long runs of periodicals and journals of electrical and allied bodies. Although in the early days, the finances of the Society of Telegraph Engineers were rather tight the germ of a library was established and by the end of 1882 the annual vote for the purchase of books, binding of periodicals, and library staff had risen to £75 per annum. Earlier in the year the librarian had been paid an honorarium of £15 'in recognition of his services in completing an Index to the first ten volumes of the *Journal*'. On the occasion of the Jubilee of the Telegraph subscriptions were invited in support of the library and an amount of £162 17s. 6d. obtained.

In 1905 the library of the Institution was reorganized and it was hoped that 'it should be the most perfect library of electrical literature in this country if not in the whole world'. The Council appointed

W. Duddell, T. Mather and Silvanus Thompson to form a small committee to investigate the literary contents of the library and to make recommendations for additions. They drew up a list of 1,623 books which, with their joint knowledge of scientific and technical literature, they considered should be added to the existing collection and within four years over 1,400 had been acquired. During the next few years the library was the subject of special attention by the Council who, in addition to the normal grant of £75 per annum, authorized an expenditure of £650 to be spread over three years. There were no lending facilities in the library at this time and in spite of efforts made by Mr. C. H. Wordingham and others no such addition could be faced on account of the cost. It was estimated that a capital expenditure of £690 would be required with an additional annual cost of £200 and it was not until 1914 that this valuable feature was added.

In addition to the main library, which today contains practically every publication of electrical engineering interest, there are two special collections received as the result of bequests. The Ronalds Library was collected at his residence at Battle in Sussex by the Sir Francis Ronalds, who has been mentioned several times in previous chapters, and this was presented to the Society of Telegraph Engineers after his death in 1873. It consists of some 2,000 volumes and 4,000 pamphlets. The Silvanus Thomson Memorial Library was purchased in 1917 with the support of members of the Institution and particularly of old students of Finsbury Technical College. It contains 900 rare books, a number of interesting manuscripts, and 2,500 scientific books of the nineteenth century, besides many thousands of periodicals, proceedings, and pamphlets. The Institution Library is a gold-mine of electrical literary treasures including, for instance, a fourteenth-century manuscript of Peter Peregrinus's *De Magnete*, and the printed edition of 1558, as well as each of the three rare editions of William Gilbert's work, dated 1600, 1628 and 1633. Much of Faraday's correspondence is included, and of more recent interest, many manuscripts, notebooks and textbooks of Oliver Heaviside.

As may be expected the *Journal* and *Proceedings of The Institution of Electrical Engineers* embrace the whole field of electrical engineering knowledge as contributed by its members and others. The *Journal*, which has grown continuously in size and variety of contents from the early days, was in 1941 divided into three parts: Part I, General; Part II, Power Engineering; Part III, Radio and Communication Engineering. Eight years later the name *Proceedings* was adopted accompanied by a new type of *Journal* giving matters of topical interest. In 1955, owing to rapid developments in the subject, a further major change became necessary and the present form of *Proceedings* was adopted, consisting of the *Journal* and the *Proceedings*, in Parts A, B and C. The *Journal*, published monthly, contains special articles based on Institution papers or describing current electrical engineering events or developments, summaries of lectures, and short reviews of papers published in Parts A and B of the *Proceedings*. It also covers a wide range of announcements of interest to members. Papers are now allocated to Parts A and B of the *Proceedings* on the basis of content rather than, as formerly, on the meeting at which they were read. Part A is for power engineering, that is supply and utilization, including all related measurement and control section papers, while Part B contains electronics and communications section papers including those papers read in the measurement and control section which are of interest to light current engineers. Part C, Monographs, contains papers which, because they are highly specialized or for some other reason are likely to be required by only a small minority of members, is published twice a year.

The membership of the Institution of Electrical Engineers in 1961 was 47,500 and it is one of the largest societies of its kind in the world. Most countries have bodies with similar objects as, for instance, the American Institute of Electrical Engineers, the Engineering Institute of Canada, Institute of Engineers, Australia, Société des Ingenieurs Civils de France, Verband Deutsche Ingenieure, as in some of the Commonwealth countries, while in the United States, radio engineering is looked after separately by the Institute of Radio Engineers.

At an early stage in its history the Institution gave attention to the requirements to be met by candidates for its membership. Before the granting of the Royal Charter in 1921 care was increasingly exercised as to the level of attainment in electrical engineering education and the level of electrical engineering responsibility undertaken by a candidate before he was admitted. Following the grant of the Royal Charter, and the privilege which ensued for corporate members to describe themselves as Chartered Electrical Engineers, the Council have recognized their responsibility to the public in determining those who may claim this distinction and regularly review the requirements for membership to ensure that these keep abreast of the growing range and complexity of electrical technology and its impact on the national life.

The preceding brief account of the Institution's history charts the development of its present responsibility which may be classified under three heads: first, to act as a learned society providing opportunities by its meetings and publications for the exchange of information and ideas on electrical engineering and all that this connotes; secondly, as a qualifying body, through admission to its various categories of membership, as has just been described; and thirdly, as the regulating body for the profession, determining, through its bye-laws, standards of professional conduct which its members agree to observe.

The successful growth of a profession of electrical engineering is one of the most striking developments throughout the whole history of electrical engineering and the Institution of Electrical Engineers, by its wide range of activities and its extensive ramifications at home and abroad, has achieved a recognition which places it at the highest level of professional engineering bodies.

REFERENCES

1. *J.I.E.E.* (Sec. Tel. Engrs.), vol. 9, p. 400, 1880.
2. Rollo Appleyard, *History of the Institution of Electrical Engineers*, 1871–1931.

CHAPTER XX

Social and Historical Background

In tracing the history of any movement in a specialized field it is often desirable to place it in the perspective of the surrounding scene. Events which at first sight do not appear to be closely connected with the narrative often assume greater relevance and increased interest when viewed in this way. So it is with the history of electrical engineering. The political, social and economic surroundings into which it was born, the influence of these surroundings on the individuals prominent in the story, and the ultimate influence of the development itself on human progress all constitute a backcloth for the act without which its full significance cannot be appreciated.

Within a few weeks of the young bookbinder's apprentice, Faraday, making those historical notes of Davy's lecture at the Royal Institution in the spring of 1812, Napoleon collected the Grand Army, a mighty host of soldiers from France and many other countries, armed and provisioned for a massive onslaught on Russia. As the defenders retired they laid waste the country before the advancing horde but Napoleon pressed on. At Borodino, where 100,000 dead were left on the battlefield, the last major obstacle was overcome and by the end of the summer Napoleon was in Moscow. This, however, proved to be a trap. Notwithstanding all his blandishments he failed to secure the co-operation of the Muscovites and, with the winter approaching only retreat was possible. Then followed one of the most tragic marches in history. Without food or shelter, the soldiers killed and ate their horses and died as they slept, covered by snow. Weapons and booty were abandoned and in the middle of December less than 20,000 ragged emaciated stragglers returned, all that remained of the 600,000 who had set out as a disciplined army six months before. During this same period Faraday had bound up his notes of the four lectures and submitted them to Davy with a request to be found a post as assistant at the Royal Institution. As the stragglers from Moscow were re-entering France Faraday received a letter from Davy making an appointment which led to that long association with the Royal Institution during which he made so many fundamental contributions to electrical engineering. Napoleon and Faraday! What an interesting comparison and what an indictment of man's ingratitude! Napoleon, the creator of misery and death for hundreds of thousands, rests in solemn grandeur in Les Invalides; Faraday, the great benefactor who, as founder of electrical engineering, did so much for human progress, lies in a modest grave at Highgate cemetery known to few.

Considerable light is thrown on the situation in Europe at this period by a record which Faraday left. Soon after his appointment at the Royal Institution Davy decided on carrying out a 'journey of scientific enquiry'. His wife and a maid were to accompany him and he took Faraday, then twenty-two, to assist him in experiments to be carried out on the way and to act as secretary-valet. They were to be away for eighteen months, at a time between the Retreat from Moscow (1812) and the Battle of Waterloo (1815), when England was at war with France. They were to visit many cities in France, Italy and Germany which, with our modern experiences of wartime security, seems incredible. Nor were they, Englishmen in a hostile country, apparently restricted in any way. They openly watched military preparations and went anywhere seeing whom they pleased. The journey from London to Paris occupied eleven days. Two days by road brought them to Exeter and the next day to Plymouth. After waiting two days for the wind, they sailed to Morlaix, sighting a French privateer on the way but without incident. Customs examination at the French port seems to have been unduly troublesome as, among other items, they carried boxes of scientific apparatus including 'glass tubes, small receivers, retorts, and capsules, a blow-pipe apparatus, a small pneumatic trough, a delicate balance, and a few other necessary articles'. They reached Paris by carriage via Rennes and Versailles and stayed three months, during which time they were treated in a friendly way by the French, being admitted to museums and libraries any day of the week while the ordinary public were restricted to two days a week.

Between accompanying Davy in his talks with Ampère and other scientists, Faraday had opportunities for seeing the sights. He describes the Emperor's (Napoleon) visit to the Senate in full state, 'sitting in one corner of his carriage covered and almost hidden from sight by an enormous robe of ermine and his face shaded by a tremendous plume of feathers that descended from a velvet hat . . . his carriage was very rich and fourteen servants stood upon it in various parts. An enormous crowd surrounded him. . . . No acclamation was heard where I stood, and no comments.' Faraday also described the view over Paris from Montmartre where he observed the Chappé semaphore telegraphs, some of them operating all day long. The electric telegraph was waiting for Oersted's discovery seven years later.

On the way to Nice many observations were made, extinct volcanoes, a house on fire extinguished by sealing up all cracks and excluding the air—'very philosophical,' comments Faraday—and, at Genoa, carrying out experiments on the torpedo fish.

Contributions to the electrical age have come from all civilized countries but, for various reasons, it was in Britain that the new ideas fell early on the most fertile soil. The telegraph of Cooke and Wheatstone in 1837 resulted, a few years later, in the establishing of the submarine-cable industry on the banks of the Thames where, up to the present day, long-distance cables have been made and shipped to all parts of the world. In this country, too, the invention of the steam engine and the availability of coal and iron produced the earliest example of industrial society. Consequently, among fringe factors affecting the history of electrical engineering, the story of the Industrial Revolution during the latter part of the eighteenth century and the first half of the nineteenth is of paramount interest.

While Galvani was experimenting in Bologna with his frogs' legs, France and England were in the throes of two quite distinct national movements which were to shape human destiny for all time. In the case of France it is possible to give a date, for it was on 5 May 1789 that the meeting of the States-General held at Versailles brought to a head the deep discontent with the regime and ushered in those painful years of the French Revolution which ultimately led to a new constitution. In England the process was a more gradual one and while, at times, it involved misery for many, there was no large-scale bloodshed. The Industrial Revolution did not spring from discontent or political pressure but rather from a realization of opportunities for exploiting natural resources and inventions. Britain already had the freedom which the French were demanding and were thus one step ahead, although the population of France was nearly three times that of England—twenty-six millions to our nine millions. The last of the serfs had disappeared by the beginning of the seventeenth century and, at the period we are discussing, the inhabitants were able to move about the country more freely than any other people in the world.

On the accession of George III in 1760 agriculture was the principal industry, and there was a small cotton industry, centred in Manchester and Bolton, with an export value of only £50,000 per annum,[1] and a small wool industry in Yorkshire. At this time cotton-spinning and weaving were carried out in the homes of agricultural workers as a part-time occupation, the 'spinsters' of the family sitting around with their distaffs in hand supplying the one loom. It required six to eight spinners to keep a loom supplied.

The development during the next few decades assumed a complex pattern of interrelated factors. The increasing scarcity of wood for fuel turned attention to coal, but flooding made deep mining impracticable. The state of the roads, which were mostly suitable for packhorses only, made communication difficult and transport of goods almost impossible. Newcomen's rudimentary steam pump invented in 1712 offered a solution to the coal-mining problem and many were installed. Attempts to save labour in the cotton industry, already scarce, resulted in a small amount of mechanization by water power on streams in hilly country. Even these small efforts emphasized the transport problem and a comprehensive canal system resulted. The first canal was built by the Duke of Bridgwater at Manchester and completed in 1761 and by 1830 there were over 3,000 miles of navigable inland waterways. As more coal became available, an iron industry was established, the raw material usually being found near the coal beds. Various processes, however, depended on hand-operated or water-powered devices such as blowers for the forge, tilt-hammers and ventilation of the mine, while horse winches were used for hauling the minerals.

The arrival of James Watts' invention of the rotary steam engine changed everything. Engines on Newcomen's condensing principle had been built and installed at costs of about £1,000 each in many parts of Northern England and Scotland, but were extremely slow and wasteful owing to the cooling down of the cylinders at every stroke for condensing the steam. By introducing the separate condenser with automatic valve gear in 1765 Watt reduced the coal bill to one-quarter and increased the output of a given size of cylinder many times.

As the application of the steam engine increased, many inventions appeared in the textile industry. In the first place these were applied in the domestic sphere, but the advantages of a closer control of the

PLATE XLVII

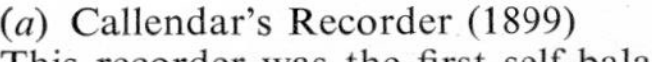

(*a*) Callendar's Recorder (1899)
This recorder was the first self-balancing potentiometer recorder ever made and could be used for temperature measurement both with resistance pyrometers and with thermocouples. It achieved a world-wide reputation.
(*Photo: Cambridge Instrument Co. Ltd.*)

(*b*) Darwin's Thread Recorder
Sir Horace Darwin invented the thread recorder as an economic alternative to Callendar's instrument. In this modification the end of the moving galvanometer pointer was periodically depressed so that it brought an ink impregnated thread into momentary contact with the chart. The record was thus in the form of a line of dots.
(*Photo: Cambridge Instrument Co. Ltd.*)

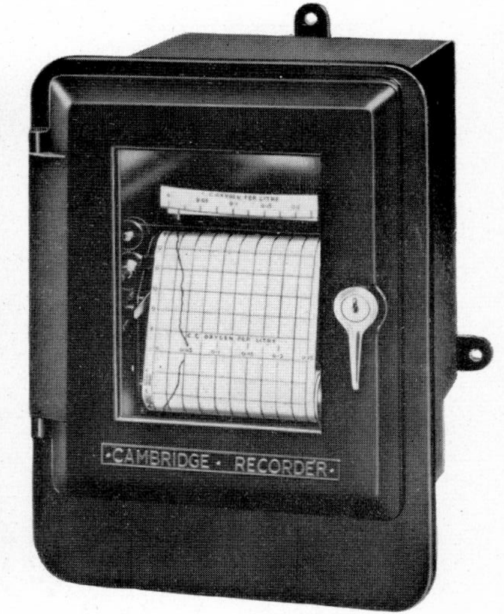

(*c*) Commercial Type Recorder
The thread recorder has been developed for many purposes. This instrument is one example employed for recording the results of gas analysis.
(*Photo: Cambridge Instrument Co. Ltd.*)

PLATE XLVIII

(*a*) The First Electrocardiograph, 1909
Einthoven, of Holland, first invented the cardiograph in 1903, and it was made at Cambridge in a commercial form from 1909 onwards.
(*Photo: Cambridge Instrument Co. Ltd.*)

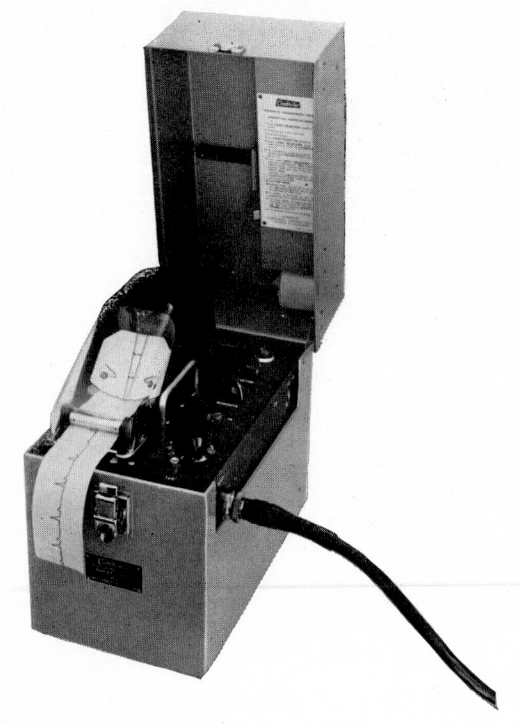

(*b*) Portable Cardiograph, 1960
This instrument has been developed to meet the need for a portable instrument to replace the complicated outfit shown above. Electrodes applied to the patient with special conducting grease now obviate the need for immersion baths.
(*Photo: Cambridge Instrument Co. Ltd.*)

operations soon resulted in bringing together groups of work people in mills. The application of cast iron to the construction of machinery followed and a rapidly expanding demand for the products at home and overseas attracted capital for building and equipping factories on an ever-increasing scale. Until the output of the machines caught up with the demand, and the increase in population in the nineteenth century had its effect, there was a serious shortage of labour in spite of an influx of Irish immigrants and the dispossession of peasant farmers by the Enclosure Acts. Notwithstanding this problem, Britain withstood the costly Napoleonic Wars and the secession of the United States. It emerged as the greatest manufacturing country in the world, as well as the world's greatest banker and carrier.

During the year 1800 weavers became so scarce, and so much trade was being lost by having to send the spun thread abroad to be woven, that special consideration was given to the possibilities of increasing the output of the loom. The result was a great stimulus to invention and by 1835 there were over 100,000 looms engaged on the weaving of cotton alone.

An interesting illustration of the development of one industry from another at this time arose from the chemical requirements of the textile industry. The old method of bleaching cotton by steeping in sour milk and exposing it to the air for several months was altogether inadequate for the increasing trade and the interest of the chemists was secured with the result that several bleaching agents were produced, the most satisfactory one being chlorine, proposed by the French chemist Berthollet. This process quickly became a complete industrial success. Dyeing and printing followed and many new chemical works were established. These in turn called for an extension of the machinery, steam engine, and metallurgical industries.

The beginning of the electrical era coincided with most depressing conditions for the poorer classes in this country. The seventeenth-century textile industry had been based on the home worker. Parents and children had carried out the spinning and weaving as a family, the raw material and woven cloth being brought to them and taken away by the merchants who paid for the labour into the family pool. To secure better supervision labour was then brought to centres, a movement encouraged by the application of small water-wheels, and later to mills specially built for the purpose and supplied with power from steam engines. People were at first reluctant to forego the family relationship but gradually succumbed to the economic conditions. The process of crowding into industrial centres fed on itself as ancillary industries grew up centred on the coal mines for steam power and the iron mines for the raw material of machinery.

Before the eighteenth century there had been devastating epidemics of plague, typhus and smallpox, but plague, at any rate, was becoming rarer, some claims being made that this was due to the adoption, about 1800, of burying the dead in coffins. The congestion brought about by the movement of populations, however, produced a new and worse situation. There was no control over building and such features as the back-to-back house became prevalent while many people lived in crowded cellars. There was no main drainage or disposal of house refuse even in crowded areas. Filth overflowed into the streets and wells and it was unusual to find a clean water supply. Typhus and smallpox became rampant, and, added to this, the facilities for disposal of the dead became hopelessly inadequate. Under these conditions it is not surprising that most people lived very short lives. Knowles[1] states that about 1840 the average age at death for different classes in Preston was as follows:

Gentry	47·39 years
Tradesmen	31·63 years
Operatives	18·28 years.

While home conditions and factory hours were major factors in the degradation of the people it was probably the condition of women and children in coal mines which made the strongest appeal to reformers at the beginning of the nineteenth century. In this connection the engineer also made his contribution as, for instance, in applying steam power for underground haulage which in 1837 released children from this work. Up to a still later date women were employed in large numbers carrying coal up ladders out of the coal mines on their backs. Even after the Watt's engine became available for this purpose the hempen ropes used were costly and wore out quickly. Women were cheaper. It was only when the steel rope was invented that women were released from this labour.

Looking back today on the development of the steam engine and its many applications we cannot avoid the impression that following the original invention of Watt the subsequent progress was unduly slow. There seems to have been no dearth of financial backing. England was rich and inventors never appeared to lack support. In fact it was not unusual for a particular technical requirement, as for instance the early manipulation and dyeing of fibres, to be advertised in the hope that some inventor would produce the answer.

The influence of the founding of the Bank of England at the end of the seventeenth century had resulted in the accumulation of capital in the country. A century later the activities of the young Chancellor of the Exchequer, William Pitt, fresh from studying Adam Smith's *Wealth of Nations* as an undergraduate at Cambridge, had restored the national finances depleted by the American War.

The factor which did hold back progress was the lack of men skilled in the design and manufacture of engineering devices in metal. The early mill machinery was largely constructed of wood but when iron became available as a stronger substitute, it was necessary to build up an entirely new technique. There was no tradition of mechanical fitting and only poor lathes with a few drilling machines.

In 1800, the year of Volta's discovery of the steady current, that is 25 years after Watt's invention of the rotary steam engine, there were less than 300 in use with a total output of only about 4,500 horse power. Serious defects such as leaking pistons, non-circular cylinders and flywheels, valves which failed and crude alignment of parts were never overcome until the machine-tool industry was started early in the nineteenth century and better lathes, planing machines, steam hammers, etc., were made available. With our modern recognition of the need for precision in rotating machinery it seems incredible that some of the early heavy cast-iron engine flywheels, many feet in diameter, were made and ran for years without the use of a lathe and with no more balancing than that obtainable by the adjustment of two sets of four wedges on a square shaft. Once the machine tool was established the steam engine, locomotive and all kinds of auxiliary plant became widely available under the direction of great mechanical engineers like Boulton, Paul, Whitworth and others.

During the nineteenth century living conditions in this country steadily improved. The situation in which most people lived at starvation level and only the very few ever saw such minor luxuries as tea, currants, sultanas, coffee and other common commodities of today was changed by the application of steam to navigation. Food, raw rubber and other articles of commerce arrived in the ever-increasing number of steamers built in our own shipyards and these returned to provide our colonies with the machinery and other products of our workshops. At home the extension of the railway made coal generally available both for factory and domestic use. The use of fire grates and gas lighting spread rapidly. Important advances were made in medicine and Lister's application of Pasteur's epoch-making discoveries in micro-biology revolutionized hospital practice.

This was the matrix then into which the seed was sown by Volta in 1800, from which the crystal of electrical engineering has been growing for one hundred and sixty years and which still shows no sign of abating either in dimensions or variety of aspect.

It is not easy to place the various contributions of electrical engineering in the order in which they have benefited humanity, owing to the fact that some most fundamental effects produce their results in an indirect manner while others are immediately evident. The improvement in our standard of living due to the adoption of the electric drive in factories and workshops for instance, is not so immediately obvious as that due to the instant availability of the electric light in our homes, nor has the day-by-day use of the electric telegraph such a dramatic appeal as the various developments which have sprung from the telegraph, in the telephone, radio and television. There is moreover extensive interweaving of branches: electric traction for example, involves telegraphs, lighting and power, yet as a social factor it has its own direct repercussions. Again, electrical instrumentation without particular application, except in such examples as computers and medicine, touches every branch of electrical engineering.

On broad lines for the purpose of this chapter, then, we may use the classifications Lighting, Power, Communications and Transport. A brief reference to each of these fields will show how stupendous has been the effect of electrical engineering on material and, indeed, on intellectual progress. Consider first of all lighting. It has transformed life throughout the civilized world. From time immemorial up to only about two centuries ago the setting sun brought the normal activities of the day to an end even during the shortest days of winter. The oil wick lamp, in which a piece of wick lay in a container of oil, had of course been used for centuries throughout the East where oil was available. But in European countries the very poor spent the hours before bedtime sitting in the fitful gleams of a wood fire. A few turned to the home-made rush light and, as the scale of prosperity improved, tallow candles made from mutton fat became general with the more expensive beeswax candles for the rich. The relative costs of these three methods of lighting was probably in the order of 1 to 5 to 20. One candle would have to serve the needs of the whole household and an ingenious device enabled close work such as needlework to be carried out during the long evenings by several members of the family. This consisted of a stand carrying a number of hollow glass balls filled with water and arranged in a circle around the candle. The needlewomen took up positions so as to catch the concentrated rays of light, which appears to have enabled them to continue the finest work for hours on end.[2]

When it came to illuminating large buildings by means of candles and wick lamps the problems involved were quite serious, but there were noteworthy examples in which brilliant effects were achieved. Faraday's diary contains an entry which suggests that on occasions he visited the circus for entertainment. After the successful conclusion of his experiments in electromagnetic rotation at Christmas 1821, he was so delighted with his discovery that he proposed to his brother-in-law, George Barnard, a visit to the theatre to commemorate the event. 'Oh let it be Astley's to see the horses,' said George and to Astleys they went.[3] It is known that in 1780 the arena of this famous circus was lit by 28 candelabra with a total of 1,200 float wicks burning in coloured glass containers.

The condition of the atmosphere in a theatre lit by such large numbers of lamps or candles must have been very injurious to the health of both performers and audience. It is said that Lavoisier, the great chemist, in his experiments on oxygen, came to the conclusion that, at one theatre he investigated, the candles themselves consumed at least one-quarter of the oxygen available for breathing by the occupants.

Bearing in mind that matches, which could be struck anywhere to light either fire or candle, did not come into general use until about 1830, candles and indeed oil lamps later were very inconvenient both for home and public use. Boswell, the biographer of Dr. Johnson, describes in a poignant fashion the dilemma he was in while writing at 2 o'clock in the morning. He accidentally snuffed out the candle which was the sole source of illumination in his cold and fireless attic. This may indicate that he had not even the benefit of the snuffless candle, the wick of which curls outwards and snuffs itself. Whatever the reason, he was in the dark anxious to continue his writing but unable to find his tinder box. Hearing the night watchman in the street below he descended stealthily, lest his landlord should shoot him in the dark mistaking him for a burglar. Here he 'relumed' his candle at the horn lantern then carried so generally on the streets at night.

The gap between the candle age and the electric era was bridged by competing developments of the oil lamp and gas lighting. Before the end of the eighteenth century the Argand oil lamp with a flat wick, bent into a tubular form with a central air duct, had been used by the wealthy, and gas appeared on the scene during the first few decades of the nineteenth century. In 1838 experiments were carried out in the House of Commons with a view to replacing the existing system of candle lighting by gas light. At first the heat was intense and only after the introduction of special ventilation were the gas pipes with their hundreds of small burning holes accepted.

In the lighting of theatres there was obviously a great advantage to be obtained from gas in the facilities for control from one central point. The battle for supremacy was soon taken up by the electric light following the work of Edison and Swan, and in spite of the Welsbach gas mantle, the advantages of cleanliness and adaptability very soon placed electricity in the position it is today, the recognized method of illumination for homes of all classes, for factories, streets, road, rail and air vehicles. The link-boy with his smoky, smelly torch dripping tar and dispelling gloom for only a few feet ahead of the pedestrian has been replaced by the general lighting of our streets so that we now see our way clearly. On long journeys by night our coach, even our own private vehicle, throws a powerful beam of light ahead, illuminating brightly the surface of the road and every obstacle—a far cry from the tiny pool of candlelight which travelled slowly along with the vehicle a century ago.

In a less dramatic way, so far as the general public is concerned, electric power has revolutionized our lives as much as electric light. It is not easy to select one example from this field to epitomize this civilizing influence of electrical engineering. From the spectacular machines driving steel rolling mills and the winding gear in collieries to the fractional horse-power motors used by the hundred thousand in so many ways, there is a continuous range of size and application. In factories noisy and dangerous belts and shafting went out when the individual motor drive came in. Not only have these improved the amenities in a factory but, for certain classes of engineering work, have so added to the facilities of speed control and machine tool design that machines employed only ten or twenty years ago are already archaic in comparison with up-to-date devices which possess astonishing outputs and precision in performance.

In the domestic sphere electric power is being extensively employed and continually extended. The vacuum cleaner, the washing machine, small fans, mixers for the kitchen, shoe polishers and hair dryers all ease labour and provide opportunities for a fuller life for the housewife, while the husband's interest is catered for by the availability of comprehensive tool sets assembled around a small electric motor which enable him to carry out a wide range of 'do it yourself' jobs from polishing a board to drilling a hole in a brick wall to hang up a picture. The versatility of the electric motor is one of its most outstanding characteristics.

Turning to the transport field, electrical engineering has not had everything its own way as in lighting and power. Only five years after Oersted had discovered that there was a connection between electricity and magnetism, and long before the practical electric motor was devised, the Stockton and Darlington Railway was opened to use steam locomotion (in 1825) and from then on vast networks of railways have been constructed in this and other countries employing the steam locomotive inaugurated by Stephenson. Only today (1960) are we witnessing the final withdrawal of this magnificent enterprise and seeing electrical engineering, admittedly sharing the honours with the Diesel train, once more making a great contribution in high-voltage traction.

As we have seen in Chapter XI, there have been two distinct movements in electric traction, the sudden growth in popularity of the electric tramcar about 1900, which however waned before the Second World War, and the urban and suburban railways which, in London, started in 1890 with the opening of the City and South London tube. The tramway, with a few notable exceptions, ultimately gave way to the competition from the motor bus with its individual power unit but in its day it certainly made a great contribution to city life. At a time when the tempo of industrial development was at its height, tramcars enabled people to avoid congested cities and live in the suburbs. It also helped those who congregated in and near the factories to seek recreation and fresh air in the surrounding country. Particularly in Britain and, to some extent, abroad, the electric tram is rapidly becoming an anachronism and in spite of the success of the trolley bus, it certainly would seem that in this direction electrical engineering has no promising future. On the other hand suburban electric railways have developed extensive systems such as that in the south of London without which it would seem to be impossible to accommodate the vast number of city workers who commute daily. All too brief references have also been made in previous chapters to the development of high-voltage traction to main lines. This in itself is an epoch-making contribution of electrical engineering which will influence long-distance transport for many decades to come.

In the rapid development of air transport there is no immediate prospect of electricity providing the power for propulsion but, in lighting and the operation of auxiliary equipment, in navigation and communication, electrical engineering has already established itself as an indispensable adjunct.

To put the magnificent contributions of electrical engineering through communications into perspective we may first of all recall the experiments of Ronalds already described in Chapter V. In 1816 he offered to the Admiralty what was clearly a working electric

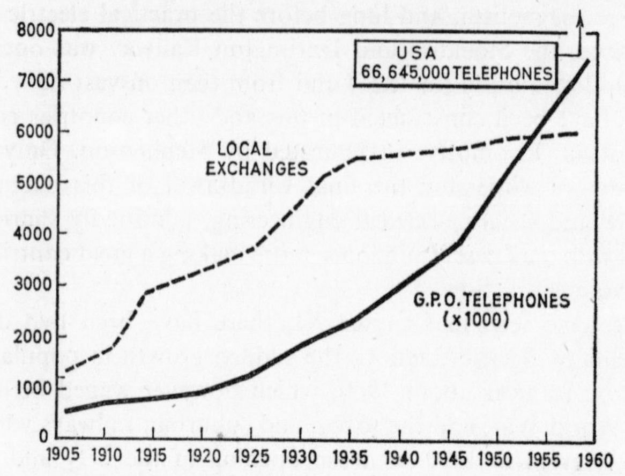

Fig. 71. The Growth in Telephones as a measure of progress in electrical engineering. (*Acknowledgments to Postmaster-General.*)

telegraph and was told in reply that 'telegraphs were wholly unnecessary'. It was not until the dramatic capture of a murderer twenty-three years later through the use of the electro-magnetic telegraph that the persistence of Cooke and Wheatstone was rewarded and telegraphy was established as a public service.

The following year (1840) saw the introduction of the universal penny post and the use of stamps in place of collection of dues on delivery of the letters.[5] In 1847 the Queen's Speech for the first time was transmitted by telegraph and, within the next decade, continental news was being received within hours instead of days as had been usual formerly. By the time the Atlantic cable was established telegraphs had been set up all over the country. In the year 1869–70 the Post Office revenue from the electric telegraph was £100,000, with a profit of about 40 per cent, and by 1900 the revenue had reached £3,350,000, with the cost of running the service considerably in excess of the receipts.

Soon after its introduction the telephone assumed an important position in the commercial world. In 1889 the three principal companies which had carried through the pioneer work were amalgamated to form the National Telephone Company with some 28,000 lines. From 1892 the Post Office came into the telephone field and in 1896 all trunk lines came within official control. The final stage in the development came in 1911 when the Post Office took over the whole telephone system of Britain. About this time the state of the telephone service in this country in comparison with conditions abroad is shown by the following table:

*Inhabitants per Telephone Station**

	1910[1]	1958[7]
United Kingdom	68	6·9
Germany	65	10·5
Sweden	30	3·1
Denmark	30	4·6
Norway	39	5·2
France	186	11·9
Russia	1,022	Not available
United States	11	2·6

As the telephone systems of the world grew, radio communication came into being, first as radio telegraphy and later as radio telephony, with all their well-known contributions to safety at sea and convenience in bridging long distances for domestic and commercial use. World networks linked the continents but a further development which has proved to have even greater influence on the everyday life of hundreds of millions of people is, of course, broadcasting. Other branches of electrical engineering have contributed enormously to man's material comfort and well-being but no other field has had such an influence on mind and spirit as the programmes of infinite variety which are now broadcast, throughout most hours of every day, from hundreds of centres throughout the world. In broad culture, in literature, music, politics, science and every field of mental and moral activity broadcasting is the most powerful educational influence ever known. It extends the voice and influence of one man far beyond the wildest dreams of the ancient orators as we saw in the subjugation of prewar Germany, a civilized state, by a demoniac Hitler. The Platonic idea of the size of a city being decided by the range of a man's voice, today brings the whole of mankind into one

* Total number of instruments of all kinds including extensions, having access to the general telephone network.

community through radio and broadcasting. National boundaries are overcome, so much so that jamming is a recognized procedure both to stifle truth and to kill propaganda. Thus, on the one hand, dictatorship is helped in the suppression of individual liberty by the services of radio but, on the other hand, the spread of ideas should result in an increase of individual freedom and democratic institutions. News now travels fast and far. Disasters like the sinking of a ship at sea, or a terrible earthquake with its heavy toll of dead and homeless, brings help from many nations within a few hours while there is reason to hope that the far-reaching discussion of problems involved by the people concerned may reduce the risk of future wars.

Strengthening of its effectiveness by permitting the speaker, performer or spectacle to be seen in millions of homes simultaneously through the extension of the device to television has indeed provided a potential power which not one of the pioneers could ever have anticipated.

We reach here as nowhere else the summit of achievement. Gilbert, Franklin and others who groped in the dark, ill-equipped with all but serene mental vision, wondered at the glimmer of light and did not understand. Volta found the current on which Oersted and Ampère built so well. The great Faraday opened a new gateway with ever-widening vistas which Kelvin, Hopkinson, Edison and Swan followed. Enthusiastic visionaries and practical men joined in. From Cambridge came J. J.'s electron and Maxwell's clear enunciation of basic theory which illuminated the track so many were treading, while finally that inscrutable man Oliver Heaviside brought his crock of gold, so difficult to accept yet so necessary and so perfect. These men all aimed at the good, and today we hold the treasure resulting from their thought and work. The measure of good which can result from its use is infinite but from its misuse could come the greatest of all world disasters. Man must decide.

REFERENCES

1. L. C. A. Knowles (Mrs.), *Industrial and Commercial Revolutions*. Routledge, 1941.
2. Green, *History of the English People*.
3. Rollo Appleyard, *A Tribute to Michael Faraday*. Constable, 1931.
4. Silvanus P. Thomson, *Michael Faraday. His Life and Work*. 1821.
5. G. R. Porter, *The Progress of the Nation*. Methuen, 1912.
6. W. T. O'Dea, *The Social History of Lighting*. Routledge, 1958.
7. Figures furnished by the G.P.O.

Chronological Table

1269 Peter Peregrinus describes a mariner's compass in case with 360° scale.
1492 Christopher Columbus observes variation of magnetic declination in crossing the Atlantic.
1544 Hartmann discovers magnetic dip.
1576 Robert Norman establishes a factory at Wapping to make compass needles.
1600 William Gilbert published *de Magnete*.
1660 Otto von Guericke made first frictional machine.
1665 *Philosophical Transactions* first published.
1687 Newton published *Principia*.
1712 onwards. Use of Newcomen and Savery's steam engine for mine pumping spread rapidly.
1720 Von Guericke. First insulated lines.
1745 Cunaeus discovered the Leyden Jar.
1747 Stephen Gray transmitted discharge from Leyden Jar across the Thames.
1750 Benjamin Franklin invented the lightning conductor.
1752 Benjamin Franklin erected a lightning rod on his home.
1753 'CM' wrote famous letter predicting electric telegraph.
1761 Kinnersley investigated heating of wires by electric current.
1762 Sulzer discovered potential between dissimilar metals.
1764 Hargreaves invented the spinning jenny.
1765 Watt invented steam engine with separate condenser.
1768 Arkwright's spinning machine.
1769 St. Paul's Cathedral equipped with lightning conductor.
1772 Cavendish investigated electrical conduction through sea water.
1774 Lesage uses frictional electricity to transmit signals.
1775 Volta invented the electrophorus.
1785 Coulomb invented the torsion balance electrometer.
1786 Galvani observed electrical convulsion in the legs of dead frogs.
1787 Bennet invented the electroscope.
1794 Chappé's semaphore telegraph sent messages 150 miles in 2 minutes.
1797 Pearson observed electrical decomposition of water.
1800 Volta announced the Voltaic Pile.
1800 Nicholson and Carlisle investigated electrolysis of water.
1801 Humphry Davy appointed to the Royal Institution.
1801 Moyes first to observe electric arc between carbon rods.
1806–7 Davy's Bakerian Lectures on electro-chemical effects.
1808 Cruickshank makes large practical electrical battery.
1808 Davy produces spectacular arc at Royal Institution lecture.
1809 Sömmering's telegraph and insulated underground wires.

Chronological Table

1811 Sömmering and Schiller send current through submarine cable across River Isar.
1812 Faraday as a boy attends Royal Institution lecture by Davy.
1813 Faraday appointed laboratory assistant at the Royal Institution.
1816 Ronalds. Practical telegraph offered to and rejected by the Admiralty.
1818 Institution of Civil Engineers founded.
1820 Oersted discovers deflection of a magnetic needle by current in a wire.
1820 Ampère communicates important analysis to academy of Science.
1820 Schweigger invents the 'Multiplier', the first practical galvanometer.
1821 Faraday produces magnetic rotation of conductor and magnet. First electric motor.
1824 Arago announced his disc experiment.
1825 Sturgeon produced practical electromagnets.
1826 Poggendorff adopted mirror in galvanometer movement.
1828 Henry produced silk covered wire and more powerful electromagnets.
1828 Kemp and Sturgeon amalgamated zinc plates in batteries.
1830 dal Negro obtained rotary motion by ratchet and pawl.
1831 Faraday discovered electromagnetic induction, ring and core experiments. Also magnet and rotating disc.
1832 Sturgeon made a magnetic engine.
1832 dal Negro produced a primitive dynamo with reciprocating parts.
1832 *c.* Pixii announced first rotating dynamo.
1832 Schilling's electromagnetic telegraph.
1833 Saxton's dynamo.
1833 Faraday laid down the laws of electrolysis.
1833 Gauss and Weber operated electric telegraph commercially.
1834 Snow Harris invented the attracted disc electrometer.
1835 Watkins' primitive electric motor.
1835 Sturgeon adopts the two-part commutator.
1836 Sturgeon uses suspended coil in galvanometer.
1836 Clark–Saxton dynamo priority dispute.
1836 Morse devised simple relay.
1836 Daniell devised first self-depolarizing cell.
1836 Moncke's rudimentary telegraph at Heidelberg.
1836 Electric light from batteries shown at Paris Opera.
1837 Page's beam engine motor.
1837 Morse and Vail produced the Morse Code.
1837 Gauss and Weber invented telegraph in Germany.
1837 Cooke and Wheatstone first practical electric telegraph on L. & N.W. Railway London-Camden Town.
1837 Pouillet invented the tangent galvanometer.
1837 Submarine cable industry started on the Thames.
1837 Davidson invented 'paddle wheel' motor.
1838 Jobart produced light by heating carbon rod in vacuo.
1838 Clarke's magneto-generator.
1838 Davenport made small magnetic engines.
1838 Jacobi propelled a boat by electric motor.
1839 Grove cell invented.
1839–1845 *c.* Gauss and Weber developed the basis of the electromagnetic system of units.
1839 O'Shaughnessy Brooke laid submarine cable across the Hoogly.
1840 Davidson's electric car.
1840 Wheatstone invented ABC telegraph.
1840 Bunsen and Smee cells appeared.
1841 Cooke-Wheatstone arbitration case.

841 Weber invented electro dynamometer.

841 Poggendorff devised the potentiometer.

842 Elias devises concentric ring motor.

842 Morse transmits signals across a river through opposing pairs of plates.

842 Slough murder case establishes the electric telegraph.

843 Wheatstone announces the Wheatstone Bridge.

843 William Siemens brings plating patent to London.

843 Morse transmits signals Baltimore to Washington.

844 Hand operated arc lamps at Paris Opera House.

844 Woolrich invents multipolar electromagnetic machine for plating.

844 Froment makes 'paddle wheel' motors.

845 Morse opened public telegraph service.

845 Matteui invented mica condenser.

845 Starr produced light by heating electrically a plumbago rod in vacuo.

845 *c.* Weber's reflecting galvanometer.

846 Bain invented automatic telegraph.

847 Gutta-percha became available for cable insulation. Siemens made 3,000 miles.

848 Swan started experiments on filament lamps.

850 Nollet invented dynamo for lighthouse illumination.

850 First channel cable laid England to France.

851 First commercially successful cable England to France.

851 Great Exhibition, Crystal Palace, London.

852 Farmer proposes multiplex telegraph working.

854 Hughes invented his printing telegraph.

855 Kelvin produced his precision attracted disc voltmeter.

1856 Varley's moving coil pointer galvanometer.

1856 Werner von Siemens introduced the shuttle armature.

1856 Kelvin developed theory of submarine cable.

1856 Formation of Atlantic Telegraph Company.

1857 Holmes' dynamo accepted for lighthouses.

1857 First attempt to lay Atlantic cable.

1858 Holmes' first electric lighthouse at South Foreland.

1858 Kelvin invented mirror galvanometer.

1858 Steamship 'Great Eastern' launched.

1858 Second Atlantic cable completed and first message sent.

1858 Firm of Siemens Bros. established.

1859 Du Moncel made carbon lamp filaments of sheepskin.

1859 Hooper introduced rubber for cable dielectrics.

1859 Staite and Moleyns used glowing platinum to produce light.

1859 Henley built his main cable factory at North Wooland on the Thames.

1860 Planté produced a lead storage battery.

1860 Atlantic cable enquiry held.

1860 *c.* Roschenschold observed rectification in semi-conductors.

1860 Pacinotti invented the ring armature.

1861 Western Union established telegraph service New York to San Francisco.

1861 Electrically fired mines used in American Civil war.

1861 British Association set up Standards Committee, Chairman Lord Kelvin.

1861 Weber suggested a fundamental system for electrical and magnetic measurements.

1862 Varley adopted curbing condensers on ocean cables.

1863 Wilde patented separately excited dynamo.

1863 Reis transmitted sound electrically.
1863 Clerk Maxwell determines the ohm.
1864 Clerk Maxwell postulated electromagnetic waves.
1865 Attempt to lay Atlantic cable with 'Great Eastern' abandoned.
1866 'Great Eastern' recovers end of 1865 cable, lays complete new one and establishes ocean telegraphy.
1866–7 Werner von Siemens, Wheatstone and Varley all announce self excitation of dynamos.
1867 Holmes produces improved form of automatic arc lamp.
1867 Kelvin invents the siphon recorder.
1867 *c.* Public exhibition of arc lighting in London, Paris and Philadelphia.
1868 Already 16,000 miles of telegraph line in Britain.
1868 Leclanché invents his famous cell.
1868 Wheatstone knighted on completion of the automatic telegraph.
1868 Wilde announces synchronization of A.C. machines.
1870 Rubber came into common use as an electrical insulator.
1870 British Post Office takes over the telegraph system.
1870 Gramme re-invents the ring armature.
1871 Society of Telegraph Engineers (later I.E.E.) founded.
1872 Dynamos in wide use for electro-plating.
1872 *The Telegraphic Journal and Electrical Review* (later *Electrical Review*) founded.
1872 Stearn originates the artificial cable balancing line for telegraphs.
1873 Maxwell produces treatise on Electricity and Magnetism.
1873 Identical machines used in Vienna as dynamos and motors.
1873 Latimer Clark produces his standard cell.
1873 Gramme first transmitted electric power over three-quarters o a mile at Vienna Exhibition.
1873 Hefner Alteneck adopts th drum type of armature.
1874 Baudot invents an automati telegraph.
1874 Lodyguine made 200 lamp with incandescent carbon rod in vacuo.
1874 Wallace produced the firs American dynamo.
1874 William Siemens built the firs cable ship *Faraday*.
1875 Electricity used for furnaces.
1875 Graham Bell invented the tele phone.
1875 Farmer 'subdivided the electri light' by parallel operation.
1875 Carey made the first proposa for a mosaic for television.
1875 Crookes investigated discharg in gases at low pressure.
1876 Kohlrausch invented movin iron galvanometer.
1876 Varley invented compound fiel winding.
1877 Brooks introduced an oil-fille underground cable system.
1877 Crompton lighting ironwork with arc lamps.
1877 Edison invented carbon tele phone transmitter.
1877 William Siemens lectured i public on A.C. and D.C.
1877 Telephone exchanges intro duced, New Haven, U.S.
1877 Jablochkoff candles installed i Avenue de l'Opera.
1877 Gaiety Theatre London ex terior illuminated by arc lamps
1877 Edison Electric Light Co formed.
1877 Telephones produced com mercially.
1878 Hughes demonstrates the micro phone.
1878 Swan demonstrates a practica incandescent filament lamp.
1878 Crompton invents arc lam with overhead mechanism.
1878 Jablochkoff candles on th Thames Embankment.
1878 Brush invents the open-co dynamo.

1878 *The Electrician* founded.
1878 Crompton making Burgin dynamos in quantity.
1879 Ayrton and Perry produce the first practical portable ammeter.
1879 Siemens makes first arc furnace.
1879 A French farmer, Chrétien, ploughed by electricity.
1879 First telephone exchange switchboard in England.
1879 Siemens and Halske exhibited an electric railway in Berlin.
1879 Bailey demonstrated a rotating electromagnetic field.
1879 Crompton uses arc lamps in his house at Porchester Gardens.
1879 Edison announced the effect of occluded gases in filaments.
1879 City and Guilds college founded at Finsbury.
1879 Edison produced his first dynamo for the Jeannette expedition.
1880 *c.* Siemens and Halske commercialized Weber's electrodynamometer.
1880 Edison produced his electric pen, the first commercial motor.
1880 Society of Telegraph Engineers added 'and Electricians' to its title.
1881 The Swan Electric Lighting Co. formed.
1881 Savoy Theatre, London, completed equipped with filament lamps.
1881 Faure preformed accumulator grids with red lead paste.
1881 Niagara first power station brought into use.
1881 Lord Kelvin's house in Glasgow lit by Swan lamps.
1881 Paris Conference. Electrical terminology formulated.
1881 Edison at Paris Exhibition. Exhibited his Jumbo dynamo.
1881 Edison constructed his first electric power station at Pearl St. N.Y.
1881 Edison and Swan joined forces.
1881 Brighton electric supply inaugurated.
1881 Ferranti at the age of 17 starts manufacturing electrical equipment.
1881 Callender introduces vulcanized bitumen cables.
1881 Potier suggested use of electrostatic voltmeter as wattmeter.
1881 Early hydro-electric plant on River Wey, Godalming.
1881 Rayleigh checked resistance standards by revolving coil.
1881 Ayrton and Perry invented direct reading ohmmeter.
1882 Crompton receives future King Edward VII at Crystal Palace Exhibition.
1882 D'Arsonval reflecting moving coil galvanometer.
1882 Edison dynamos installed at Holborn Viaduct, at Pearl St., New York and Milan.
1882 Ayrton and Perry (later Aron) pendulum type supply meter.
1882 Compound winding of field coils introduced.
1882 Crompton lights the new Law Courts.
1882 Ayrton invented the ohmmeter.
1882 First Electric Lighting Act in England.
1882 First permanent electric supply, Brighton.
1882 Gaulard and Gibbs arranged transformer in series.
1882 Berthoud-Borel Company formed to make lead covered cables.
1882 Tentative electric tramway at Leytonstone, London.
1882 35 mile 2000 volt A.C. transmission at Munich.
1882 Torsion type dynamometer appears.
1882 Early water-power electricity generating station, Cragside, Northumberland.
1882 *c.* Hopkinson and Edison independently invented 3-wire D.C. system.
1882 First application of electricity underground, Forest of Dean.
1883 Volk installed experimental electric railway at Brighton.
1883 Thomson–Houston Co. formed, Lynn, Mass.
1883 Overhead trolley railways started at Portrush and Richmond (Va.)

1883 Grosvenor Gallery 100 volt public supply system inaugurated.
1883 *c.* Step down transformers come into use.
1883 Crompton installed largest electric light system, Opera House, Vienna.
1883 Charing Cross Electric Supply Company formed.
1883 Ayrton and Perry invent solenoid type of ammeter.
1883 Laminated cores and carbon brushes introduced.
1883 Ferranti introduces mercury type energy meters.
1883 Cardew invented the hot-wire voltmeter.
1883 Electric tramways at Kew and Giants Causeway.
1883 Edison observes the blackening of a carbon filament lamp.
1884 Parsons constructs the first parallel flow reactor turbine.
1884 Alexander Siemens invents the variable speed A.C. commutator motor.
1884 Willans constructs his vertical high-speed central valve engine.
1885 Dynamos of 100 kW in production.
1885 Fleming and Crompton introduce commercial potentiometer.
1885 Frères invents recording ammeter.
1886 Ferranti appointed in charge of Grosvenor Gallery system.
1886 Hopkinson published famous paper 'Dynamo-electric machinery'.
1886 Kelvin developed the ampere balance standard.
1886 Crompton inaugurates Kensington Court system.
1886 Paddington G.W.R. system with Gordon alternators inaugurated.
1886 Metropolitan Electric Supply Co. formed with £500,000 capital.
1886 Wright invented his voltage recorder.
1886 Impregnated lead-covered cable manufactured in U.S.
1886–7 Hughes, Preece and Thompson separately laid foundations of telephone transmission theory.
1887 Hertz demonstrates electromagnetic waves.
1887 Parsons made 10 turbo-sets up to 32 kW.
1887 Elihu Thomson introduced electric welding.
1887 Hookham invents motor meter with eddy current brake.
1887 Kelvin constructed a 10,000 volt quadrant electrometer.
1887 Hellesen produced the first commercial dry cell.
1887 Tesla patented 2- and 3-phase alternators and induction motors.
1887 London Electric Supply Corporation formed. Capital £1,000,000.
1888 Ferrari and Tesla independently produced rotating fields by 2-phase current.
1888 Experiments with condenser discharge confirmed Maxwell's prediction.
1888 Amendment of 1882 Lighting Act.
1888 Parsons built a 75 kW turbine for Newcastle.
1888 Schallenberger meter first used in England.
1888 Kensington and Knightsbridge Company formed.
1888 Weston invented the direct-reading portable moving-coil ammeter.
1889 'Dry core' paper telephone cable introduced.
1889 Strowger proposed the automatic telephone exchange.
1889 Instititution of Electrical Engineers incorporated under its present title.
1889 Formation of National Telephone Co.
1889 C. A. Parsons and Co. established.
1889 General Electric Co. formed in England.
1889 Atherton forms British Insulated Wire Co. to make impregnated paper cables.

1889 Ferranti's Deptford station under steam at 10,000 volts.
1889–98 John Hopkinson Professor at King's College.
1889 Evershed and Goolden invent the ohmmeter.
1889 Kennedy devised underground conduit system.
1890 Wright invented shaded pole meter.
1890 Dobrowolsky transmitted 100 H.P. by 3-phase current.
1890 Kelvin invented the 'carriage lamp' voltmeter.
1891 Branly invented the 'coherer'.
1891 City and South London tube railway opened.
1891 Crompton introduces the term 'load factor'.
1891 Ferranti lays 10,000 volt Deptford cable.
1891 Mordey introduced laminated transformer cores.
1891 *Lightning*, later the *Electrical Times*, began publication.
1891 Thomson–Houston motor designed for street cars.
1891 First submarine telephone cable England to France.
1891 Leeds corporation installs electric tramway system.
1891 National Physical Laboratory founded.
1891 Parsons built 100 kW set with first condenser.
1891 Langdon Davies produces his induction motor.
1891 Ferranti 112 kW transformers for London Electric Supply Co.
1891 Board of Trade sets up Standards Committee.
1891 Von Miller transmits 3-phase current 110 miles at 25,000 volts, Frankfurt.
1892 Heaviside publishes his Electrical Papers.
1892 Blondel suggested the oscillograph.
1892 Electrical Exhibition, Crystal Palace.
1892 Sprague builds a 1,000 H.P. electric locomotive.
1892 Weston constructs his standard cell.
1892 White invents the 'solid back' telephone transmitter.
1892 Hayes proposed the Hayes telephone exchange circuit.
1893 Patchell introduced superheated steam in power stations.
1893 Electrical Congress, Chicago.
1893 Wright invents Maximum Demand Indicator.
1894 Order in Council established British electrical standards and Cardew current-balance as legal standard for current.
1894 Semenza's criticism of British electric supply stations.
1894 Parsons 350 kW turbine installed Manchester Square, London.
1894 Silvanus Thompson's important British Association paper.
1894 *c.* Extensive doubling and twisting of single wire telephone circuits.
1894 Oliver Lodge transmits dots and dashes by radio at Royal Institution.
1895 Marconi transmitted signals over one kilometre in Boulogne.
1895 Niagara Falls Co. installed 10 5,000 H.P. 2-ph. 2,200 volt water turbines.
1895 Rotary automatic telephone exchange switch introduced.
1895 Sprague invents multiple train control.
1896 First Annual Convention of Incorporated Municipal Electrical Association.
1896 British Thomson, Houston, Co. formed in U.K.
1896 Parsons installs 120 kW single-phase turbo generator at Pittsburgh.
1896 Campbell invents resonant reed type of frequency meter.
1896 Post Office takes over trunk telephone lines.
1897 Duddell invented practical oscillograph.
1897 Brown invented oil-immersed circuit breaker.
1897 Marconi transmits radio signals across the Bristol Channel.

1897 Elihu Thomson invents commutator motor type of meter.
1897 Braun demonstrates a cathode ray tube.
1897 Bastion invents the electrolytic meter.
1897 J. J. Thomson discovers the electron.
1897 Lodge introduces syntonic wireless or tuning on frequency.
1898 *c.* Nernst invents the Nernst lamp.
1898 *c.* Hartmann and Braun produce the sagging hot wire voltmeter.
1898 Schallenberger invents the energy meter.
1898 Duddell invents the electromagnetic oscillograph.
1898 *c.* 22 automatic exchanges operating in the U.S.A.
1899 Bremer patented the flame arc.
1899 Charles Merz set up as consulting engineer and later, joined by William McLellan formed the world renowned organization.
1899 Pupin proposed the telephone loading coil.
1899 Marconi transmitted radio signals from South Foreland to Wimereux.
1900 Wireless first installed on merchant ships.
1900 Parsons installed 4,000 volt sets at Elberfeld.
1900 Central London Railway opened.
1900 First public 6,600 volt 3-phase supply at Newcastle.
1900 Electrification of District Railway, London, first section opened.
1900 *c.* British Post Office takes over telephone subscribers' lines.
1900 First central battery telephone exchange in England opened, Bristol.
1900 Shaped conductor 3-core cables installed, Wood Lane, London.
1900 Wright invented an electrolytic meter.
1900 Addenbrooke used electrostatic wattmeter to measure dielectric loss.
1900 Cuenod and Thury proposed high voltage constant current D.C. system.
1900 Cooper Hewitt invented mercury arc lamp.
1901 Giorgi proposes the MKS system of units.
1901 Cathode ray oscillograph in use.
1901 Murray automatic telegraph system developed.
1902 Marconi transmitted signals across the Atlantic.
1902 Richardson's fundamental publication on thermionic emission.
1902 Loading coils used commercially.
1902 Parsons installed first revolving field generator at Newcastle.
1902 Cooper Hewitt invented mercury arc rectifier.
1903 Paul invents the 'Unipivot' instrument.
1903 Einthoven announced the string galvanometer.
1903 A 5,000 kW vertical Curtis turbine set installed at Chicago.
1903 Poulsen generated undamped oscillations by enclosed arc.
1903 Birkeland fixed nitrogen electrically.
1904 Fleming invented the thermionic diode.
1904 N.E. Railway, Newcastle–Tynemouth electrified.
1905 Tantalum lamps introduced.
1905 Braun produced the high-emission cathode ray tube.
1905 Niagara Falls 60,000 volt transmission over 200 miles.
1905 First power station system control room in Great Britain at Carville.
1905 Clothier and Price made the first withdrawable metal clad high voltage switch.
1906 Lee de Forest invented the triode.
1906 Tungsten filament lamps introduced.
1906 International Electrotechnical Commission founded.
1906 The last large reciprocating engine sets (3,500 kW) installed

for L.C.C. tramways at Greenwich.
1906 Hewlett introduced high voltage suspension insulator.
1906 Moutiers–Lyon D.C. transmission at 60 kV.
1907 Campbell constructs his mutual inductance standard.
1908 Electrical Standards adopted internationally.
1908 Coolidge produced tungsten filaments.
1908 Lancashire–Heysham Railway.
1908 Lots Road station built for London underground railways.
1908 Edison invented the nickel alkaline battery.
1909 Drysdale constructs his A.C. potentiometer.
1909 First electrocardiograph.
1910 First loaded submarine telephone cable, Dover–Calais.
1910 Institution of Electrical Engineers moved into Savoy Place.
1910 Order in Council brought electrical standards up to date.
1911 Trolley-buses first run in Leeds and Bradford.
1911 British Post Office completed takeover of telephone system.
1911 British Electrical and Allied Industries Association formed.
1911 Hunter invented the split conductor protective gear.
1912 Clinker valve.
1912 Epsom, first automatic telephone exchange in England.
1912 Parsons installed 25,000 kW set at Chicago.
1912 *Titanic* disaster emphasizes need for continuous alert on ship's radio.
1913 Kaplan invented adjustable blade water turbine.
1913 *c.* Coolidge made tungsten lamp filaments.
1913 Langmuir substituted inert gas in filament lamp.
1913 *c.* Franklin suggested the feedback principle.
1913 Continuously loaded submarine cable laid to Vancouver Island.
1914–18 Valve amplifiers introduced on telephone circuits.
1914–18 22,000 volt underground cable a commercial proposition.
1915 Commercial telephone service established New York–San Francisco.
1916 *c.* Duddell Mather developed wattmeter for voltages up to 10,000.
1917 A 7,000 kW 3,000 r.p.m. turbo-alternator installed at Willesden.
1917 First high pressure boilers (475 lb. per sq. in.) adopted at North Tees.
1918 Radio transmission Eiffel Tower.
1918 English Electric Co. formed.
1918 *c.* Hochstadter introduced screening on high voltage cables.
1918 Steam pressure of 475 lb. per sq. in. in use at North Tees power station.
1919 Armstrong invented the 'superheterodyne' circuit.
1919 Electricity (Supply) Act.
1919 British Electrical Development Association formed.
1919 Beard invented the compensated pilot protective gear.
1919 Air cooling of alternators introduced at Blaydon Burn.
1919 Merchant Shipping (Wireless Telegraph) Act makes it compulsory for ships to carry radio officers.
1920 Electricity Commissioners appointed.
1920 Vallauri formulates mathematical operation of triode.
1920 Radio broadcasting instituted.
1920 Advisory Committee sets pattern of railway electrification.
1920 Radio Research Board established.
1920 Emanueli introduced oil-filled cable.
1920 Fisher and Atkinson's paper on discharges in high voltage cables.
1920 14,000 kW Lungström turbine at Vasteras.
1921 A Metrovick 12,500 kW set at work in Liverpool.

1921 Institution of Electrical Engineers Royal Charter.
1922 British Broadcasting Company formed.
1922 Cathode ray oscilloscope.
1923 Photo-electric cells.
1923 Coltano (Italy) continuous wave radio station established.
1924 First meeting of World Power Conference.
1924 Commercial selenium rectifiers.
1924 Lossev applies the 'cat's whisker' in radio.
1924 Parsons 50,000 kW set installed at Chicago.
1924 Bennett invents the oilostatic high voltage cable system.
1925 Star Quad telephone cable introduced.
1925 Barking Power Station opened by King George V.
1925 *c.* 132,000 oil impregnated cables manufactured.
1925 Weir Report on electricity supply.
1925 London Electricity, two Acts of Parliament.
1926 Electricity (Supply) Act. Standard 132 kV grid established and standard 50 cycle frequency.
1926 Round invented the tetrode.
1926 Tilleger and Holst invented the pentode.
1926 Busch invented the electro-magnet lens for the electron microscope.
1926 *c.* Abraham and Edgecumbe electrostatic voltmeter for high voltages.
1926 Copper-oxide rectifiers developed.
1927 Television established.
1927 British Broadcasting Corporation succeed British Broadcasting Co.
1928 Mercury arc rectifiers commercialized.
1928 First 33,000 volt turbo-alternator installed at Brimsdown.
1929 First section of British 132 kV grid operating.
1929 Reyrolles set up short circuit testing station.
1929 Associated Electrical Industries formed.
1929 Gas-filled grid-controlled thyratrons.
1929 A.C. Multiplex 12 Channel telegraph system in operation.
1929 First pentode valve.
1930 Moullin invented electronic voltmeter.
1931 Zworykin invents the Iconoscope.
1931 Wilson wrote his classical exposition on semi conductors.
1932 Ministry of Transport Order standardizes electric traction system of England.
1932 Knoll and Ruska produce a practical electron microscope.
1932 McGee and Tedham develop a photosensitive mosaic for a television camera.
1933 Battersea A station commissioned with steam pressure 600 lb. per sq. in. and 105,000 kW turbo set.
1933 London–Brighton line electrified.
1934 Coiled coil filament lamp introduced.
1934 First parabolic reflector microwave radio link, Dover–Calais.
1936 First pumpless steel clad mercury rectifiers.
1936 Overall thermal efficiency of turbo-alternators at Battersea reached 29 per cent.
1936 McGee and Lubszynski develop sensitive Super-Emitron.
1936 Television broadcasting inaugurated in U.K.
1936 Boulder Dam hydro-electric scheme with 115,000 H.P. turbines.
1936 Electricity Supply (Meters) Act.
1937 12-channel carrier telephone circuits established by British Post Office.
1937 First coaxial telephone cable in U.K. London–Birmingham.
1938 Hydrogen cooling of large alternators introduced.
1939 Repeaters become normal on long distance telephone cables.
1939 Power station boiler pressures

seldom exceed 600 lb. per sq. in.

1939–45 German magnetic mines defeated by degaussing.

1940 Oliphant, Randall and Boot invent the cavity magnetron.

1941 1,500 kW wind turbo-generator at Vermont.

1941 Grand Coulee dam constructed.

1942 British Electricity Authority planned 275 kV grid.

1942 First self sustaining atomic pile in operation at Chicago. 200 watts.

1943 Uranium graphite pile produces 1,000 kW at Oakridge.

1943 First submerged telephone repeater, Holyhead Isle of Man.

1944 Parsons built first long life gas turbine 500 H.P.

1945 Battersea B station in commission. Steam pressure 1,350 lb. per sq. in.

1946 Central Electricity Board interconnected 180 electricity companies, 354 municipal undertakings and 9 joint electricity authorities.

1947 Electricity Act nationalized British electric supply.

1947 N.P.L. formally promulgated the Paris (1946) decision on absolute electrical units.

1948 Barden and Brattain produced the first practical transistor.

1948 Cambridge Instrument Co. developed the direct writing cardiograph.

1949 First hydrogen cooled alternator installed at Littlebrook.

1949 Littlebrook B station, steam pressure 1,235 lb. per sq. in.

1950 Loch Sloy hydro-electric scheme 45,000 Francis turbines.

1950 International Electrotechnical Commission adopted the rationalized M.K.S. system.

1950 *c.* Turbo-alternators officially standardized 30,000 kW (600 lb. per sq. in.) 60,000 kW (900 lb. per sq. in.) with exceptions permitted.

1952 288,000 kW transmitted 600 miles at 380,000 volts in Sweden.

1952 Portobello station 60,000 kW set. Steam pressure 1,360 lb. per sq. in. Efficiency attained 32 per cent.

1952 First part of British 275,000 volt grid completed.

1953 Steam pressure of 1,500 lb. per sq. in. and temperature of 1050°F. adopted at Stourport.

1953 Steam pressure of 2,350 lb. per sq. in. and temperature of 1,100°F. adopted at New Jersey.

1955 National Physical Laboratory installs 3.2 million volt Ferranti impulse generator.

1956 First transatlantic submarine telephone cable completed.

1956 British Transport Commission adopt 25 kV for railway electrification.

1956 First atomic power station in operation at Calder Hall.

1959 Kariba hydro-electric scheme on Zambesi with 100 MW sets and 330 kV lines working.

1959 325 MW turbo-generator commissioned at Fort Rouge, Detroit.

1961 Membership of Institution of Electrical Engineers exceeds 48,000.

Index